Advances in integrated weed management

It is widely recognised that agriculture is a significant contributor to global warming and climate change. Agriculture needs to reduce its environmental impact and adapt to current climate change whilst still feeding a growing population, i.e. become more 'climate-smart'. Burleigh Dodds Science Publishing is playing its part in achieving this by bringing together key research on making the production of the world's most important crops and livestock products more sustainable.

Based on extensive research, our publications specifically target the challenge of climate-smart agriculture. In this way we are using 'smart publishing' to help achieve climate-smart agriculture.

Burleigh Dodds Science Publishing is an independent and innovative publisher delivering high quality customer-focused agricultural science content in both print and online formats for the academic and research communities. Our aim is to build a foundation of knowledge on which researchers can build to meet the challenge of climate-smart agriculture.

For more information about Burleigh Dodds Science Publishing simply call us on +44 (0) 1223 839365, email info@bdspublishing.com or alternatively please visit our website at www.bdspublishing.com.

Related titles:

Integrated weed management for sustainable agriculture
Print (ISBN 978-1-78676-164-4); Online (ISBN 978-1-78676-166-8, 978-1-78676-167-5)

Managing soil health for sustainable agriculture Volume 1: Fundamentals
Print (ISBN 978-1-78676-188-0); Online (ISBN 978-1-78676-191-0, 978-1-78676-190-3)

Managing soil health for sustainable agriculture Volume 2: Monitoring and management
Print (ISBN 978-1-78676-192-7); Online (ISBN 978-1-78676-195-8, 978-1-78676-196-1)

Measuring soil health
Print (ISBN 978-1-78676-426-3); Online (ISBN 978-1-78676-428-7, 978-1-78676-429-4)

Improving soil health
Print (ISBN 978-1-78676-670-0); Online (ISBN 978-1-78676-672-4, 978-1-78676-673-1)

Chapters are available individually from our online bookshop:
https://shop.bdspublishing.com

climate•SMART•publishing
IN AGRICULTURAL SCIENCE

BURLEIGH DODDS SERIES IN AGRICULTURAL SCIENCE
NUMBER 113

Advances in integrated weed management

Edited by Professor Per Kudsk, Aarhus University, Denmark

burleigh dodds
SCIENCE PUBLISHING

Published by Burleigh Dodds Science Publishing Limited
82 High Street, Sawston, Cambridge CB22 3HJ, UK
www.bdspublishing.com

Burleigh Dodds Science Publishing, 1518 Walnut Street, Suite 900, Philadelphia, PA 19102-3406, USA

First published 2022 by Burleigh Dodds Science Publishing Limited

Library of Congress Control Number: 2021947122

British Library Cataloguing in Publication Data
A catalogue record for this book is available from the British Library

ISBN 978-1-78676-745-5 (Print)
ISBN 978-1-78676-748-6 (PDF)
ISBN 978-1-78676-747-9 (ePub)
ISSN 2059-6936 (print)
ISSN 2059-6944 (online)

DOI 10.19103/AS.2021.0098

Typeset by Deanta Global Publishing Services, Dublin, Ireland

Contents

Series list

Title	Series number
Achieving sustainable cultivation of maize - Vol 1 From improved varieties to local applications Edited by: Dr Dave Watson, CGIAR Maize Research Program Manager, CIMMYT, Mexico	001
Achieving sustainable cultivation of maize - Vol 2 Cultivation techniques, pest and disease control Edited by: Dr Dave Watson, CGIAR Maize Research Program Manager, CIMMYT, Mexico	002
Achieving sustainable cultivation of rice - Vol 1 Breeding for higher yield and quality Edited by: Prof. Takuji Sasaki, Tokyo University of Agriculture, Japan	003
Achieving sustainable cultivation of rice - Vol 2 Cultivation, pest and disease management Edited by: Prof. Takuji Sasaki, Tokyo University of Agriculture, Japan	004
Achieving sustainable cultivation of wheat - Vol 1 Breeding, quality traits, pests and diseases Edited by: Prof. Peter Langridge, The University of Adelaide, Australia	005
Achieving sustainable cultivation of wheat - Vol 2 Cultivation techniques Edited by: Prof. Peter Langridge, The University of Adelaide, Australia	006
Achieving sustainable cultivation of tomatoes Edited by: Dr Autar Mattoo, USDA-ARS, USA and Prof. Avtar Handa, Purdue University, USA	007
Achieving sustainable production of milk - Vol 1 Milk composition, genetics and breeding Edited by: Dr Nico van Belzen, International Dairy Federation (IDF), Belgium	008
Achieving sustainable production of milk - Vol 2 Safety, quality and sustainability Edited by: Dr Nico van Belzen, International Dairy Federation (IDF), Belgium	009
Achieving sustainable production of milk - Vol 3 Dairy herd management and welfare Edited by: Prof. John Webster, University of Bristol, UK	010
Ensuring safety and quality in the production of beef - Vol 1 Safety Edited by: Prof. Gary Acuff, Texas A&M University, USA and Prof. James Dickson, Iowa State University, USA	011
Ensuring safety and quality in the production of beef - Vol 2 Quality Edited by: Prof. Michael Dikeman, Kansas State University, USA	012
Achieving sustainable production of poultry meat - Vol 1 Safety, quality and sustainability Edited by: Prof. Steven C. Ricke, University of Arkansas, USA	013
Achieving sustainable production of poultry meat - Vol 2 Breeding and nutrition Edited by: Prof. Todd Applegate, University of Georgia, USA	014

Introduction

Weed management continues to face many challenges, including herbicide resistance, invasive species, climate change and how best to deploy the range of non-chemical control methods available. To tackle these challenges, integrated weed management (IWM) needs to evolve to embrace a more holistic, landscape-based and agroecological approach.

This volume provides an authoritative review of the latest developments in IWM, including the changes in understanding the complex ways weeds interact with their environment and with each other, as well as how some weed species may contribute to ecosystem services such as soil health. The book is split into three parts. Chapters in Part 1 focus on weed ecology, Part 2 chapters examine intelligent weed control technologies and Part 3 provides five case study chapters that focus on the use of IWM in various settings.

Part 1 Weed ecology

The first chapter of the book discusses advances in understanding the contribution of weeds to the functioning of agroecosystems. Chapter 1 first reviews key aspects of weed ecology, focusing on areas such as weed diversity, weed functional traits and ways of accounting for intraspecific variation in weeds. It also highlights the use of a response-effect model to assess weed multi-functionality and trade-offs between negative and positive effects of weeds. The chapter includes a case study showing how farmers can manage weeds beneficially, followed by a summary on how important implanting effective IWM is to food security in the future.

Chapter 2 focuses on advances in understanding the dynamics of weed communities and their responses to different IWM approaches. The chapter assesses the role of a functional trait-based approach able to capture both the complexity of weed communities and the ways they might react to different combinations of IWM techniques. Rather than weed eradication, which might be neither feasible or environmentally beneficial, such an approach can potentially lead to a more functionally diverse weed community that is less competitive in any given crop. Adopting this more holistic approach will allow IWM to create both more productive and more sustainable cropping systems.

Moving on from Chapter 2, Chapter 3 discusses advances in managing arable weed propagules which can have a major impact on weed survival and spread. The chapter first describes the ways by which weed propagules have been historically managed. It then discusses advances in managing weed propagules with a special focus on inactive propagules i.e. those that are not germinated or sprouted. Ways of managing inactive propagules reviewed in the chapter includecrop harvest (weed seed crushing and milling), weed seeds on the soil surface (weed seed predation), weed propagules in the soil matrix (weed seed decay and mechanical destruction of ramets) and the process chains around arable farming (managing manure or crop biomass transport and processing). Finally, the chapter suggests new avenues for research.

Chapter 4 considers advances in allelopathic interactions between weeds and crops. The chapter begins by highlighting allelochemical classes and plant defence, how allelochemicals can be produced in plants and the use of a rhizosphere model for belowground microbial interactions in allelopathy. It also illustrates allelochemical interactions in wheat, rice, buckwheat and sorghum, reviews experimental methodology and allelopathic trait selection and provides a case study on the weed-suppressive effect of buckwheat. A section on using allelopathy as a future component of IWM is also included, focusing on the development of new herbicides based on allelochemical templates, the use of allelopathic crops and breeding for allelopathic traits in crops. The chapter then summarises how allelopathy can potentially be used in the future for IWM practices.

The final chapter of Part 1 focuses on advances in understanding invasive characteristics in weed species. Chapter 5 first examines how genetic modifications in plants can be considered a factor in invasiveness. It then goes on to discuss the four main epigenetic modifications that effect invasiveness: DNA methylation, histone modifications, chromatin configuration and actions of non-coding RNA species that affect messenger RNA availability. The chapter concludes by emphasising how both genetic and epigenetic modification analysis is important in understanding invasiveness and weediness.

Part 2 Intelligent weed control technologies

Part 2 opens with a chapter that reviews modelling the effects of cropping systems on weed dynamics, focusing on the how best to manage the transition from process analysis to practical decision support. Chapter 6 first assesses three contrasting models which quantify the effect of a cropping system on weed dynamics: a single-equation static model, matrix-based models and a model built from process-based sub-models. The chapter moves on to discuss ways of limiting the modelled system for more practical applications, focusing on temporal, spatial and species scales. It then reviews modelling approaches, first focusing on empirical versus mechanistic models, then discussing stochastic versus deterministic models. It also considers how to bridge the gap between process analysis and decision support before concluding with an overview of why models are essential in managing weeds and selecting the optimum approach to IWM.

Chapter 7 discusses developing decision support systems (DSS) for weed management. The chapter begins by reviewing how DSS can be used in weed management to set thresholds for implementing an IWM strategy. It then examines the role of decision support systems in reducing herbicide use, as well as how these systems can be used to prevent herbicide resistance for effective, low-cost weed control. The chapter also highlights how DSS can be used for long-term management of a wide range of weed species and how the adoption of weed management DSS by farmers is slowly increasing. The chapter concludes by highlighting how research into using DSS for weed management is developing.

The next chapter focuses on advanced detection technologies for weed scouting. Chapter 8 starts by highlighting the current techniques that can be used to make weed management more efficient, such as on-ground and remote-sensing methods for weed detection. The chapter then goes on to show how more precise weed scouting can contribute to implementing and assessing the effects of different IWM techniques and the ways they can be combined. These methods range from more targeted spraying, use of cultural techniques such as more competitive crop cultivars, tillage and rotation practices, through to better assessment of weed competitiveness and resistance in response to IWM strategies. The chapter concludes by highlighting the importance of improving detection technologies for weed scouting in the future.

The subject of Chapter 9 is advances in precision application technologies for weed management. The chapter begins by reviewing advances in precision weed control systems, including more precise herbicide application techniques (such as off and online patch spraying) for site-specific weed management. It also looks at advances in areas such as camera-guided mechanical weed control and robotic weeding. The chapter then examines emerging technologies such as improvements in image processing and weed identification, the use of genetic modification, signalling compounds, topical and systematic markers to help distinguish crops more easily from weeds. It also assesses the potential of nanotechnology in such areas as non-markers and sensors. A section on herbigation – the application of herbicides through an irrigation system – is also provided, followed by a discussion on tracking spatial distribution patterns of weeds for improved pre-emergence management. A summary on why new developments in precision weed management are essential to improving IWM practices closes the chapter.

Expanding on topics previously touched upon in Chapter 9, Chapter 10 focuses on advances in mechanical weed control technologies. The chapter first discusses the principles of mechanical weed control, then goes on to examine the three main types of mechanical weed control, starting with full-width cultivators then discussing inter-row and intra-row cultivation. The chapter looks, for example, at how vision and global navigation satellite system (GNSS) technologies can improve guidance systems for mechanical intra-row weed control possible, opening the possibility of automatic intra-row weeding to revolutionize weed management in direct-sown row crops

Part 3 Case studies

The first chapter of Part 3 assesses on-farm implementation of integrated weed management. Chapter 11 reviews the cognitive, social and individual dispositional factors which help to explain the lack of IWM adoption by farmers. It assesses factors such as lack of available knowledge on IWM, limited evidence for its efficacy, reliability and cost-effectiveness of IWM. The chapter also discusses the challenges associated with trade-offs against other attributes of cropping systems and the increased complexity involved in implementing an IWM strategy. The chapter reviews the infrastructure needed to support learning by farmers to change existing beliefs of farmers and resistance to change. The chapter includes a case study on understanding the decision-making processes for on-farm IWM amongst European farmers.

Chapter 12 looks at optimising integrated weed management in narrow-row crops. The chapter uses the IWMPRAISE framework which focuses on the five pillars of IWM. It first discusses cropping system diversification, then moves on to examine cultivar choice and establishment, field and soil management and direct control tactics. The chapter includes four separate case studies on IWM programmes for cereals in the United Kingdom, France, Slovenia and Denmark. The chapter then assesses the relative success of each programme to identify those approaches worth exploring in future research.

The next chapter reviews the current status of integrated weed management for grasslands. Chapter 13 first describes the weed management toolbox for grasslands, focusing on prevention, cultural, physical and biological control. It then moves on to review how weed management practices can be integrated in grasslands, supported by case study. The chapter also addresses how multiple transitions in the weed's life cycle can be dealt with, looks at the vertical and horizontal integration of weed management practices and the integration of grazing and mowing practices. A section on use of invertebrates and pathogens for weed control in combination with other management practices is also provided. The chapter concludes with an outlook for further improving IWM in grasslands.

Chapter 14 focuses on integrated weed management in perennial woody crops. The chapter discusses two case studies. The first of these is on olive orchards in Spain, focusing on strategies such as the use of soil management systems, tillage, no tillage with chemical control, inert cover with plant residue mulches, as well as use of spontaneous and cultivated cover crops. The second case study focuses on vineyards in the UK. The case study reviews soil management systems and, in particular, a NIAB EMR integrated weed management experiment, as well as the influence of weed management on canopy development, yield and grape quality. The chapter concludes by highlighting how the most suitable integrated weed management strategy can be influenced by factors such as location, topography, soil type, crop features, farmer preferences and climatic conditions.

The final chapter of the book reviews evaluating the economics of integrated weed management. Chapter 15 first looks at the various approaches to economic evaluation, then moves on to provide a case study on the economic performance of IWM for winter wheat production in Denmark. It focuses on comparing current weed management practices using crop rotations with alternative IWM strategies. The chapter compares the economics of different IWM strategies and describes the different approaches that can be used to assess the economics of IWM.

Preface

Weeds are ubiquitous and cause substantial yield and quality losses across all arable and horticultural systems and are thus a major concern to farmers. In many countries, weeds outnumber pests and disease in terms of potential impact on crop production. Some weeds are toxic and their presence in grassland may be a threat to grazing animals. Weeds also creates problems in recreational areas and the pollen of some weeds can cause allergenic reactions in humans.

Since the introduction of organic chemical herbicides shortly after World War 2, farmers in the developed world have relied heavily on the use of chemical herbicides. Mechanical weed control and cultural practices, which can prevent or reduce weed infestation, and which were widely practiced before the introduction of chemical herbicides, were given up. This change was perhaps most clearly reflected in the adoption of less diverse crop rotations. The blanket use of chemical herbicides became a standard practice in many crops even when the use of insecticides and fungicides against pests and disease was, at least partly, based on reaching certain thresholds for use and more targeted application. One reason why many farmers have been unwilling to skip the use of herbicides, even in fields with low weed infestations, is the long-term implications of surviving weeds on the soil seed bank.

In recent years, the intensive use of chemical pesticides has come under increased scrutiny and interest in alternative control measures has increased. In conventional farming, this renewed interest in non-chemical weed control measures has very much been triggered by the steadily increasing cases of herbicide resistance. At the same time, particularly in the EU, pesticide legislation has been tightened, and the criteria required for a pesticide to receive authorization have become stricter than ever. This is expected to lead to a reduction in the number of pesticides (including herbicides) available to EU farmers. The 2009 Sustainable Use of Pesticides Directive (SUD) sought to further reduce reliance on chemical control in favour of integrated pest management (IPM). Recently, the EU agreed on the 'Green Deal', at the heart of which is the Farm to Fork strategy that includes a 50% reduction in the use of pesticides by 2030.

One of the requirements of the SUD is that EU farmers should adopt IPM and follow the eight principles laid out in one of its annexes. IPM in entomology can be traced back to the 1900s while integrated weed management (IWM) is more recent. The backbone of IWM is more diverse crop rotations, requiring the trend towards less diverse crop rotation associated with the introduction of effective chemical herbicides to be reversed. Furthermore, non-chemical methods and biological control methods need to be developed, optimized and implemented to reduce the reliance on herbicides. However, most non-chemical tools are less effective or less reliable than herbicides. They cannot be considered stand-alone methods but need to be combined with other methods to provide an IWM strategy, making effective weed control more complex. The high level of complexity of IWM partly explains why it has not received the same attention as integrated management of pests and diseases. On the other hand, if the complexity of IWM can be resolved successfully, it could inspire others to proceed and develop integrated crop management solutions, which should be the final goal.

The present book addresses some of the issues that need to be resolved to reach this goal. This book complements the book 'Integrated weed management for sustainable agriculture' edited by Prof. Bob Zimdahl but also present results from the EU Horizon 2020 project 'IWMPRAISE', the first EU research and innovation project focusing solely on IWM. Hopefully the book will be of inspiration to the reader and promote both IWM research and uptake by end-users.

Acknowledgements

Cover image acknowledgements

- Top left: Frank Poulsen, Vision Weeding, Denmark.
- Top right: Urs Schaffner, CABI, Switzerland.
- Bottom left: Federico Leoni, Scuola Superiore Sant'Anna (SSSA), Italy.
- Bottom right: Maurizio Sattin, CNR, Italy.

Part 1

Weed ecology

Chapter 1

Advances in understanding the contribution of weeds to the functioning of agroecosystems

Sandrine Petit, Séverin Yvoz, Alexandre Ploteau, Camille Zuccolo and Stéphane Cordeau
Agroécologie, AgroSup Dijon, INRAE, Univ. Bourgogne, Univ. Bourgogne Franche-Comté, Dijon, France

1 Introduction

Weeds are an essential component of the agroecosystem. They are one of the main biotic factors limiting crop productivity (Oerke, 2006), as they compete with the crop for sunlight, water and nutrients (Bastiaans et al., 2000). Their primary producer status also places them at the base of the agroecosystem food web (Pocock et al., 2012). The vast array of interactions weeds have with diverse biotic components found in cultivated fields can modulate ecological processes occurring above and below the ground in the agroecosystem (Marshall et al., 2003; Petit et al., 2011). As such, weeds are part of the functional biodiversity, defined as the biotic components that stimulate the ecological processes driving the agroecosystem and that provide ecosystem services (Blaix et al., 2018).

Research describing the functional role of plants in driving ecological processes is well developed in many ecosystems (e.g. grasslands, see Manning et al., 2015), but it is relatively recent in arable ecosystems where the focus has mostly been on processes underpinning food production (Moonen and Barberi, 2008; Martin and Isaac, 2015). However, over the last two decades, a number of studies have attempted to quantify the contribution of individual weed species and weed communities to a wide range of processes. One of the

http://dx.doi.org/10.19103/AS.2021.0098.01

rationales was to assess the potential ecological consequences of the general decline in weed diversity observed in many parts of the world (Storkey and Neve, 2018). Another objective was to enhance our capacity to identify farming management strategies that can ensure crop productivity and enhance weed biodiversity and associated ecological processes, while being economically sustainable (Petit et al., 2015; Adeux et al., 2019a).

Weeds are primarily considered as pests (e.g. Shennan, 2008) and the outcome of weed-crop competition has been the topic of numerous studies and syntheses (Oerke, 2006). Losses in crop yield due to weeds are highly variable and affected by (i) the characteristics of the crop and of the species composing the weed community (Adeux et al., 2019b), (ii) the environmental conditions and crop management (Milberg and Hallgren, 2004), and (iii) the methodological approach implemented to relate weeds to crop yield (Colbach et al., 2020). The contribution of weeds to other agroecosystem services has received much less attention, although their role as trophic resource providers has been highlighted early on (Palmer and Maurer, 1997; Norris and Kogan, 2000; Marshall et al., 2003). In a recent review, Blaix et al. (2018) identified 129 studies describing weed contribution to ecological processes underpinning regulation services. Weeds were found to contribute to nutrient cycling and were shown to improve the soil's physical properties. The review highlighted knowledge gaps concerning the benefits of weeds for crop pollination and natural pest control. In the latter case, many studies only provided evidence that the presence of weeds increases the abundance or diversity of natural pest enemies, with no quantification of the positive feedback on crop yield.

Several key issues need to be addressed in order to improve our ability to predict the potential services and disservices provided by weeds and to identify farming management strategies that could reconcile crop productivity and the provision of regulation services. We need to better understand the role of weed diversity in the functioning of the agroecosystem. Some advances are also required in the development of functional approaches, that is, identifying key functional traits and accounting for their intraspecific variability. There is also an urgent need to implement functional approaches linking farming management to weed traits and to the multiple functions provided by weeds.

2 How key issues of weeds are addressed

2.1 The role of weed diversity

Agricultural intensification, including increased use of tillage, fertilisers and herbicides, on top of the simplification of crop rotations, has led to a widespread decline of weed diversity in many parts of the world over the last decades (e.g. Sutcliffe and Kay, 2000; Fried et al., 2009; Cirujeda et al., 2011). Although field edges (Fig. 1) still harbour higher weed diversity than field interiors (e.g. Fried

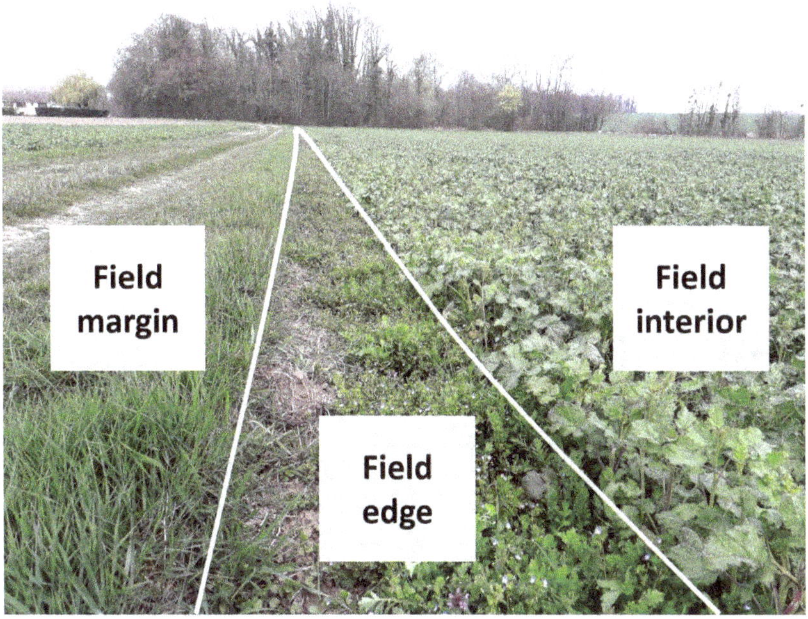

Figure 1 Field edges and field interiors harbour different weed communities.

et al., 2009), this loss of weed diversity is a concern, because it threatens the delivery of multiple functions and services in agroecosystems. The consequences of weed decline on higher trophic levels are quite easy to grasp. For example, at the national scale in UK, the decline in the population of farmland birds was partly explained by a reduction in the frequency and cover of bird food plants in arable fields (Smart et al., 2000). Similarly, the decline of bumblebee forage plants at a national scale over the last decades was identified as the likely principal cause of decline in bumblebee species across the UK (Carvell et al., 2006).

It is also increasingly suggested that in-field weed diversity could alleviate weed-crop competition, notably because it could protect the weed community from being dominated by a few highly competitive and/or herbicide-resistant weeds (Storkey and Neve, 2018). The idea that a diverse weed community will be less competitive is supported by several studies (Poggio and Ghersa, 2011; Cierjacks et al., 2016). More recently, through a detailed analysis of the effect of weed communities on several components of crop yield in a multi-year and multi-site field experiment, Adeux et al. (2019b) demonstrated that high levels of weed diversity were always associated with low weed biomass and reduced interference with the crop. The authors observed a positive relationship between the evenness of weed communities (evenness represents the similarity of contribution of the different weed species to the community and ranges from 0 to 1, a value of 1 meaning that all species in the community have an equal contribution) and crop productivity at all the critical crop stages, that is, stem elongation, heading, grain filling and maturity (Fig. 2). Besides the effect of weed diversity/evenness, the composition of weed communities was also a main factor explaining the variations in the degree of interference with the crop, with higher yield losses when competitive trait values were high at the community level.

2.2 Adopting weed functional approaches

Approaches based on functional traits have allowed a shift in perspective that better reflects the ecological processes that drive weed communities. Similarly, functional trait diversity, rather than the diversity of species *per se*, is the dimension of biodiversity most directly related to ecosystem functioning. Relevant functional traits can inform our understanding of plant responses to environmental and management factors (response traits). They can also have an effect on ecosystem processes underlying ecosystem service delivery (effect traits). Trait-based approaches have been widely applied in semi-natural ecosystems, yet their application to agriculture could help better identify the mechanisms underlying the role of agrobiodiversity in providing services. In agricultural systems, research on effect traits has initially focused on grasslands (e.g. see Manning et al., 2015). Lately, much effort has been devoted to arable systems and the identification of weed traits that are key for processes underlying the provision of agroecosystem services (Navas, 2012; Gaba et al., 2017; Cordeau et al., 2017). In parallel, weed mean trait values have become increasingly accessible in databases such as TRY (Kättge et al., 2011), LEDA Traitbase (Kleyer et al., 2008) or BiolFlor (Klotz et al., 2002).

Functional approaches accounting for the pattern of weed productivity and weed competitive ability and the resulting impact on crop yield have been the focus of several studies (Storkey, 2006; Adeux et al., 2019b). A low

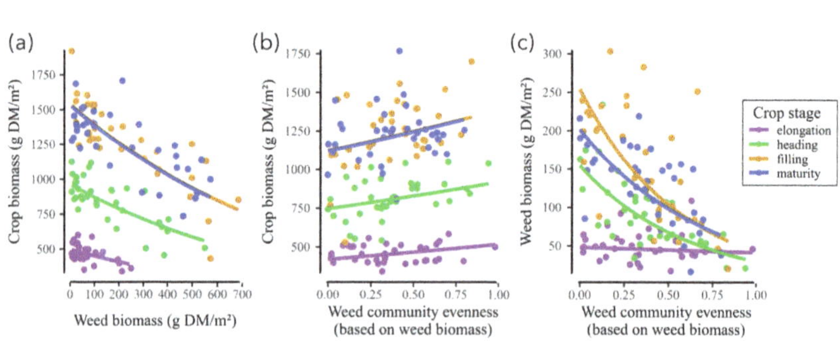

Figure 2 Relationships between (a) weed and crop dry biomass (b) weed community evenness (calculated from biomass data and crop biomass) and (c) weed community evenness (calculated from biomass data) and weed biomass at four crop stages in un-weeded winter cereals (Source: Adeux et al., 2019b).

competitive ability with the crop, a limited seed production capacity and short seed longevity are key factors, which, when combined, ensure suitable weed management over the long run in cropped fields (Table 1). Functional traits underlying the competitive ability of crops and weeds are several (Gunton et al., 2011), as they relate to different processes, for example, seed size, early growth rate, soil resource uptake and light interception. A comprehensive review can be found in Gaba et al. (2017).

Weed functional effect traits can also help elucidate the complex interactions between weeds and other trophic levels that underlie the delivery of services such as pollination or natural pest control (Table 1).

Here, analyses of large datasets composed of animal and plant records, closely matched in time and space, can provide valuable insights. For example, Brooks et al. (2012) established robust links between the functional traits of weeds and those of seed-eating carabid taxa and their trophic interactions. Autumn-germinating and small-seeded weeds were associated with small and spring-breeding carabids more specialised in seed feeding, whereas spring-germinating and large-seeded weeds were associated with a range of large autumn-breeding omnivorous carabids. Using a comparable approach, Storkey et al. (2013) modelled the variations in the abundance of phytophagous

Table 1 Examples of weed functional effect traits related to certain organs (column) that are relevant to assess the contribution of weeds to (dis)services (row), that is, crop production (yield loss), pollination (resource provision to pollinators) and pest control (resource provision to natural enemies such as parasitoids and seed-eating birds)

		Plant organs		
		Stems and leaves	Flowers	Seeds
Ecosystem service	Crop production	Relative growth rate Specific leaf area Leaf dry matter content Nitrophily Height		Size Quantity Longevity
	Pollination		Colour UV reflectance Size and symmetry Corolla form and size Nectar: quantity, sugar type(s) Pollen: quantity, protein content	
	Natural pest control	Extra-floral nectar	Colour Type of inflorescence Corolla depth Nectar: quantity, sugar type(s)	Size Lipid content Energy content

invertebrate groups using key functional traits of the co-occurring weed species. They found that more ruderal communities supported proportionally more invertebrates. Similarly, adult farmland birds, complementing their diet with seeds in autumn and winter when arthropods become scarce, express preferences for some weed species based on seed traits such as seed size, seed energy content (Gibbons et al., 2006), lipid content (Greig-Smith and Wilson, 1985) or protein content (Valera et al., 2005).

Alternatively, the identification of relevant functional effect traits can be derived from in-depth knowledge of the ecological process at play. Gardarin et al. (2018) provide a comprehensive review of the contribution of a trait-based approach to understand plant–arthropod interactions and improving the service of natural pest control. They highlighted the importance of some key traits related to resource provision. For example, the longevity of parasitoids, and thus their efficiency as biocontrol agents, can be enhanced by plant floral nectar in the field (Heimpel and Jervis, 2005). This resource must be abundant, accessible (matching of nectar depth in the corolla to the size of arthropod mouthparts) and available for long periods throughout the year, to increase the growth rates of specialists during pest outbreaks (Welch and Harwood, 2014). Similarly, Petit and Bohan (2017) provide a detailed description of factors and weed species characteristics that affect weed seed predation by carabids, which are important biocontrol agents in arable fields. Carabid preferences are linked to the seed lipid content (Gaba et al., 2019), morphology and coat thickness (Lundgren and Rosentrater, 2007). Seed size can also be a limiting factor. For example, a seed is rarely consumed by carabids if it is heavier than 3 mg (Petit et al., 2014). The plant traits that affect their interaction with pollinators are quite similar to those affecting other guilds of arthropods using floral resources, namely traits affecting flower attractiveness (flower colour (Backhaus, 1992), UV reflectance, flower symmetry and flower size). The form and the size of the corolla determine its accessibility. The quantity and the quality of the floral rewards in terms of pollen (Hass et al., 2019) and/or nectar (Pamminger et al., 2019) are also important factors.

Such knowledge is highly valuable to assess the contribution of individual species and weed communities to different services. For example, Mézière et al. (2015) developed an indicator estimating the contribution of weed communities to the maintenance of generalist insect predators that was based on traits, that is, seed size and seed lipid content of individual weed species (Table 1), and used outputs of the FlorSys model, namely the number of seeds produced per plant, weighted by the relative abundance within the community. Following a similar approach, Ricou et al. (2014) estimated the value of individual weed species for different types of pollinators using available knowledge, describing not only the characteristics of flowers affecting their attractiveness but also the quantity and quality of resources available and their accessibility to pollinators (Table 1).

2.3 Accounting for intraspecific variation in trait values

Trait values available in databases are mostly mean trait values, that is, available trait values are averaged over multiple populations and/or over a wide range of habitats, with a sampling effort that is not necessarily representative of dominant land uses. Trait values can, however, vary strongly according to the growing conditions of individual plants. These variations can have important implications for the outcome of ecological interactions (Bolnick et al., 2011). Because it is common to collate data derived from a large number of studies, it is sometimes possible to broadly assess intraspecific trait variability in existing databases, although the factors that might explain the variability are not apparent. For example, Kazakou et al. (2014) found that for most traits, interspecific variability was higher than intraspecific variability, and species ranking was conserved across different datasets and spatial scales. However, they also detected important differential responses in terms of intraspecific trait variability, depending on the trait examined. The specific leaf area (SLA), leaf dry matter content (LDMC), seed mass, seed N concentration and onset of flowering were rather stable traits, whereas leaf chemical traits and reproductive plant height were more flexible traits.

A number of empirical studies have highlighted the impact of the growing environment on weed traits that are key to the provision of ecological functions. For example, Wulff et al. (1999) demonstrated that the seed germination rate of *Chenopodium album* varied significantly in response to the maternal and the grand-maternal nutrient environment. The growing environment will thus impact not only the potential harmfulness of this weed to the crop (as an early germination date means rapid growth and thus better ability to compete with the crop at a later stage) but also the period over which non-germinated seeds remain available to seed predators. SLA, which is marker of plant resource-use strategy, was also shown to differ between plants located in field edges and those located in field interiors for common weed species such as *Fallopia convolvulus*, *Veronica hederifolia*, *Veronica persica* and *Viola arvensis* (Perronne et al., 2014). This high intraspecific variability in SLA values was also demonstrated to respond to the crop type, possibly due to differences in crop canopy closure, as SLA is sensitive to this factor (Borgy et al., 2016).

The impact of weed plant growing conditions on weed functions, and notably on their capacity to provide floral and seed resources to other organisms, was recently examined (Yvoz et al., 2020a). The authors conducted a large-scale survey of weed traits to estimate the intraspecific variation in floral and seed provision for 30 weed species in response to the within-field location (field edge vs. field interior) and the crop grown in the field (six crop types). They found that the flowering and fruiting success of most species were higher in field edges than in field interiors and lower in cereal crops than in other crops, as was the amount of flowers produced (Fig. 3). Moreover, they showed that weeds flowered and fruited earlier and that the flowering period was longer in field edges than in field interiors. Hence, this study demonstrated that within individual weed species, plants growing at field edges potentially contributed more to pest control and pollination services than their field-interior counterparts.

These examples illustrated that the relationships between traits and ecological functions are context-dependent and will vary with management practices. Linking farming management to weed (dis)services through functional approaches thus requires integrating intraspecific variations in trait values. Many authors recommended using trait values derived from measures conducted under conditions that are consistent with the research context rather than using mean trait values extracted from global databases (Lavorel and Garnier, 2002; Cornelissen et al., 2003). This situation calls for conducting ambitious campaigns of field measurement of trait values under contrasting conditions (Kazakou et al., 2014; Wood et al., 2015).

2.4 Implementing the response-effect model

Functional traits respond to environmental and management factors (response traits) and affect ecosystem processes underlying ecosystem service delivery (effect traits). Trait-based approaches thus provide a robust framework for evaluating and predicting the impact of farming management on services provided by weeds through the original 'response-effect model' trait linkages of plants. Such linkages are built on the overlap or correlations between the

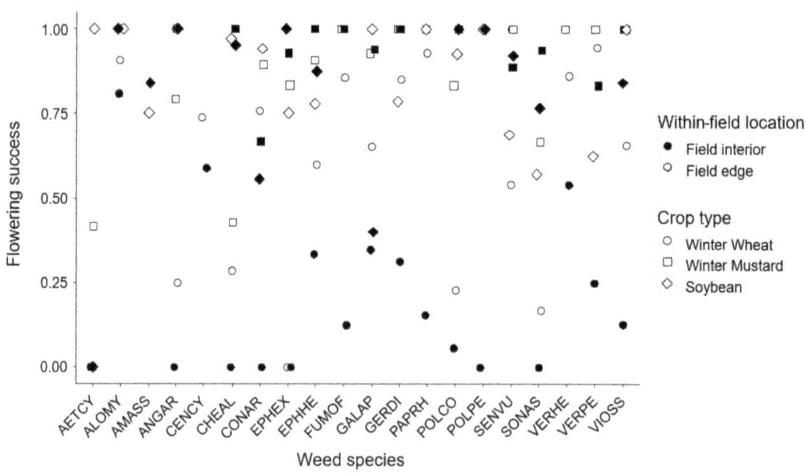

Figure 3 Flowering success of 20 weed species in different combinations of crop type and within-field location (Source: Yvoz et al., 2020a).

'response traits' that determine how the functional diversity of a community responds to an environmental factor and 'effect traits' that determine service delivery. Such extensions of the response-effect model to capture indirect effects of environmental change on ecosystem services have been developed in grasslands (Lavorel et al., 2013). This framework could further be used to develop particular trait-based management strategies that can be implemented in farming systems to increase multiple ecosystem services, as well as to manage trade-offs among ecosystem services in agriculture (Wood et al., 2015). This response-effect framework has been tested and successfully applied in grasslands to understand how the different intensity levels of land-use impacted ecosystem services through changes in vegetation (Gross et al., 2008; Minden and Kleyer, 2011). In cultivated systems, Kazakou et al. (2016) developed and tested such a model to assess how soil management practices in vineyards impacted the functional characteristics of weed communities and how, in turn, weeds affected grapevine production. Although there was a high variability among study sites, their results suggested that intensive management practices, such as tillage, filter for weed species traits favouring high growth rates, rapid nutrient mobilization and rapid decomposition and mineralization which increased resource availability and thus improved grapevine production.

Further application of the framework to weeds appears promising. There is a substantial overlap between competitive traits that can be considered both as effect traits and as response traits (Andrew et al., 2015). Regarding weed interactions with other taxa, Storkey et al. (2013) demonstrated a strong overlap between the weed traits that respond to disturbance and those that affect the abundance of phytophagous invertebrates and the diet of farmland birds. They showed that species with high SLA, a classic plant response trait to habitat disturbance, harboured a higher diversity of invertebrates. In another study, Mézière et al. (2015) used a trait-based approach to model the dynamics of weed communities in response to contrasting cropping systems and to predict the impact of the simulated weed communities on a set of weed (dis)services. In this modelling study, trait overlap included, for example, seed lipid content, which affected both the response of weeds to cropping systems (rate/speed of weed germination) and the value of weed species for insects.

2.5 Towards assessments of weed multi-functionality

Because it is now better established that weeds can potentially contribute to a large number of ecological functions (Blaix et al., 2018), some studies have attempted to quantify multiple services delivered by weed communities. The added value of such approaches is not only the capacity to analyse potential synergies or antagonisms between services but also to identify weed communities that would deliver interesting trade-offs between services. However, such studies remain scarce, to date. We present here published results as well as a case study.

Gaba et al. (2020) conducted in-field measurements of several indicators for three functions (pollination, pest control and soil fertility). They covered more than 180 fields and analysed how these functions were related to observed weed communities. They found that weed abundance and/or weed richness were strongly correlated with individual functions, that is, weed diversity was a strong contributor to ecosystem multifunctionality. Although the number of fields sampled was impressive, this approach remains correlative and the causality between weed communities and the diverse functions estimated remains to be tested.

Mézière et al. (2015) adopted quite a different approach, which aimed at identifying farming management strategies that would deliver interesting trade-offs between weed harmfulness to crops and the positive contribution of weeds to pest control and pollination. They simulated contrasting farming management strategies using the FlorSys model and derived the 'functions' from the resulting weed communities using indicators. Despite a general tendency of antagonism between biodiversity and production, some simulated weed communities provided substantial biodiversity services and caused no negative impacts on crop production.

2.5.1 Case study: linking farming management strategies to weed services

The identification of farming management strategies that would deliver weed flora to minimise disservices and bring added value in terms of biodiversity and services is one of the topics of the H2020 IWMPRAISE project. The analysis was conducted at the scale of a small territory in the arable region south of Dijon, Burgundy, France. It was built on a long-term dataset describing the annual management practices since 2004 in 97 fields on 22 farms where INRAE has been recording the weed flora for six consecutive years (2008–2013).

The analysis was developed according to a general framework that describes the links between long-term agricultural management practices implemented by each farmer on his fields, the resulting weed flora and the associated (dis)services that weeds are expected to contribute (Fig. 4). One assumption is that the weed flora in the interior of arable fields would differ from the weed flora present at the field edge (i) in their response to farming management and (ii) in their contribution to the provision of (dis)services. The implementation of the framework involved four successive steps, which are described as follows.

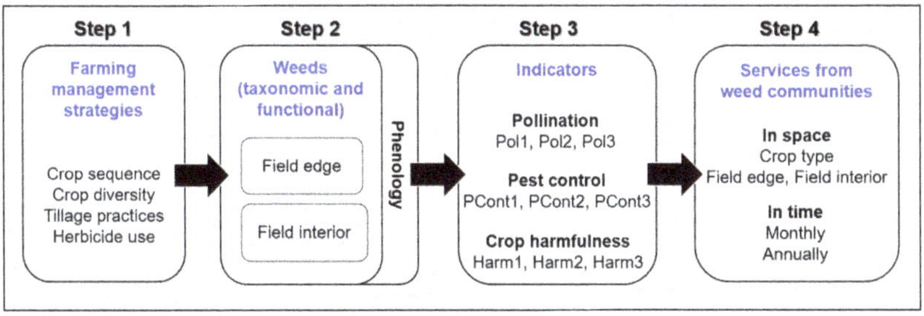

Figure 4 Framework linking farming management strategies to the contribution of weeds to (dis)services in four consecutive steps.

Step 1: Identifying farming management strategies

The first step in implementing the framework consisted in translating the detailed information on agricultural operations, collected through farmers' interviews over the last 15 years within each field of the study area, into farming management strategies. Some indicators of farming intensity (e.g. treatment frequency index) were converted into ratios that translated the value of the practice for a specific field compared to the mean value of all the fields cultivated with the same crop in the same year in the study area. Based on 14 indicators, eight contrasted farming strategies were identified within the study area. These strategies differed primarily in terms of crop diversity within the crop sequence, and then in terms of tillage and herbicide regime (for a full report, see Yvoz et al., 2020b).

Step 2: Quantifying the effect of farming management strategies on the weed flora

Overall, 155 weed taxa were identified in the study area during the 2008–2013 period and among those, 46 taxa were only found in field edges, supporting the assumption that these habitats provide refuge for weed species in intensive agricultural landscapes (Fried et al., 2009). Mean annual weed species richness was significantly higher in field edges than in field interiors, regardless of the crop type. The taxonomic and functional weed composition in field interiors and in field edges were clearly different. Field edges harboured species with ecological strategies associated with field interiors, such as ruderal (i.e. short life cycle with rapid growth in habitat with high levels of nutrient and low levels of light, producing a lot of small seeds) and competitor (low SLA and LDMC, but high seed mass, height, Ellenberg N and Ellenberg L values). However, field edges also harboured species with a conservative strategy (slow growth rate with persistent leaves) which could be explained by a spill-over from the close field margin. The number of species within each strategy was consistently higher in the field edges, suggesting a 'refugia role' of field edges for field interior species. Farming management impacted field-edge communities, though to a lesser extent, than field-interior communities, whereas the functional difference between the two habitats was less marked when management intensity was lower.

Step 3: Developing service indicators for weed species in different growing conditions

Six indicators of ecosystem services were developed that accounted for the contribution of weeds to the maintenance of pollinators (Pol1 for bees, Pol2 for bumblebees and Pol3 for hoverflies) and to the maintenance of natural pest enemies (PCont1 for birds, PCont2 for carabids and PCont3 for parasitoids). Three additional indicators of ecosystem disservices accounted for weed harmfulness (Harm1 for competition to crop, Harm2 for harvesting problems and Harm3 for soil seedbank built up).

The contribution of each weed species to different (dis)services depended on their functional characteristics (functional effect traits), as well as on their success and phenology, which varied according to the growing conditions of each plant. Such response to growing conditions was measured in the study area for 30 weed species (see Section 4 and Yvoz et al., 2020a). In order to be able to extrapolate to the 155 species recorded in the study area, the phenology and the flower and seed production of missing species were imputed by using the most comparable species through a clustering methodology using functional traits.

The computation of these nine indicators revealed that some species in some growing conditions were important contributors to pollination and pest control but did not cause much harmfulness to the crop (Table 2). Among those, *Cyanus segetum* offered an interesting bundle of (dis)services in many crop types, although there was a risk of seed bank build-up in winter oilseed rape. Several species of the *Geranium* genus were also among the top contributors.

Step 4: Assessing the contribution of weed communities to multiple services

The contribution of each weed species in each growing condition was used to estimate the nine indicators for the weed communities recorded in the study area between 2008 and 2013. The weed community's contribution to services was the sum of contributions of individual species weighted by its abundance in the community. Indicators related to pest control and pollination could be estimated several times over the crop growing season by using data collected every two weeks during the phenological survey (Yvoz et al., 2020a). Indicators for weed harmfulness to the crop were evaluated yearly.

Table 2 Ranking of the 20 combinations of weed species x growing conditions that provide the most interesting combinations of (dis)services (i.e. high services and low harmfulness)

Species	Growing conditions	Pol1	Pol2	Pol3	PCont1	PCont2	PCont3	Harm1	Harm2	Harm3
Cyanus segetum	S (edge)	0.018	0.019	0.075	0.660	0.660	0.073	0.284	0.001	0.001
Geranium columbinum	WO (interior)	0.108	0.127	0.127	0.080	0.089	0.046	0.403	0.001	0.596
Erodium cicutarium	WO (edge)	0.086	0.107	0.091	0.142	0.130	0.024	0.395	0.001	0.671
Kickxia spuria	WM (edge)	0.219	0.105	0.578	0.660	0.660	0.315	0.132	0.001	0.001
Geranium dissectum	WO (interior)	0.081	0.102	0.438	0.044	0.046	0.023	0.329	0.001	0.718
E. cicutarium	WM (edge)	0.064	0.082	0.062	0.086	0.082	0.016	0.482	0.001	0.740
Geranium rotundifolium	WO (interior)	0.130	0.150	0.152	0.099	0.096	0.044	0.430	0.001	0.669
Taraxacum officinale	WO (interior)	0.103	0.177	0.051	0.239	0.224	0.299	0.162	0.001	0.741
Veronica hederifolia	WO (edge)	0.075	0.081	0.104	0.091	0.101	0.058	0.431	0.001	0.782
Echium vulgare	SB (edge)	0.034	0.025	0.443	0.077	0.083	0.120	0.435	0.001	0.644
E. cicutarium	SB (edge)	0.143	0.161	0.161	0.236	0.222	0.068	0.406	0.001	0.574
C. segetum	WO (interior)	0.005	0.008	0.033	0.009	0.008	0.005	0.557	0.001	0.912
C. segetum	WO (edge)	0.003	0.005	0.030	0.005	0.003	0.002	0.557	0.001	0.923
Thlaspi perfoliatum	WM (edge)	0.248	0.243	0.183	0.115	0.109	0.097	0.230	0.001	0.793
K. spuria	WO (interior)	0.310	0.162	0.654	0.660	0.660	0.398	0.102	0.001	0.001
Chaenorhinum minus	WM (edge)	0.343	0.274	0.381	0.660	0.660	0.460	0.148	0.001	0.001
T. perfoliatum	WO (edge)	0.298	0.286	0.229	0.171	0.161	0.130	0.185	0.001	0.734
Potentilla reptans	WW (edge)	0.257	0.603	0.173	0.660	0.660	0.159	0.314	0.001	0.001
T. officinale	WO (edge)	0.161	0.245	0.084	0.302	0.293	0.369	0.168	0.001	0.677
P. reptans	WB (edge)	0.254	0.602	0.169	0.660	0.660	0.155	0.347	0.001	0.001

For each service, values represent the relative contribution, ranging from 0 (best) to 1 (worst).
WO, winter oilseed rape; WM, winter mustard; WB, winter barley; WW, winter wheat; SB, spring barley.

The dynamics of the indicators of pest control and pollination during the course of the crop growing season in the field edge and the field interior are presented in Fig. 5. As expected, the contribution of weeds occurred later in the season in spring barley and in soybean than in winter crops. The important finding here has more to do with the estimated differences in the timing of the contribution of weeds of field edges and of field interiors within the same crop type. In many instances, field-edge weed communities significantly increased the duration of the contribution to services, either by contributing for a longer period (PCont1 in winter mustard, Pol1 in winter barley) and/or by contributing before (Pol2, Pol3 in soybean) or after (Pol3, Pcont3 in winter oilseed rape) weed communities located in the field interior. This result suggests a complementarity between field-edge and field-interior weed communities in the functioning of the agroecosystem.

The indicator profile of weed communities at the annual scale is presented in Fig. 6, per location within the field and per crop type. Overall, the contribution of field-edge weed communities was interesting, with high values for pollination and pest control, but also expressed high values for harmfulness to the crop. For all services considered, indicators did not vary much among crop types, although the direct competition with the crop (Harm1) was slightly lower in winter oilseed rape and winter mustard than in the other crops. Conversely, the profile of field-interior weed communities varied between crop types. Cereals presented the least interesting profile, with low values for pollination and pest control, and low harmfulness to the crop. Soybean had the highest contribution to pollination and pest control, but expressed the highest contribution to harmfulness to the crop. Besides, in this crop type, weed contribution of communities from field interior was quite similar to the profile of field edge communities. Winter oilseed rape and winter mustard expressed the most interesting field interior profile with intermediate contribution to pollination and pest control, but low contribution to harmfulness (quite similar to cereal crops).

3 Conclusion

There is a critical need to ensure future food security, and increasing emphasis will be placed on greater crop productivity while reducing environmental impact and the reliance on chemical use in modern agriculture. This context, in addition to policy-driven changes in herbicide use, is fostering the emergence of farming management strategies that are less reliant on herbicides and follow the principles of integrated weed management (Swanton and Weise, 1991), including the biological control of weeds (Petit et al., 2018). This move away from farming relying solely on herbicides will trigger changes in the diversity and the composition of weed communities. These changes can be an opportunity to enhance the provision of weed functions that are beneficial for agriculture, such as the natural control of crop pests or crop pollination. They can also potentially represent a threat to crop production, through enhanced crop yield loss. For farmers to adopt alternatives to herbicides to manage weeds, it will be necessary to demonstrate that these management options can provide enhanced, stable and resilient ecological functions and services, with little or no additional risk to either crop yield or farm productivity in the long run. However, to date, there is little available evidence that demonstrates the reality of such a win-win situation. Moreover, the management principles that would result in weed communities offering interesting trade-offs between services and disservices remain largely unexplored. Strengthening our capacity to assess weed multi-functionality over a large range of agronomic contexts is thus of primary importance if we are to identify management options that best reconcile the positive and negative aspects of weeds.

Research can contribute to this overarching goal. The role of weed diversity in the functioning of agroecosystems is being revisited. Functional approaches are offering a robust framework to quantify and understand the underlying mechanisms that link farming management to weed functions. Recent studies have provided strong evidence that weeds are truly multi-functional, and this considerably enlarges the scope of future research, far beyond the sole topic

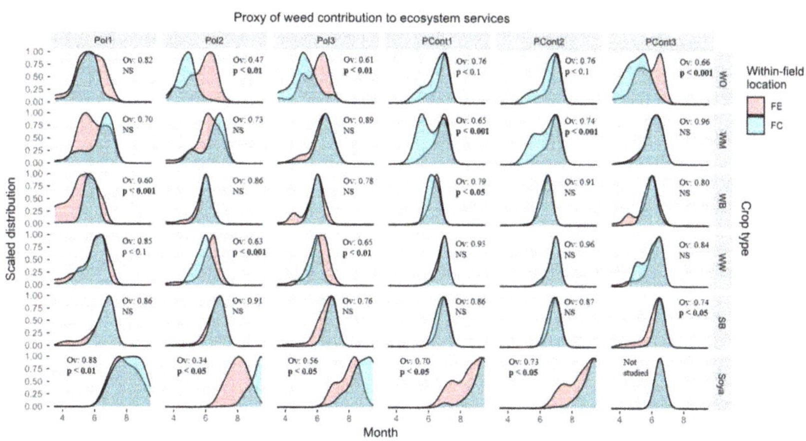

Figure 5 Distribution of the six indicators of services over time in the field edge (FE) and the field interior (FC) and per crop type (WO, winter oilseed rape; WM, winter mustard; WB, winter barley; WW, winter wheat; SB, spring barley). Density distributions are scaled per within-field location. Overlap values (*OverlapTrue* function) and test against a null hypothesis using a one-tailed direct test of significance for non-random distribution (10 000 randomizations).

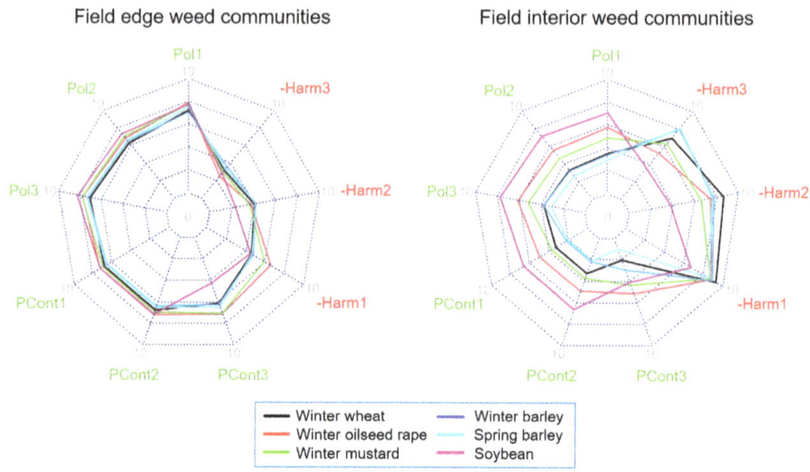

Figure 6 The mean contribution of the six main crop types to the nine indicators of (dis) services. The nine indicators were rescale between 0 and 10 and Harm1, Harm2 and Harm3 were reversed for graphical purpose (10 means low harmfulness to the crop).

of weed-crop interaction. In recent years, the understanding of the role of key responses and effect functional traits in weed communities and how they relate to specific functions has significantly progressed. We have started to uncover the sometimes-predominant role of intraspecific trait variability, its cause and its effects. The implementation of the response-effect model linking management to functions via weed functional traits has been initiated, although it remains to be further tested. Research aiming at quantifying weed multi-functionality will necessarily require a combination of approaches, from *in situ* measures of functions and also of weed traits in a wide range of contexts, to the development of indicators of functions and services derived from weed traits and the use of models, notably to predict weed community shifts in response to contrasting farming management strategies. Such a combination of approaches is necessary to generate the knowledge required to guide farming management strategies, by demonstrating that there are interesting weed function trade-offs that can be achieved that are cost-effective and credible for farmers to adopt.

4 Future trends in research

Currently, we know that some management options seem to deliver interesting trade-offs in terms of weed functions and reconcile the negative and positive aspects of weeds. Advancing our understanding of weed functions in response to farming management strategies will require additional research in at least three directions.

First, the assessment of weed functions is often indirect and based on indicators developed using weed functional traits. The development of these indicators is often expert-based and efforts should be devoted to validate these under field situations. For example, Ricou et al. (2014) developed indicators of pollination values for weeds and provided a conclusive validation through in-field measurements of pollinator visits on weed communities. Such an approach remains to be developed to validate other indicators based on weed effect traits, for example, the value of weeds for natural enemies of crop pests such as some parasitoid wasps. Recent advances in the determination of the diet of organisms in agroecosystems through molecular analysis of the gut content of weed consumers (Frei et al., 2019) can certainly be mobilised alongside field surveys to progress on this front. Besides, if weed functional traits underpinning ecological functions are relatively well understood for some ecological functions, they remain poorly documented for others. For example, investigating the impact of weed functional traits on crop quality (grain size, protein, oil and metabolite composition) has been identified as a promising research avenue (Gibson et al., 2017).

Second, weeds contribute to multiple functions, and when studies focus on several weed functions, each function is usually considered as independent from the others. Yet, there are interactions between functions, and agroecosystems shelter networks of services (Bohan et al., 2016). For example, weed seeds removed by seed-eating organisms will no longer be available to others but will also not contribute to the seed bank build-up. More generally, the contribution of weeds to enhanced pollination and pest control services should translate into a positive feedback on crop yield, although the strength of such feedback can be highly variable (Tamburini et al., 2019). These interactions are not necessarily easy to account for in the evaluation of weed multi-functionality, but current progress in the scientific field of ecosystem services assessment could be mobilised in the near future.

Finally, a key issue is the identification of management options that can deliver weed communities offering interesting trade-offs. Here, multiple spatial scales could be considered (Duru et al., 2015). Within farmed fields, available knowledge suggests that field edges are key habitats to maintain weed diversity and strongly contribute to the provision of floral and seed resources for mobile organisms that provide services to agriculture. This calls for revisiting the principles of field-edge management and functional approaches could help anticipate whether a relaxed field-edge management to enhance the provision of resources would lead to in-field weed infestations. At the scale of farmed landscapes, interesting weed trade-offs could be achieved by increasing the total length/area of field edge habitats or by increasing the diversity of crop types and farming management strategies, as such diversification options would enable the expression of complementary weed functions within a small landscape (Colbach et al., 2018).

5 Where to look for further information

A number of key papers presenting reviews or syntheses on the application of plant functional approaches to weeds are provided in this chapter.

The topic of weed functions is studied not only by large organisations devoted to ecology (British Ecological Society; Ecological Society of America) but also by organisations involving agronomists (European Society of Agronomy). Discussions of weed services can be found across a variety of international societies and their meetings. Primary amongst these are the European Weed Research Society--notably the Weeds and Biodiversity Working Group (http://www.ewrs.org). The recent review by Blaix et al. (2018) on weed services originates from this group.

6 References

Adeux, G., Munier-Jolain, N., Meunier, D., Farcy, P., Carlesi, S., Barberi, P. and Cordeau, S. (2019a). Diversified grain-based cropping systems provide long-term weed control while limiting herbicide use and yield losses. *Agron. Sustain. Dev.* 39(4), 42.

Adeux, G., Vieren, E., Carlesi, S., Barberi, P., Munier-Jolain, N. and Cordeau, S. (2019b). Mitigating crop yield losses through weed diversity. *Nat. Sustain.* 2(11), 1018–1026.

Andrew, I. K., Storkey, J. and Sparkes, D. L. (2015). A review of the potential for competitive cereal cultivars as a tool in integrated weed management. *Weed Res.* 55(3), 239–248.

Backhaus, W. (1992). Color vision in honeybees. *Neurosci. Biobehav. Rev.* 16(1), 1–12.

Bastiaans, L., Kropff, M. J., Goudriaan, J. and van Laar, H. H. (2000). Design of weed management systems with a reduced reliance on herbicides poses new challenges and prerequisites for modeling crop-weed interactions. *Field Crops Res.* 67(2), 161–179.

Bohan, D. A. and Quintessence consortium (2016). Networking our way to better ecosystem service provision. *Trends Ecol. Evol.* 31(2), 105–115.

Bolnick, D. I., Amarasekare, P., Araújo, M. S., Bürger, R., Levine, J. M., Novak, M., Rudolf, V. H., Schreiber, S. J., Urban, M. C. and Vasseur, D. A. (2011). Why intraspecific trait variation matters in community ecology. *Trends Ecol. Evol.* 26(4), 183–192.

Borgy, B., Peronne, R., Kohler, C., Grison, A.-L., Amiaud, B. and Gaba, S. (2016). High importance of intraspecific functional variability in weed response. *Weed Res.* 56(2), 102–113.

Blaix, C., Moonen, A. C., Dostatny, D. F., Izquierdo, J., Le Corff, J., Morrison, J., Von Redwitz, C., Schumacher, M. and Westerman, P. R. (2018). Quantification of regulating ecosystem services provided by weeds in annual cropping systems using a systematic map approach. *Weed Res.* 58(3), 151–164.

Brooks, D. R., Storkey, J., Clark, S. J., Firbank, L. G., Petit, S. and Woiwod, I. P. (2012). Trophic links between functional groups of arable plants and beetles are stable at a national scale. *J. Anim. Ecol.* 81(1), 4–13.

Carvell, C., Roy, D. B., Smart, S. M., Pywell, R. F., Preston, C. D. and Goulson, D. (2006). Declines in forage availability for bumblebees at a national scale. *Biol. Cons.* 132(4), 481–489.

Cierjacks, A., Pommeranz, M., Schulz, K. and Almeida-Cortez, J. (2016). Is crop yield related to weed species diversity and biomass in coconut and banana fields of northeastern Brazil? *Agric. Ecosyst. Environ.* 220, 175–183.

Cirujeda, A., Aibar, J. and Zaragoza, C. (2011). Remarkable changes of weed species in Spanish cereal fields from 1976 to 2007. *Agron. Sustain. Dev.* 31(4), 675–688.

Colbach, N., Cordeau, S., Garrido, A., Granger, S., Laughlin, D., Ricci, B., Thomson, F. and Messéan, A. (2018). Landsharing vs landsparing: how to reconcile crop production and biodiversity? A simulation study focusing on weed impacts. *Agric. Ecosyst. Environ.* 251, 203-217.

Colbach, N., Petit, S., Chauvel, B., Deytieux, V., Lechenet, M., Munier-Jolain, N. and Cordeau, S. (2020). The pitfalls of relating weeds, herbicide use and crop yield: don't fall into the trap! a critical review. *Front. Agron.* 2, 615470. Available at: https://www.frontiersin.org/articles/10.3389/fagro.2020.615470/.

Cordeau, S., Ryan, M. R., Bohan, D. A., Reboud, X. and Chauvel, B. (2017). Which traits allow weed species to persist in grass margin strips? *Weed Sci.* 65(3), 381–394.

Cornelissen, J. H. C., Lavorel, S., Garnier, E., Díaz, S., Buchmann, N., Gurvich, D. E., Reich, P. B., Steege, Ht, Morgan, H. D., Heijden, M. G. A. van der, Pausas, J. G. and Poorter, H. (2003). A handbook of protocols for standardised and easy measurement of plant functional traits worldwide. *Aust. J. Bot.* 51(4), 335–380.

Duru, M., Therond, O., Martin, G., Martin-Clouaire, R., Magne, M. A., Justes, E., Journet, E. P., Aubertot, J. N., Savary, S., Bergez, J. E. and Sarthou, J. P. (2015). How to implement biodiversity-based agriculture to enhance ecosystem services: a review. *Agron. Sustain. Dev.*, *35*(4), 1259-1281. https://doi.org/10.1007/s13593-015-0306-1

Frei, B., Guenay, Y., Bohan, D. A., Traugott, M. and Wallinger, C. (2019). Molecular analysis indicates high levels of carabid weed seed consumption in cereal fields across Central Europe. *J. Pest Sci.* 92(3), 935–942.

Fried, G., Petit, S., Dessaint, F. and Reboud, X. (2009). Arable weed decline in Northern France: crop edges as refugia for weed conservation? *Biol. Cons.* 142(1), 238–243.

Gaba, S., Perronne, R., Fried, G., Gardarin, A., Bretagnolle, F., Biju-Duval, L., Colbach, N., Cordeau, S., Fernández-Aparicio, M., Gauvrit, C., Gibot-Leclerc, S., Guillemin, J.-P., Moreau, D., Munier-Jolain, N., Strbik, F. and Reboud, X. (2017). Response and effect traits of arable weeds in agro-ecosystems: a review of current knowledge. *Weed Res.* 57(3), 123–147.

Gaba, S., Deroulers, P., Bretagnolle, F. and Bretagnolle, V. (2019). Lipid content drives weed seed consumption by ground beetles (Coleopterea, Carabidae) within the smallest seeds. *Weed Res.* 59(3), 170–179.

Gaba, S., Cheviron, N., Perrot, T., Piutti, S., Gautier, J.-L. and Bretagnolle, V. (2020). Weeds enhance multifunctionality in arable lands in South-West of France. *Front. Sustain. Food Syst.* 4, 71.10.3389/fsufs.2020.00071.

Gardarin, A., Plantegenest, M., Bischoff, A. and Valantin-Morison, M. (2018). Understanding plant–arthropod interactions in multitrophic communities to improve conservation biological control: useful traits and metrics. *J. Pest Sci.* 91(3), 943–955.

Gibbons, D. W., Bohan, D. A., Rothery, P., Stuart, R. C., Haughton, A. J., Scott, R. J., Wilson, J. D., Perry, J. N., Clark, S. J., Dawson, R. J. G. and Firbank, L. G. (2006). Weed seed resources for birds in fields with contrasting conventional and genetically modified herbicide-tolerant crops. *Proc. Biol. Sci.* 273(1596), 1921–1928.

Gibson, D. J., Young, B. G. and Wood, A. J. (2017). Can weeds enhance profitability? Integrating ecological concepts to address crop-weed competition and yield quality. *J. Ecol.* 105(4), 900–904.

Greig-Smith, P. W. and Wilson, M. F. (1985). Influences of seed size, nutrient composition and phenolic content on the preferences of bullfinches feeding in ash trees. *Oikos* 44(1), 47–54.

Gross, N., Robson, T. M., Lavorel, S., Albert, C., Le Bagousse-Pinguet, Y. and Guillemin, R. (2008). Plant response traits mediate the effects of subalpine grasslands on soil moisture. *New Phytol.* 180(3), 652–662.

Gunton, R. M., Petit, S. and Gaba, S. (2011). Functional traits relating arable weed communities to crop characteristics. *J. Veg. Sci.* 22(3), 541–550.

Hass, A. L., Brachmann, L., Batáry, P., Clough, Y., Behling, H. and Tscharntke, T. (2019). Maize-dominated landscapes reduce bumblebee colony growth through pollen diversity loss. *J. Appl. Ecol.* 56(2), 294–304.

Heimpel, G. E. and Jervis, M. A. (2005). Does floral nectar improve biological control by parasitoids? In: Wäckers, F., van Rijn, P., Bruin, J. (Eds) *Plant-Provided Food For Carnivorous Insects: A Protective Mutualism and Its Applications*. Cambridge: Cambridge University Press, pp. 267–304.

Kattge, J., Diaz, S., Lavorel, S., Prentice, I. C., Leadley, P., Bönisch, G., Garnier, E., Westoby, M., Reich, P. B., Wright, I. J., Cornelissen, J. H. C., Violle, C., Harrison, S. P., Van Bodegom, P. M., Reichstein, M., Enquist, B. J., Soudzilovskaia, N. A., Ackerly, D. D., Anand, M., Atkin, O., Bahn, M., Baker, T. R., Baldocchi, D., Bekker, R., Blanco, C. C., Blonder, B., Bond, W. J., Bradstock, R., Bunker, D. E., Casanoves, F., Cavender-bares, J., Chambers, J. Q., Chapin, F. S., Chave, J., Coomes, D., Cornwell, W. K., Craine, J. M., Dobrin, B. H., Duarte, L., Durka, W., Elser, J., Esser, G., Estiarte, M., Fagan, W. F., Fang, J., Fernández-méndez, F., Fidelis, A., Finegan, B., Flores, O., Ford, H., Frank, D., Freschet, G. T., Fyllas, N. M., Gallagher, R. V., Green, W. A., Gutierrez, A. G., Hickler, T., Higgins, S. I., Hodgson, J. G., Jalili, A., Jansen, S., Joly, C. A., Kerkhoff, A. J., Kirkup, D., Kitajima, K., Kleyer, M., Klotz, S., Knops, J. M. H., Kramer, K., Kühn, I., Kurokawa, H., Laughlin, D., Lee, T. D., Leishman, M., Lens, F., Lenz, T., Lewis, S. L., Lloyd, J., Llusià, J., Louault, F., Ma, S., Mahecha, M. D., Manning, P., Massad, T., Medlyn, B. E., Messier, J., Moles, A. T., Müller, S. C., Nadrowski, K., Naeem, S., Niinemets, Ü, Nöllert, S., Nüske, A., Ogaya, R., Oleksyn, J., Onipchenko, V. G., Onoda, Y., Ordoñez, J., Overbeck, G., Ozinga, W. A., Patiño, S., Paula, S., Pausas, J. G., Peñuelas, J., Phillips, O. L., Pillar, V., Poorter, H., Poorter, L., Poschlod, P., Prinzing, A., Proulx, R., Rammig, A., Reinsch, S., Reu, B., Sack, L., Salgado-negret, B., Sardans, J., Shiodera, S., Shipley, B., Siefert, A., Sosinski, E., Soussana, J. -F., Swaine, E., Swenson, N., Thompson, K., Thornton, P., Waldram, M., Weiher, E., White, M., White, S., Wright, S. J., Yguel, B., Zaehle, S., Zanne, A. E. and Wirth, C. (2011). TRY – a global database of plant traits. *Glob. Change Biol.* 17(9), 2905–2935.

Kazakou, E., Violle, C., Roumet, C., Navas, M. L., Vile, D., Kattge, J. and Garnier, E. (2014). Are trait-based species rankings consistent across data sets and spatial scales? *J. Veg. Sci.* 25(1), 235–247.

Kazakou, E., Fried, G., Richarte, J., Gimenez, O., Violle, C. and Metay, A. (2016). A plant trait-based response-and-effect framework to assess vineyard inter-row soil management. *Bot. Lett.* 163(4), 373–388, 10.1080/23818107.2016.1232205.

Kleyer, M., Bekker, R. M., Knevel, I. C., Bakker, J. P., Thompson, K., Sonnenschein, M., Poschlod, P., van Groenendael, J. M., Klimeš, L., Klimešová, J., Klotz, S., Rusch, G.

M., Hermy, M., Adriaens, D., Boedeltje, G., Bossuyt, B., Dannemann, A., Endels, P., Götzenberger, L., Hodgson, J. G., Jackel, A., Kühn, I., Kunzmann, D., Ozinga, W. A., Römermann, C., Stadler, M., Schlegelmilch, J., Steendam, H. J., Tackenberg, O., Wilmann, B., Cornelissen, J. H. C., Eriksson, O., Garnier, E. and Peco, B. (2008). The LEDA Traitbase: a database of life-history traits of the northwest European flora. *J. Ecol.* 96(6), 1266–1274.

Klotz, S., Kuhn, I. and Durka, W. (2002). Biolflor: eine Datenbank zu Biologisch-Okologischen Merkmalen der Gefaßpflanzen in Deutschland. Schriftenreihe fur Vegetationskunde 38. *Bundesamt fur Naturschutz 334 S*, Bonn, Germany.

Lavorel, S. and Garnier, E. (2002). Predicting changes in community composition and ecosystem functioning from plant traits: revisiting the Holy Grail. *Funct. Ecol.* 16(5), 545–556.

Lavorel, S., Storkey, J., Bardgett, R. D., de Bello, F., Berg, M. P., Le Roux, X., Moretti, M., Mulder, C., Pakeman, R. J., Díaz, S. and Harrington, R. (2013). A novel framework for linking functional diversity of plants with other trophic levels for the quantification of ecosystem services. *J. Veg. Sci.* 24(5), 942–948.

Lundgren, J. G. and Rosentrater, K. A. (2007). The strength of seeds and their destruction by granivorous insects. *Arthropod Plant Interact.* 1(2), 93–99.

Manning, P., de Vries, F. T., Tallowin, J. R. B., Smith, R., Mortimer, S. R., Pilgrim, E. S., Harrison, K. A., Wright, D. G., Quirk, H., Benson, J., Shipley, B., Cornelissen, J. H. C., Kattge, J., Bönisch, G., Wirth, C. and Bardgett, R. D. (2015). Simple measures of climate, soil properties and plant traits predict national-scale grassland soil carbon stocks. *J. Appl. Ecol.* 52(5), 1188–1196.

Marshall, E. J. P., Brown, V. K., Boatman, N. D., Lutman, P. J. W., Squire, G. R. and Ward, L. K. (2003). The role of weeds in supporting biological diversity within crop fields. *Weed Res.* 43(2), 77–89.

Martin, A. R. and Isaac, M. E. (2015). REVIEW: plant functional traits in agroecosystems: a blueprint for research. *J. Appl. Ecol.* 52(6), 1425–1435.

Mézière, D., Petit, S., Granger, S., Biju-Duval, L. and Colbach, N. (2015). Developing a set of simulation-based indicators to assess harmfulness and contribution to biodiversity of weed communities in cropping systems. *Ecol. Indic.* 48, 157–170.

Milberg, P. and Hallgren, E. (2004). Yield loss due to weeds in cereals and its large-scale variability in Sweden. *Field Crops Res.* 86(2–3), 199–209.

Minden, V. and Kleyer, M. (2011). Testing the effect-response framework: key response and effect traits determining above-ground biomass of salt marshes. *J. Veg. Sci.* 22(3), 387–401.

Moonen, A. C. and Barberi, P. (2008). Functional biodiversity: an agroecosystem approach. *Agric. Ecosyst. Environ.* 127(1–2), 7–21.

Navas, M.-L. (2012). Trait-based approaches to unravelling the assembly of weed communities and their impact on agro-ecosystem functioning. *Weed Res.* 52(6), 479–488.

Norris, R. F. and Kogan, M. (2000). Interactions between weeds, arthropod pests, and their natural enemies in managed ecosystems. *Weed Sci.* 48(1), 94–158.

Oerke, E.-C. (2006). Crop losses to pests. *J. Agric. Sci.* 144(1), 31–43.

Palmer, M. W. and Maurer, T. A. (1997). Does diversity beget diversity? A case study of crops and weeds. *J. Veg. Sci.* 8(2), 235–240.

Pamminger, T., Becker, R., Himmelreich, S., Schneider, C. W. and Bergtold, M. (2019). The nectar report: quantitative review of nectar sugar concentrations offered by bee-visited flowers in agricultural and non-agricultural landscapes. *PeerJ* 7, e6329.

Perronne, R., Gaba, S., Cadet, E. and Le Corre, V. (2014). The interspecific and intraspecific variation of functional traits in weeds: diversified ecological strategies within arable fields. *Acta Bot. Gal* 161(3), 243–252.

Petit, S., Boursault, A., Guilloux, M., Munier-Jolain, N. and Reboud, X. (2011). Weeds in agricultural landscapes: a review. *Agron. Sustain. Dev.* 31(2), 309–317.

Petit, S., Boursault, A. and Bohan, D. A. (2014). Weed seed choice by carabid beetles: linking field measurements and laboratory diet assessments. *Eur. J. Entomol.* 111(5), 615–620.

Petit, S., Munier-Jolain, N., Bretagnolle, V., Bockstaller, C., Gaba, S., Cordeau, S., Lechenet, M., Mézière, D. and Colbach, N. (2015). Ecological intensification through pesticide reduction: weed control, weed biodiversity and sustainability in arable farming. *Environ. Manag.* 56(5), 1078–1090.

Petit, S. and Bohan, D. A. (2017). The use of insects in integrated weed management. In: Zimdahl, R. (Ed). *Integrated weed management for sustainable agriculture. Burleigh Dodds Series in Agricultural Science*. ISBN-13: 9781786761644, 480p.

Petit, S., Cordeau, S., Chauvel, B., Bohan, D., Guillemin, J. and Steinberg, C. (2018). Biodiversity-based options for arable weed management: a review. *Agron. Sustain. Dev.* 38(5), 48.

Pocock, M. J. O., Evans, D. M. and Memmott, J. (2012). The robustness and restoration of a network of ecological networks. *Science* 335(6071), 973–977.

Poggio, S. L. and Ghersa, C. M. (2011). Species richness and evenness as a function of biomass in arable plant communities. *Weed Res.* 51(3), 241–249.

Ricou, C., Schneller, C., Amiaud, B., Plantureux, S. and Bockstaller, C. (2014). A vegetation-based indicator to assess the pollination value of field margin flora. *Ecol. Indic.* 45, 320–331.

Shennan, C. (2008). Biotic interactions, ecological knowledge and agriculture. *Philos. Trans. R. Soc. B Biol. Sci.* 363(1492), 717–739.

Smart, S. M., Firbank, L. G., Bunce, R. G. H. and Watkins, J. W. (2000). Quantifying changes in abundance of food plants for butterfly larvae and farmland birds. *J. Appl. Ecol.* 37(3), 398–414.

Storkey, J. (2006). A functional group approach to the management of UK arable weeds to support biological diversity. *Weed Res.* 46(6), 513–522.

Storkey, J., Brooks, D., Haughton, A., Hawes, C., Smith, B. M. and Holland, J. M. (2013). Using functional traits to quantify the value of plant communities to invertebrate ecosystem service providers in arable landscapes. *J. Ecol.* 101(1), 38–46.

Storkey, J. and Neve, P. (2018). What good is weed diversity? *Weed Res.* 58(4), 239–243.

Sutcliffe, O. L. and Kay, Q. O. N. (2000). Changes in the arable flora of central southern England since the 1960s. *Cons* 93(1), 1–8.

Swanton, C. J. and Weise, S. F. (1991). Integrated weed management: the rationale and approach. *Weed Technol.* 5(3), 657–663.

Tamburini, G., Bommarco, R., Kleijn, D., van der Putten, W. H. and Marini, L. (2019). Pollination contribution to crop yield is often context-dependent: a review of experimental evidence. *Agric. Ecosyst. Environ.* 280, 16–23.

Valera, F., Wagner, R. H., Romero-Pujante, M., Gutiérrez, J. E. and Rey, P. J. (2005). Dietary specialization on high protein seeds by adult and nestling serins. *Condor* 107(1), 29–40.

Welch, K. D. and Harwood, J. D. (2014). Temporal dynamics of natural enemy pest interactions in a changing environment. *Biol. Control* 75, 18–27.

Wood, S. A., Karp, D. S., DeClerck, F., Kremen, C., Naeem, S. and Palm, C. A. (2015). Functional traits in agriculture: agrobiodiversity and ecosystem services. *Trends Ecol. Evol.* 30(9), 531–539.

Wulff, R. D., Causin, H. F., Benitez, O. and Bacalini, P. A. (1999). Intraspecific variability and maternal effects in the response to nutrient addition in Chenopodium album. *Can. J. Bot.* 77(8), 1150–1158.

Yvoz, S., Cordeau, S., Zuccolo, C. and Petit, S. (2020a). Crop type and within-field location as sources of intraspecific variations in the phenology and the production of floral and fruit resources by weeds. *Agric. Ecosyst. Environ.* 302, 107082.

Yvoz, S., Petit, S., Biju-Duval, L. and Cordeau, S. (2020b). A framework to type farming strategies within production situations to improve the comprehension of weed communities. *Eur. J. Agron.* 115, 126009.

Chapter 2

Advances in understanding the dynamics of weed communities in integrated weed management systems

Jonathan Storkey, Sustainable Agriculture Science, Rothamsted Research, UK

1 Introduction

1.1 The need for a new approach in weed science

Integrated weed management (IWM) represents a paradigm shift in approach to protecting crop yield from weed competition and presents a new challenge to weed scientists seeking to predict the response of weed populations and communities to management. The current, dominant weed control paradigm is built around chemical control and began with the introduction of herbicides in the 1960s. The increasing availability of cheap, broad-spectrum herbicides (along with insecticides and fungicides) had the indirect effect of simplifying cropping systems. Without the need for diverse crop rotations to disrupt the life cycles of pests, weeds and diseases, the most profitable crops could now be grown more frequently (sometimes continuously) and crop breeding focussed on traits that optimise yield in conditions free of weeds, pests and disease (Storkey et al., 2017). Chemical crop protection products have been an important factor, therefore, in realising the potential gains of these high yielding crop genotypes globally (Oerke, 2006) and, along with increased mechanisation and inputs of inorganic fertilisers, have underpinned the dramatic rises in food production associated with the Green Revolution (Evenson and Gollin, 2003).

http://dx.doi.org/10.19103/AS.2021.0098.02

In this scenario, a farm manager plans his cropping system around short-term economic decisions, largely driven by commodity prices and weed control is purely *reactive*. The simplified, intensively managed cropping systems associated with the Green Revolution represent a narrow ecological niche for weeds and have selected a small number of species that are well adapted to the system. These species tend to be nitrophilous and with phenology that allows the competition of their life cycle between disturbance events. This has effectively created a list of competitive, 'target' weeds associated with different cropping systems; for example, *Alopecurus myosuroides* in winter wheat and *Echinochloa crus-galli* in maize dominated systems. These species have become the focus of efforts to develop new herbicides and, for the farmer, weed control becomes a purely technical exercise in determining the timing and dose of the appropriate active ingredient to kill weed *a* in crop *x*. This has been reflected in an emphasis in the weed science literature in the latter half of the twentieth century on understanding the biology of the target weed species that are well adapted to these simplified cropping systems and optimising their control. As such, it has been argued that weed science should primarily be an applied discipline – delivering solutions to emergent problems (Moss, 2008). In this paradigm, success is measured as reduced weed abundance with the ultimate goal of weed-free fields.

The reactive approach to weed control, facilitated by the steady flow of new, cheap herbicide active ingredients in the 1970s and 1980s has led to multiple, well documented, negative unintended consequences (Pingali, 2012). These include declines in farmland biodiversity and the pollution of water courses but, despite the introduction of legislation, it is unlikely that these environmental concerns on their own would have prompted the paradigm shift in crop protection we are now witnessing. Rather it has been the rapid evolution of herbicide resistance that is challenging the dominant weed control paradigm that relies mainly or solely on herbicides (Hawkins et al., 2019). The evolution of herbicide resistance has led to an arms race between weeds and the agrochemical industry, as new products are sought that can control weed populations that have evolved resistance to existing chemistry. It could be argued that, as the frequency of product discovery lengthens and cross-resistance to multiple modes of action becomes more widespread, the weeds are winning. The battle has, perhaps, reached its nadir in the stacking of genes in genetically modified herbicide-tolerant crops to confer tolerance to multiple herbicides as weeds have evolved resistance to glyphosate. The unsustainability of this approach has rightly been highlighted (Mortensen et al., 2012).

The widespread evolution of resistance to herbicides has led to important advances in the science of predicting the impact of management on weed populations and communities. Specifically, it has catalysed interdisciplinary approaches that combine an understanding of evolutionary (Neve et al., 2009) and epidemiological (Comont and Neve, 2020) processes with agronomy to predict the response of weeds at a genomic level. However, in two important respects, the approach to studying the response of weeds to management has remained largely the same. Firstly, the focus remains on controlling a few target weeds species on short time scales. It could be argued that this phenomenon has now become more acute as greater research effort is spent on a handful of weeds that have now become extremely problematic because of herbicide resistance, such as *A. myosuroides* in northwest Europe (Delye et al., 2010), *Amaranthus palmeri* in southern and central USA (Ward et al., 2013) and *Lolium rigidum* in Australia (Owen et al., 2014). Secondly, weed science remains largely a reactive discipline to emergent problems, of which herbicide resistance is currently the most acute.

1.2 The promise of functional ecology for integrated weed management

IWM has emerged as a more sustainable alternative to weed control in response to the breakdown of cropping systems that are based mainly or solely on herbicides owing to the evolution of resistance (Table 1). The principles of IWM are explored more fully elsewhere in this book; here I explore the implications for how we seek to understand the impact of this new approach on the dynamics of weed communities. The change in approach to weed control represents a

Table 1 Comparison of two paradigms of weed management: herbicide-based systems and IWM

Herbicide-oriented system	Integrated weed management
Characteristics	
Reactive to emerging problematic weeds	Prevents any weed species becoming dominant
Implemented within single cropping season	Implemented across whole cropping system
Focussed on individual weed species	Aims to manage whole weed community
Dependent on herbicides as main control tool	Uses herbicides as one option in an integrated system with multiple non-chemical options
Focus of research	
Understand biology of target species and optimise control of individual species	Understand impact of contrasting management systems on composition and impact of weed community
Measure of success	
Reduced abundance or eradication of target species	Increased functional diversity of weed flora, reduced impact of weeds in any single growing season

significant challenge to the weed science community. As opposed to studying relatively tractable systems confined to specific weed crop combinations in a single growing season, implementing IWM demands new methodologies that capture the impact of multiple management interventions over several years on weed communities that may contain hundreds of individual species. IWM can, therefore, be characterised as a 'knowledge intensive' system replacing an 'input-intensive system'. While it may be possible to quantify, model and predict the long term impact of multiple chemical and cultural interventions on a single weed species (as has been done for *A. myosuroides* (Lutman et al., 2013)), such a 'narrow deep' approach can take decades and is not tractable as the way forward for designing cropping systems that are resilient to dominant weeds. A recent review of all alternative weed control techniques identified more than 30 options for controlling weeds from stale seedbeds and cover crops to flame weeding and harvest seed destruction. The impact of combinations of options are, therefore, too numerous to capture using the conventional species-based approach even before any potential synergies or antagonisms are considered.

This challenge has necessitated a fundamental shift in thinking when developing methodologies for predicting the impact of IWM on weed populations and communities (Table 1). Specifically, it has opened up a new front in weed science that applies principles and understanding gained from studies of the assembly of plant communities in semi-natural systems to managed cropped fields (Booth and Swanton, 2002; Gaba et al., 2014; Navas, 2012). This change in thinking means more than just including biotic interactions between weed species in a traditional community ecology approach as this would still require a detailed knowledge of the autecology of every species in the community. Rather, to make the system tractable, weed scientists have begun to apply approaches developed in the field of functional ecology that seeks to understand the evolutionary constraints that have led to plants adapted to similar habitats sharing similar functional traits (Gaba et al., 2017; Violle et al., 2007); for example, seed mass, plant height and flowering time. Evolutionary trade-offs mean that functional groups can be discriminated along environmental gradients (e.g. Wright et al., 2004) and species within a group will similarly respond to perturbations and have a similar role in the ecosystem (Diaz et al., 2007).

The foundation for predicting weed responses is no longer the autecology of individual species but a fundamental understanding of the ecological processes underlying the assembly of whole plant communities defined in terms of their traits (Table 1). In this respect, the impact of specific management options associated with IWM on weeds can be understood in terms of general abiotic and biotic gradients that determine community assembly: frequency and intensity of disturbance, availability of resource and biotic interactions (competition, facilitation, parasitism and allelopathy). These factors act on the weed community as a series of 'filters' that select for species with a combination of traits that is well adapted to the resulting ecological niche (Booth and Swanton, 2002). For example, the use of a stale seedbed and post-emergence herbicides can both be understood as a form of disturbance that reduces the opportunity for weeds to complete their life cycle selecting for species with a wide germination range and early flowering (Fried et al., 2012). The focus of this functional, community-based approach is no longer the shift in the relative abundance of species at a taxonomic level (although this can be predicted *post hoc* see Section 3) but rather a prediction of the response of the functional characteristics of the weed community defined either as the dominant trait values or functional diversity (Mason et al., 2013). If the effect of the relevant traits on the ecosystem function of the weeds (either negative as crop competition or positive in supporting biodiversity) can be quantified, the impact of management change on the behaviour of the wider system can then be predicted (Gaba et al., 2017; Storkey et al., 2013). This aspect is discussed in more detail elsewhere in this book.

A final distinction between the new trait-based approach to predicting the impact of management on weed communities is how the output from such studies can be related to metrics of success (Table 1). Whereas the traditional approach to weed science has focussed on weed abundance as the most important metric and measures success in terms of the eradication of weeds, the trait-based approach determines the impact of management on the distribution of trait values in the community. For example, in a given cropping system, is a community dominated by one or two functional types or is there a greater evenness of trait values? It has been suggested that, if total weed eradication is acknowledged to be an unattainable and (for environmental reasons) an undesirable goal, the goal of weed management should be a more functionally diverse weed community that will be less competitive in any given crop (Smith et al., 2010; Storkey and Neve, 2018). This focus on the functional composition of weed communities, as opposed to absolute numbers, is in accord with the general philosophy of IWM that seeks to take a proactive approach to build systems that are resilient to any one species (or functional group) becoming dominant.

Several reviews have been written outlining the principles of the trait-based approach and how they can be applied to weed communities (Gaba et al., 2014; Gaba et al., 2017; Navas, 2012). In this chapter, case studies that all study the impact of a change in cultivation practice on weed functional traits are presented that demonstrate the potential and challenges of this approach in predicting the impact of IWM on weed communities. Secondly, the application of the trait-based approach to population dynamics modelling is discussed as a way of exploring the impact of novel IWM strategies on weed communities. Finally, recommendations are made for the future direction of research.

2 Empirical case studies

In reviewing the literature on the functional trait-based approach to weed science, an important distinction needs to be made. A body of literature is now emerging that applies methodologies developed in functional ecology to understand the fundamental ecological principles that determine 'weediness' (Bourgeois et al., 2019; Kuester et al., 2014) and explain the changes in weed communities over time (Fried et al., 2009; Storkey et al., 2010). While these studies can give an insight into future functional shifts in response to management change, they do not specifically address the impact of individual management interventions that could be part of an IWM strategy. The number of papers that apply the trait-based approach to quantifying the impact of different components of an IWM strategy are growing, including studies of the effect of crop competition (Colbach et al., 2020; Gunton et al, 2011) and soil fertility (Ryan et al., 2010). Here I compare and contrast four European studies that quantify the impact of soil tillage on weed functional traits to illustrate the potential and challenges of this approach to predicting the impact of IWM on weed communities. In all cases, response data were aboveground weed diversity (assessed as density or percentage cover) used in combination with data on traits from online databases. The intensity of soil tillage can be considered a general category that captures several specific interventions that could form part of an IWM strategy. These include rotational ploughing, stale seedbeds and inter-row hoeing.

Two of the studies, both from France, used data from regional surveys of weed floras along a gradient of soil disturbance. Fried et al. (2012) analysed weed data from 218 fields as part of a national survey for which all management data were available. Aboveground weed diversity was assessed using cover-abundance classes and tillage defined both in terms of depth and intensity (no tillage < minimum tillage < conventional tillage). Trichard et al. (2013) sampled 52 fields in the Côte d'Or department of France along a gradient of time since the adoption of conservation agriculture with reduced tillage. In this case, the response of plant traits was analysed in relation to time since the adoption of direct drilling. Two studies from Spain analysed data from long term cropping system experiments. Armengot et al. (2016) reported the results of a meta-analysis of seven experiments studying the effect of reduced tillage in low input or organic systems from across Europe. The specific tillage treatments differed between experiments but in all cases mouldboard, inversion tillage could be compared to reduced tillage. Finally, Plaza et al. (2015) analysed 24 years of weed survey data from a long-term experiment started in 1985 with contrasting tillage treatments with increasing intensity: no tillage < minimum tillage < conventional tillage. The results from the four studies in terms of the functional traits that were observed to respond to tillage are presented in Table 2. For ease of comparison, all results have been interpreted in terms of the effect of reduced tillage intensity on weed traits.

The underlying assumption of the application of trait-based approaches to predicting the response of weed communities to alternative IWM strategies is that management will impact functional traits in a consistent way in terms of the traits that respond (Gaba et al., 2017) and the direction of the effect. If this can be demonstrated to be true, the impact of IWM on weed function can be predicted regardless of species diversity or identity. A comparison of the results of the four studies only provides partial support for this assumption. In the case of life form, there is a consistent effect of reducing tillage intensity with perennials increasing in non-inversion systems. But, although other traits (including seed size) were impacted by tillage in multiple studies, the direction of the response was inconsistent. This was despite the fact that relationships between traits were similar reflecting broad ecological strategies (small-seeded, short, early flowering ruderals vs. larger seeded, tall, late-flowering competitors). Nevertheless, decreasing tillage appeared to 'push' weed communities in different directions ecologically. Although from a theoretical perspective, decreasing the intensity of disturbance should select for a more

Table 2 Summary of effect of decreasing tillage intensity (defined as either depth and/or frequency) on weed traits observed in four European studies

Traits with a significant response	Effect of decreased tillage intensity	Reference
Plant height Seed weight Flowering onset Germination range Dispersal syndrome Life form	Shallower tillage and decreased frequency of soil disturbance were associated with *taller* weeds with *larger seeds, later flowering, a narrower range for germination* that were *more likely to be wind dispersed*. Some *perennials were observed to increase*.	Fried et al. (2012)
Plant height Seed weight Nutrient affinity Life form	In comparison to mouldboard ploughing, reduced tillage was associated with *shorter* weeds *with smaller seeds, earlier flowering, a lower affinity for nitrogen* that were *more likely to be perennials*.	Armengot et al. (2016)
Life form Flowering onset Fecundity Functional group	Direct drilled systems were associated with *earlier flowering* species with *reduced seed production* and *were more likely to be grasses* and *perennials*.	Trichard et al. (2013)
Specific leaf area Plant height Seed weight Fecundity	Decreased intensity of tillage was associated with *shorter* weeds with *smaller seeds, higher specific leaf area*, and *greater seed production*.	Plaza et al. (2015)

competitive ecological strategy (as was observed in the Fried et al. study), more ruderal species were observed to increase in the two studies from Spain. There are three possible reasons for this inconsistency that provide important insights for how the trait-based approach to predicting weed community responses to IWM could be further refined.

Firstly, 'intensity of tillage' is defined differently in the different studies and is often compared using categories such as 'conventional', 'minimum' and 'no-till' (or 'direct drilling'). However, the specific characteristics of these systems will differ between studies meaning it is difficult to be sure similar systems are being compared. In discussing the conflicting results of their study compared to Fried et al., 2012, Armengot et al. (2016) identified the frequency of soil disturbance as a confounding factor. One of the features that separated the tillage categories in the Fried et al. study was the number of passes and, although tillage depth was analysed separately, it is likely that the shift from ruderal to competitive strategies was partly a response to the reduced frequency of disturbance. Trichard et al. (2013) also mention the importance of the timing of tillage (associated with the drilling date of crops) as an additional confounding factor. Indeed, sowing time has been observed to be an important *driver* of weed community functional composition in other studies (Smith, 2006). It will be necessary in the future, therefore, to more clearly define tillage in terms of the combination of factors that act on different ecological processes that may need to be analysed separately in terms of their impact on traits. For example, a comparison of tillage treatments could comprise a combination of tillage depth, degree of soil mixing, frequency and timing, which may need to be included as separate factors when analysing survey or experimental data.

The second factor that may explain uncertainty in identifying the functional response of weed communities to tillage across the four studies is the selection of functional traits used in the analysis. Each study used a different set of traits in the analysis and no one trait was common across all studies. This makes it challenging to draw common conclusions despite the fact that some traits (such as seed size and seed production per plant) will be correlated. Not including certain traits may also lead to alternative interpretations of the data. For example, dispersal syndrome was included in the Fried et al. study but not in any of the others. Species with wind-dispersed seeds, that were observed to increase in response to reduced tillage intensity, tend to also be small seeds, which may partly explain the decrease in the community weighted mean of this trait in no-till systems in comparable studies. Some traits are also indicative of multiple ecological processes that may have contrasting responses to the different components of tillage described above. An example is seed weight which is a trait that interfaces the regenerative and established phases of weed growth as well as partly determining dispersal ability (Westoby et al., 2002). Increased seed reserves in larger seeds enable them to emerge from greater depth, giving them an advantage if tillage is deeper. However, smaller seeds also tend to be more persistent enabling them to survive longer if buried at depths that induce secondary dormancy. Whether or not a change in tillage practice selects for large or small seeds will therefore be a product of the interaction of depth and frequency of tillage. This problem may partly be overcome through a clearer definition of 'tillage' that acknowledges these separate components, as discussed above, but it is also possible to include additional seed traits that more precisely reflect specific ecological processes. For example, a combination of seed mass and shape has been identified as an indicator of seedbank persistence (Thompson et al., 1993) and mortality in the soil has been related to seed coat characteristics (Gardarin et al., 2010; Villora et al., 2019).

The final challenge to predicting the functional response of weeds to tillage illustrated by these studies is that soil disturbance cannot be isolated from other parts of the system with which it interacts. Both Armengot et al. and Trichard et al. discussed the importance of soil fertility, the former identifying the effect of tillage in altering the distribution of nutrients in the soil in a low input system and the latter highlighting the effect of inorganic fertilisers as an additional management filter on the weed communities. A change in tillage practice will also often be associated with a change in cropping, which also imposes additional selection pressure on the weed communities through the timing of tillage, crop competition and in-crop weed management interventions. Because of these confounding effects of other parts of the system, it is difficult to separate out the effect of tillage from these studies and draw definitive conclusions in terms of the traits that respond and the direction of the effect.

I will return to these challenges to the trait-based approach highlighted by the comparison of these four studies in the conclusion where recommendations will be made for future work.

3 A trait-based approach to population dynamics modelling

We have seen that one of the big challenges facing the weed science community in predicting the response of weed communities and populations to IWM is coping with complexity. Quantifying the effect of different levels of a single factor on a few weed species (e.g. a dose-response to a herbicide) is relatively tractable and can be achieved through conventional experimental approaches. However, IWM involves the combination of multiple interventions with the aim of managing the whole weed community such that no weed species become dominant to the extent that the system becomes overly reliant on too few herbicide active ingredients (with the associated risk of herbicide resistance). These different interventions may have an additive, synergistic or antagonistic effect on different species in the community. One approach to managing this complexity is to use statistical models to study the impact of management at the level of functional traits in systems with contrasting weed management strategies, as discussed above. However, the parameters of the IWM system will be constrained by the management options available or included in the study system limiting the opportunity to predict the effect of novel combinations of options. Simulation models of weed populations dynamics have traditionally been used to overcome this limitation and derive hypotheses about the behaviour of novel systems that can then be challenged in empirical studies (Holst et al., 2007). However, as discussed above, although these models exist at a level of detail necessary to predict the response of a single species to IWM (Colbach et al., 2006), taking this approach for all species in the community has not been tractable because of the effort required to derive the many parameters.

In recent years, modelling groups in the UK and France have begun to apply trait-based approaches as a pragmatic way of extending the scope of their models parameterised for individual species to whole weed communities. Although the models differ, the fundamental principle remains the same: quantifying relationships between weed functional traits and model parameters to allow species to be modelled for which the model has not been parameterised. In Chapter 1, the advances that have been made in applying this approach to a complex mechanistic model of annual weed population dynamics to predict the impact of IWM on the functioning of weed communities in France is described in detail (Colbach et al., 2014). Here, I present the progress that has been made in modelling functional shifts in UK weed communities in response to management change using a simpler empirical model of population dynamics.

Empirical models of weed population dynamics are built around a series of equations that model the transition from one 'state' to another. These states are defined by the life cycle of an annual weed (Fig. 1) and underpin all population dynamics models, whether empirical or mechanistic (Holst et al., 2007): (1) seed in the seedbank, (2) emerged seedlings, (3) mature plants and (4) fresh seed. The impact of management on these transitions can be modelled by changing the values of parameters in the equations. For example, Moss (1990) predicted the impact of a change in soil cultivation by defining the proportion of the seed bank that could emerge from a shallow and a deep layer and the distribution of those seeds between the layers under different cultivation scenarios. The challenge in building these models is finding the right balance between including sufficient detail to capture the interaction of multiple management interventions while ensuring sufficient data are available to parameterise them. The model described here, therefore, aimed to only use relationships between traits and mode parameters for which trait data were widely available for all

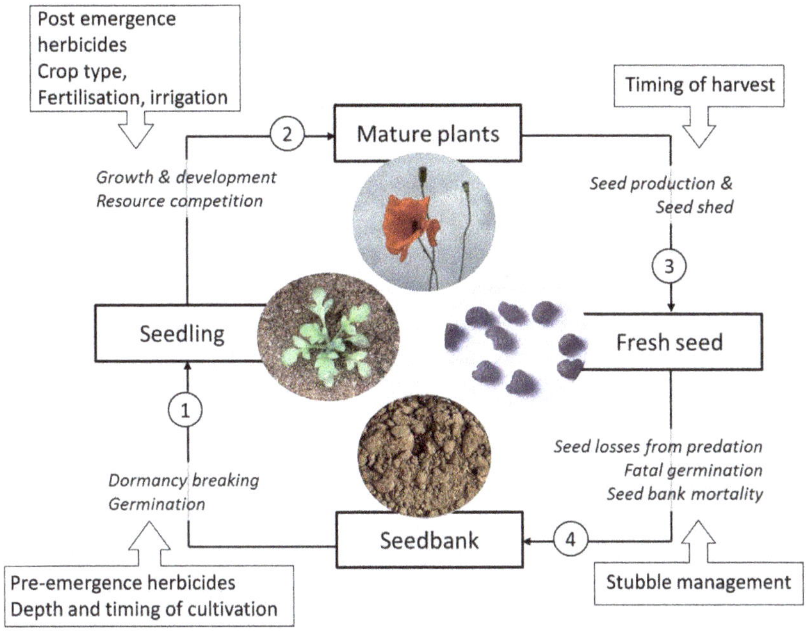

Figure 1 Representation of the life cycle of an annual weed with four 'states' (boxes) and 'transitions' (arrows) that are determined by the biological processes in italics. Examples of weed management interventions that aim to disrupt the life cycle at different points are also included.

weed species from online databases such as the Ecoflora Database for UK plants (Fitter and Peat, 1994) and the Seed database held at Kew Gardens, UK (Flynn et al., 2004). The ambition is to be able to predict the response of the whole UK annual weed flora.

The approach of combining functional traits with an empirical model of weed population dynamics was first demonstrated by Storkey et al. (2015) in a simple proof of concept using just a single equation to model each of the transitions in Fig. 1 incorporating only two relationships between traits and model parameters. Firstly, the growth and competition of weeds were related to plant height (transition #2) with seedlings of taller species being more competitive and achieving greater mature biomass. This parameter was further modified by fertiliser inputs, with small-seeded, tall species being relatively more competitive in more fertile conditions (Storkey et al., 2010). Secondly, seed production was related to seed weight (transition #3) with smaller seeded species producing more seed per unit of mature plant biomass. The power of this initial model to predict the impact of multiple IWM options was, therefore, very limited but it was used to illustrate the concept of modelling communities at the level of functional traits, deriving fitness contours that predict the population dynamics of weed ideotypes with different trait combinations under any given management scenario. Weed species can then be mapped onto this space *post hoc*. The shift in weed communities in response to increased herbicide and fertiliser use was used to validate this new conceptual approach to weed community ecology (Fig. 2).

From this simple starting point, the model has been substantially improved to include the functionality required to model the impact of many more management interventions (Metcalfe et al., 2020) including changes in tillage practice (as discussed in Section 2), crop competition and harvest weed seed control. The principles remain the same and the parameters have been derived from a relatively small number of traits for which data are available for the whole UK weed flora but the impact on a greater range of processes has been included greatly increasing the functionality of the model (Table 3). The exception is the use of herbicides. It has been hypothesised that there may be functional traits that can be used to predict the susceptibility of species to herbicides (Gaba et al., 2017) (including the characteristics of the leaf cuticle) but, because of the biochemical specificity of many active ingredients, in this case, the impact of herbicides was modelled using species-specific dose-response curves.

This new version of the model now has the functionality to explore many more combinations of IWM management interventions. For example, one of the changes to a cropping system that would be expected to have the largest impact on weed communities would be a change in the identity or diversity of crops grown in a rotation (including the use of companion or cover crops). However, specific management filters associated with this change that determine the

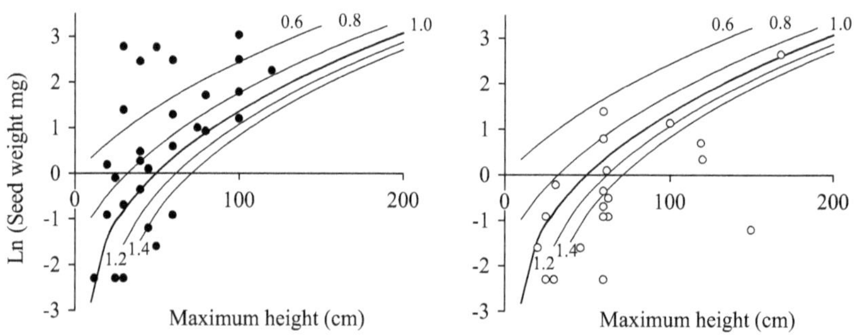

Figure 2 Output of the life cycle model for a generic annual weed, with different combinations of seed weight and maximum height, expressed as fitness contours indicating population growth (λ > 1) or decline (λ < 1) for a scenario of high herbicide mortality and high fertility that is typical of intensively managed cropping systems. Data on the height and seed weight of two sets of weeds species have been mapped onto the contour plots: left, ● 31 rare or declining arable weed species and right, ○ 22 species commonly found in UK winter crops (Storkey et al., 2015).

Table 3 Summary of weed functional traits included in the version of a trait-based empirical model of weed population dynamics published by Metcalfe et al. (2020) and the IWM interventions the model now has the functionality to predict the effect on weed communities

Transition	Management intervention	Weed response traits
Seedbank → Seedlings	Date of sowing	Emergence periodicity[a]
	Depth of cultivation	Seed weight
	Herbicide timing	Emergence periodicity[a]
	Frequency and depth of cultivation	Seedbank persistence[a]
Seedlings → mature plants	Fertilisation	Ellenberg N number[a]
	Crop competition	Maximum height + Onset of flowering
Mature plants → Fresh seed	Timing of harvest and weed seed destruction	Flowering time + seed weight

[a]Categorical functional groups.

response of the weed community are multiple and complex: changes in the tillage regime (timing, frequency and potentially depth), different in-crop weed control options (chemical and non-chemical), altered crop competition and a change in the timing of harvest. This model has the functionality to model the interaction of all of these factors on the relevant processes driving weed populations based on their functional response traits. The effect of changing a rotation (e.g. including an additional break crop) on the functional diversity and characteristics of any UK weed community (regardless of the species pool) can, therefore, be predicted.

The model has been validated on an intensively managed arable field in Suffolk, UK, for which detailed management data on cropping, soil cultivation and inputs were available for 30 years (1987–2016). These input data were used to simulate the filtering effect of long-term management on a regional pool of 101 annual weed species. Because of stochastic processes to do with dispersal and establishment, it would be surprising if the model were able to accurately predict the local species pool in the field. Indeed, the model had mixed success at predicting the species found in the study field with only 8 of the 15 annuals observed in the field ranking among the 20 most abundant species in any simulation (Metcalfe et al., 2020). However, as described in Section 1.2, the objective of the trait-based approach is not to necessarily predict the effect of IWM at the taxonomic level but rather on the functional diversity and characteristics of the emergent weed community. In this respect, the model was more successful, predicting the correct trajectory of selection for several traits, including seed size and emergence pattern (Fig. 3). The former was explained by the regular ploughing and high inputs of agro-chemicals (supporting the conclusions of Fried et al. (2012)) and the latter by the high frequency of spring-sown crops in the rotation. However, there were more generalist species

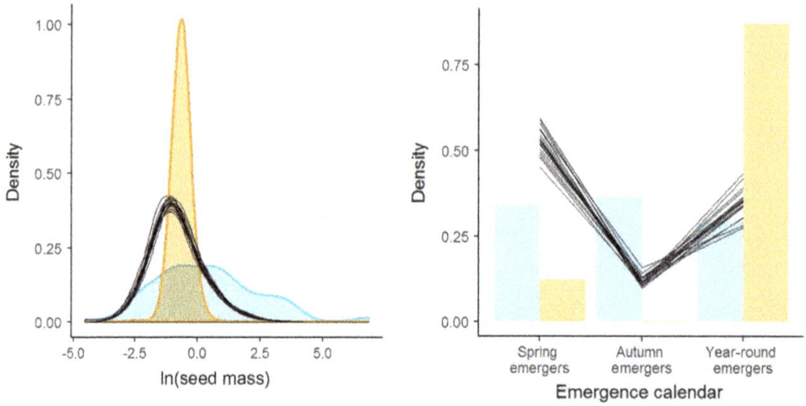

Figure 3 Results from an empirical weed population dynamics model that predicts the impact of management at the level of functional traits. The shift in the distribution of two traits is illustrated: seed mass (left) and periodicity of emergence (right). The blue area represents the frequency distribution of the traits in a regional species pool of 101 annual weed species. The orange area is the frequency distribution of traits from a local species pool sampled from a field in Suffolk, UK, for which management data were available for 30 years. Lines are model output for individual runs using the management data from the Suffolk field as input and initialising the model with equal numbers of all the species in the regional pool. The model includes some stochastic functions (e.g. the proportion of the seedbank that emerges) meaning output varies each time the model is run (Metcalfe et al., 2020).

(adapted to emerge throughout the year) in the observed local species pool than were predicted by the model.

4 Conclusions

In this chapter, I have argued that IWM assumes a philosophical shift in attitude towards weeds with important implications for weed science. The efficacy and availability of cheap herbicides created the promise of weed-free fields, the evolution of herbicide resistance has shown this promise to be false and highlighted the environmental and agronomic unsustainability of systems that are over-reliant on chemical crop protection products (MacLaren et al., 2020). In response, it can be argued that IWM represents a step towards a more ecological approach to designing sustainable cropping systems that are founded on diversification and broadening the habitat niche for weeds such that no one species becomes dominant. As such, the objective becomes less controlling target species and more managing whole weed communities. The novel trait-based approaches to predicting the response of weed communities that have emerged in response to this challenge represent a fruitful synthesis of agronomic and ecological knowledge and principles. Great progress has been made in this rapidly expanding field over the last decade, but I conclude this chapter by suggesting an important area that needs to be addressed to fully realise its potential in the future.

The review of papers studying the response of weed traits to a change in tillage practice revealed apparently conflicting results that was partly a consequence of a lack of standardisation across studies. Firstly, different sets of traits were included in each analysis with some traits only appearing in one study and no traits being common to all four. Within the plant ecology community, a great deal of effort has been spent standardising trait definitions and identifying the subset of traits that have the greatest power to predict the explain community assembly and predict the response to change (Diaz et al., 2004; Diaz et al., 2016). While individual research groups have begun to compile their own weeds traits databases, there is a need for a similar coordinated effort across the weed ecology community that identifies the key traits that should be included in a study modelling functional responses to management change and standardises the sources of data. This is particularly important because weeds tend to have high phenotypic plasticity and their growth habit in cropped fields may differ from semi-natural habitats where values for traits currently in online databases may have been measured (Kattge et al., 2011). As well as standardising information on traits, however, the example of the effect of tillage also highlighted the need to more precisely define the management filters associated with IWM. Specifically, it was argued that broad categories, such as conventional or no-till were too crude and did not capture the multiple filters that interact in these contrasting systems. What is required is a framework for defining agronomic interventions in terms of the specific perturbation effect they represent for a weed - or taking a 'weeds-eye' view of IWM (Storkey et al. 2021). If such a framework can be developed, contrasting systems or studies could be compared using a standardised set of building blocks that can be combined to derive a *gradient* of IWM. Such a framework would also be a valuable resource for weed population dynamics models as it would provide a template that could be applied to multiple models in an Ensemble approach.

As the potential for applying ecological principles to practical weed science problems grows, we are at an exciting juncture where this consolidation of knowledge will begin to deliver robust tools for predicting the response of weed communities in the more complex, sustainable cropping systems of the future.

5 Acknowledgements

Part of the material in this chapter was funded by the European Union Horizon 2020 project 'Integrated Weed Management: Practical implementation and solutions for Europe (IWMPraise)' grant agreement no. 727321.

6 Where to look for further information

A number of research groups are currently working on the topic of modelling weed community assembly in integrated weed management systems. As well as my own group at Rothamsted Research (UK), these include the groups of Matt Liebman at Iowa State University (US) and the groups of Sabrina Gaba, Sandrine Petit and Nathalie Colbach groups at INRAE (France). As well as the need to derive standardised lists of traits and sources of data (as discussed above), current questions in this research area include the importance of morphological plasticity and intra-specific trait variability in determining the persistence of weeds and outcome of community assembly. Analyses of functional traits often use a single value for each species x trait combination whereas, in reality, natural variation and adaptation to local environments result in a range of values. This can be particularly important for weeds where species traits available in online databases may have been measured in semi-natural plant communities that may be very different from the conditions in a cultivated field. Finally, there is also a need to develop user-friendly platforms that make the science of trait-based weed community dynamics accessible to practitioners for the prediction of the impact of management change on local weed communities.

7 References

Armengot, L., Blanco-Moreno, J. M., Barberic, P., Bocci, G., Carlesi, S., Aendekerk, R., Berner, A., Celette, F., Grosse, M., Huiting, H., Kranzler, A., Luik, A., Mader, P., Peigne, J., Stoll, E., Delfosse, P., Sukkel, W., Surbock, A., Westaway, S. and Sans, F. X. (2016). Tillage as a driver of change in weed communities: A functional perspective. *Agriculture, Ecosystems and Environment* 222, 276–85.

Booth, B. D. and Swanton, C. J. (2002). Assembly theory applied to weed communities. *Weed Science* 50(1), 2–13.

Bourgeois, B., Munoz, F., Fried, G., Mahaut, L., Armengot, L., Denelle, P., Storkey, J., Gaba, S. and Violle, C. (2019). What makes a weed a weed? A large-scale evaluation of arable weeds through a functional lens. *American Journal of Botany* 106(1), 90–100.

Colbach, N., Durr, C., Roger-Estrade, J., Chauvel, B. and Caneill, J. (2006). AlomySys: Modelling black-grass (*Alopecurus myosuroides* Huds.) germination and emergence, in interaction with seed characteristics, tillage and soil climate: I. *European Journal of Agronomy* 24(2), 95–112.

Colbach, N., Granger, S., Guyot, S. H. M. and Meziere, D. (2014). A trait-based approach to explain weed species response to agricultural practices in a simulation study with a cropping system model. *Agriculture, Ecosystems and Environment* 183, 197–204.

Colbach, N., Munier-Jolain, N., Dugue, F., Gardarin, A., Strbik, F. and Moreau, D. (2020). The response of weed and crop species to shading: How to predict their morphology and plasticity from species traits and ecological indexes? *European Journal of Agronomy* 121, 18.

Comont, D. and Neve, P. (2020). Adopting epidemiological approaches for herbicide resistance monitoring and management. *Weed Research* 61(2), 81-7.

Delye, C., Michel, S., Berard, A., Chauvel, B., Brunel, D., Guillemin, J. P., Dessaint, F. and Le Corre, V. (2010). Geographical variation in resistance to acetyl-coenzyme A carboxylase-inhibiting herbicides across the range of the arable weed *Alopecurus myosuroides* (black-grass). *New Phytologist* 186(4), 1005-17.

Diaz, S., Hodgson, J. G., Thompson, K., Cabido, M., Cornelissen, J. H. C., Jalili, A., Montserrat-Marti, G., Grime, J. P., Zarrinkamar, F., Asri, Y., Band, S. R., Basconcelo, S., Castro-Diez, P., Funes, G., Hamzehee, B., Khoshnevi, M., Perez-Harguindeguy, N., Perez-Rontome, M. C., Shirvany, F. A., Vendramini, F., Yazdani, S., Abbas-Azimi, R., Bogaard, A., Boustani, S., Charles, M., Dehghan, M., de Torres-Espuny, L., Falczuk, V., Guerrero-Campo, J., Hynd, A., Jones, G., Kowsary, E., Kazemi-Saeed, F., Maestro-Martinez, M., Romo-Diez, A., Shaw, S., Siavash, B., Villar-Salvador, P. and Zak, M. R. (2004). The plant traits that drive ecosystems: Evidence from three continents. *Journal of Vegetation Science* 15(3), 295-304.

Diaz, S., Kattge, J., Cornelissen, J. H. C., Wright, I. J., Lavorel, S., Dray, S., Reu, B., Kleyer, M., Wirth, C., Prentice, I. C., Garnier, E., Bonisch, G., Westoby, M., Poorter, H., Reich, P. B., Moles, A. T., Dickie, J., Gillison, A. N., Zanne, A. E., Chave, J., Wright, S. J., Sheremet'ev, S. N., Jactel, H., Baraloto, C., Cerabolini, B., Pierce, S., Shipley, B., Kirkup, D., Casanoves, F., Joswig, J. S., Gunther, A., Falczuk, V., Ruger, N., Mahecha, M. D. and Gorne, L. D. (2016). The global spectrum of plant form and function. *Nature* 529(7585), 167-71-U73.

Diaz, S., Lavorel, S., de Bello, F., Quetier, F., Grigulis, K. and Robson, T. M. (2007). Incorporating plant functional diversity effects in ecosystem service assessments. *Proceedings of the National Academy of Sciences of the United States of America* 104(52), 20684-89.

Evenson, R. E. and Gollin, D. (2003). Assessing the impact of the Green Revolution, 1960 to 2000. *Science* 300(5620), 758-62.

Fitter, A. H. and Peat, H. J. (1994). The ecological flora database. *Journal of Ecology* 82(2), 415-25.

Flynn, S., Turner, R. M. and Dickie, J. B. (2004). Seed information database (version 6.0 October 2004).

Fried, G., Chauvel, B. and Reboud, X. (2009). A functional analysis of large-scale temporal shifts from 1970 to 2000 in weed assemblages of sunflower crops in France. *Journal of Vegetation Science* 20(1), 49-58.

Fried, G., Kazakou, E. and Gaba, S. (2012). Trajectories of weed communities explained by traits associated with species' response to management practices. *Agriculture, Ecosystems and Environment* 158, 147-55.

Gaba, S., Fried, G., Kazakou, E., Chauvel, B. and Navas, M. L. (2014). Agroecological weed control using a functional approach: A review of cropping systems diversity. *Agronomy for Sustainable Development* 34(1), 103-19.

Gaba, S., Perronne, R., Fried, G., Gardarin, A., Bretagnolle, F., Biju-Duval, L., Colbach, N., Cordeau, S., Fernandez-Aparicio, M., Gauvrit, C., Gibot-Leclerc, S., Guillemin, J. P., Moreau, D., Munier-Jolain, N., Strbik, F. and Reboud, X. (2017). Response and effect traits of arable weeds in agro-ecosystems: A review of current knowledge. *Weed Research* 57(3), 123-47.

Gardarin, A., Durr, C., Mannino, M. R., Busset, H. and Colbach, N. (2010). Seed mortality in the soil is related to seed coat thickness. *Seed Science Research* 20(4), 243-56.

Gunton, R. M., Petit, S. and Gaba, S. (2011). Functional traits relating arable weed communities to crop characteristics. *Journal of Vegetation Science* 22(3), 541-50.

Hawkins, N. J., Bass, C., Dixon, A. and Neve, P. (2019). The evolutionary origins of pesticide resistance. *Biological Reviews of the Cambridge Philosophical Society* 94(1), 135-55.

Holst, N., Rasmussen, I. A. and Bastiaans, L. (2007). Field weed population dynamics: A review of model approaches and applications. *Weed Research* 47(1), 1-14.

Kattge, J., Diaz, S., Lavorel, S., Prentice, I. C., Leadley, P., Bonisch, G., Garnier, E., Westoby, M., Reich, P. B., Wright, I. J., Cornelissen, J. H. C., Violle, C., Harrison, S. P., van Bodegom, P. M., Reichstein, M., Enquist, B. J., Soudzilovskaia, N. A., Ackerly, D. D., Anand, M., Atkin, O., Bahn, M., Baker, T. R., Baldocchi, D., Bekker, R., Blanco, C. C., Blonder, B., Bond, W. J., Bradstock, R., Bunker, D. E., Casanoves, F., Cavender-Bares, J., Chambers, J. Q., Chapin, F. S., Chave, J., Coomes, D., Cornwell, W. K., Craine, J. M., Dobrin, B. H., Duarte, L., Durka, W., Elser, J., Esser, G., Estiarte, M., Fagan, W. F., Fang, J., Fernandez-Mendez, F., Fidelis, A., Finegan, B., Flores, O., Ford, H., Frank, D., Freschet, G. T., Fyllas, N. M., Gallagher, R. V., Green, W. A., Gutierrez, A. G., Hickler, T., Higgins, S. I., Hodgson, J. G., Jalili, A., Jansen, S., Joly, C. A., Kerkhoff, A. J., Kirkup, D., Kitajima, K., Kleyer, M., Klotz, S., Knops, J. M. H., Kramer, K., Kuhn, I., Kurokawa, H., Laughlin, D., Lee, T. D., Leishman, M., Lens, F., Lenz, T., Lewis, S. L., Lloyd, J., Llusia, J., Louault, F., Ma, S., Mahecha, M. D., Manning, P., Massad, T., Medlyn, B. E., Messier, J., Moles, A. T., Muller, S. C., Nadrowski, K., Naeem, S., Niinemets, Ü, Nollert, S., Nuske, A., Ogaya, R., Oleksyn, J., Onipchenko, V. G., Onoda, Y., Ordonez, J., Overbeck, G., Ozinga, W. A., Patino, S., Paula, S., Pausas, J. G., Penuelas, J., Phillips, O. L., Pillar, V., Poorter, H., Poorter, L., Poschlod, P., Prinzing, A., Proulx, R., Rammig, A., Reinsch, S., Reu, B., Sack, L., Salgado-Negre, B., Sardans, J., Shiodera, S., Shipley, B., Siefert, A., Sosinski, E., Soussana, J.-F., Swaine, E., Swenson, N., Thompson, K., Thornton, P., Waldram, M., Weiher, E., White, M., White, S., Wright, S. J., Yguel, B., Zaehle, S., Zanne, A. E. and Wirth, C. (2011). TRY: A global database of plant traits. *Global Change Biology* 17(9), 2905-35.

Kuester, A., Conner, J. K., Culley, T. and Baucom, R. S. (2014). How weeds emerge: A taxonomic and trait-based examination using United States data. *New Phytologist* 202(3), 1055-68.

Lutman, P. J. W., Moss, S. R., Cook, S. and Welham, S. J. (2013). A review of the effects of crop agronomy on the management of *Alopecurus myosuroides*. *Weed Research* 53(5), 299-313.

MacLaren, C., Storkey, J., Menegat, A., Metcalfe, H. and Dehnen-Schmutz, K. (2020). An ecological future for weed science to sustain crop production and the environment: A review. *Agronomy for Sustainable Development* 40(4), 29.

Mason, N. W. H., de Bello, F., Mouillot, D., Pavoine, S. and Dray, S. (2013). A guide for using functional diversity indices to reveal changes in assembly processes along ecological gradients. *Journal of Vegetation Science* 24(5), 794-806.

Metcalfe, H., Milne, A. E., Deledalle, F. and Storkey, J. (2020). Using functional traits to model annual plant community dynamics. *Ecology* 101(11), e03167.

Mortensen, D. A., Egan, J. F., Maxwell, B. D., Ryan, M. R. and Smith, R. G. (2012). Navigating a critical juncture for sustainable weed management. *BioScience* 62(1), 75-84.

Moss, S. R. (1990). The seed cycle of *Alopecurus myosuroides* in winter cereals: A quantitative anlysis. In: *Proceedings of the EWRS Symposium: Integrated Weed Management in Cereals*, Helsinki, Finland, pp. 27-35.

Moss, S. R. (2008). Weed research: Is it delivering what it should? *Weed Research* 48(5), 389-93.
Navas, M.-L. (2012). Trait-based approaches to unravelling the assembly of weed communities and their impact on agro-ecosystem functioning. *Weed Research* 52(6), 479-88.
Neve, P., Vila-Aiub, M. and Roux, F. (2009). Evolutionary-thinking in agricultural weed management. *New Phytologist* 184(4), 783-93.
Oerke, E.-C. (2006). Crop losses to pests. *Journal of Agricultural Science* 144(1), 31-43.
Owen, M. J., Martinez, N. J. and Powles, S. B. (2014). Multiple herbicide-resistant *Lolium rigidum* (annual ryegrass) now dominates across the Western Australian grain belt. *Weed Research* 54(3), 314-24.
Pingali, P. L. (2012). Green Revolution: Impacts, limits, and the path ahead. *Proceedings of the National Academy of Sciences of the United States of America* 109(31), 12302-8.
Plaza, E. H., Navarrete, L. and Gonzalez-Andujar, J. L. (2015). Intensity of soil disturbance shapes response trait diversity of weed communities: The long-term effects of different tillage systems. *Agriculture, Ecosystems and Environment* 207, 101-8.
Ryan, M. R., Smith, R. G., Mirsky, S. B., Mortensen, D. A. and Seidel, R. (2010). Management filters and species traits: Weed community assembly in long-term organic and conventional systems. *Weed Science* 58(3), 265-77.
Smith, R. G. (2006). Timing of tillage is an important filter on the assembly of weed communities. *Weed Science* 54(4), 705-12.
Smith, R. G., Mortensen, D. A. and Ryan, M. R. (2010). A new hypothesis for the functional role of diversity in mediating resource pools and weed-crop competition in agroecosystems. *Weed Research* 50(1), 37-48.
Storkey, J., Brooks, D., Haughton, A., Hawes, C., Smith, B. M. and Holland, J. M. (2013). Using functional traits to quantify the value of plant communities to invertebrate ecosystem service providers in arable landscapes. *Journal of Ecology* 101(1), 38-46.
Storkey, J., Bruce, T. J. A., McMillan, V. E. and Neve, P. (2017). The future of sustainable crop protection relies on increased diversity of cropping systems and landscapes. In: Lemaire, G., Carvalho, P. C. F., Kronberg, S. and Recous, S. (Eds), *Agro-Ecosystem, Diversity: Impact on Food Security and Environmental Quality*, Elsevier, London, UK, pp. 199-209.
Storkey, J., Helps, J., Hull, R., Milne, A. E. and Metcalfe, H. (2021). Defining integrated weed management: A novel conceptual framework for models. *Agronomy* 11(4), 747.
Storkey, J., Holst, N., Bojer, O. Q., Bigongiali, F., Bocci, G., Colbach, N., Dorner, Z., Riemens, M. M., Sartorato, I., Sonderskov, M. and Verschwele, A. (2015). Combining a weed traits database with a population dynamics model predicts shifts in weed communities. *Weed Research* 55(2), 206-18.
Storkey, J., Moss, S. R. and Cussans, J. W. (2010). Using assembly theory to explain changes in a weed flora in response to agricultural intensification. *Weed Science* 58(1), 39-46.
Storkey, J. and Neve, P. (2018). What good is weed diversity? *Weed Research* 58(4), 239-43.
Thompson, K., Band, S. R. and Hodgson, J. G. (1993). Seed size and shape predict persistence in soil. *Functional Ecology* 7(2), 236-41.

Trichard, A., Alignier, A., Chauvel, B. and Petit, S. (2013). Identification of weed community traits response to conservation agriculture. *Agriculture, Ecosystems and Environment* 179, 179–86.

Villora, R. A., Plaza, E. H., Navarrete, L., Sanchez, M. J. and Sanchez, A. M. (2019). Climate and tillage system drive weed communities' functional diversity in a Mediterranean cereal-legume rotation. *Agriculture, Ecosystems and Environment* 283, 10.

Violle, C., Navas, M. L., Vile, D., Kazakou, E., Fortunel, C., Hummel, I. and Garnier, E. (2007). Let the concept of trait be functional! *Oikos* 116(5), 882–92.

Ward, S. M., Webster, T. M. and Steckel, L. E. (2013). Palmer amaranth (Amaranthus palmeri): A review. *Weed Technology* 27(1), 12–27.

Westoby, M., Falster, D. S., Moles, A. T., Vesk, P. A. and Wright, I. J. (2002). Plant ecological strategies: Some leading dimensions of variation between species. *Annual Review of Ecology and Systematics* 33(1), 125–59.

Wright, I. J., Reich, P. B., Westoby, M., Ackerly, D. D., Baruch, Z., Bongers, F., Cavender-Bares, J., Chapin, T., Cornelissen, J. H. C., Diemer, M., Flexas, J., Garnier, E., Groom, P. K., Gulias, J., Hikosaka, K., Lamont, B. B., Lee, T., Lee, W., Lusk, C., Midgley, J. J., Navas, M. L., Niinemets, U., Oleksyn, J., Osada, N., Poorter, H., Poot, P., Prior, L., Pyankov, V. I., Roumet, C., Thomas, S. C., Tjoelker, M. G., Veneklaas, E. J. and Villar, R. (2004). The worldwide leaf economics spectrum. *Nature* 428(6985), 821–27.

Chapter 3

Advances in managing arable weed propagules

Bärbel Gerowitt, University of Rostock, Germany; and Barbara Baraibar, University of Lleida – Agrotecnio, Spain

1 Introduction

This chapter is about plant propagules. A propagule is a part of the plant that can detach from the rest of the plant and grow into a new plant. The survival of plants relies on these propagules which play a key role in reproduction. Propagules either arise from generative or vegetative reproduction and spread. Generatively produced propagules are seeds, while vegetative propagules are fragments of roots, stolons (stems growing along the ground) and rhizomes (stems growing underground) as well as bulbs or tubers originating from clonal growth. According to Harper (1977), in population biology, every seed is a genet, hence a genetically different organism, while fragments from clonal growth are genetically identical ramets. Together genets and ramets constitute the set of plant propagules required for the survival and spread of the plant.

The spread of seeds starts with the process of maturation which allows the seed to be released from the plant. Seeds use various mechanisms (including autochory, anemochory and zoochory, see Benvenuti 2007) to move away from the mother plant in order to maximise their chances of survival, with distances of seed spread varying widely. Arable soils carry thousands of weed

http://dx.doi.org/10.19103/AS.2021.0098.03

seeds per m^2. Intact, inactive seeds are tiny and hard to detect in the soil and thus well adapted to resist mechanical soil treatments. In contrast to seeds, ramets of perennial weeds are mainly the result of soil management processes such as tillage which break up roots and other plant material and are spread primarily by these practices. Both seeds and ramets can spread within the field or between fields. Both crops and weed plants require soil not only for root establishment to sustain above-ground growth and development but also as a refuge to survive unfavourable conditions.

In arable farming, the main medium for weed propagules is the tilled zone of the soil that extends from the soil surface through the topsoil down to the limits of arable management. Physical processes like ploughing usually reach down to 30 cm, though deep tillage practices may go deeper. Arable farmers establish crops in this soil layer which is the focus of this chapter. All arable soils also carry weeds (seeds and ramets) as a reservoir for spontaneous vegetation which means all arable soils have a soil propagule bank. The importance of the seed bank is well known in weed science and farming (Haring and Flessner 2018). However, the propagule bank resulting from vegetative reproduction has been studied less and is more complex as it includes both creeping ramets and the deep root system (Håkansson 1982). Hatcher (2017) refers to this as the bud bank. The deep root system is responsible for survival, while the creeping ramet system grows locally and spread via fragmentation.

Measuring and forecasting the magnitude of weed establishment from a seed bank in a timely and accurate manner has been widely studied (Gonzalez-Andujar et al. 2016). However, while plant propagules are ubiquitous in arable soils, there is strong variation as to whether and when seeds germinate or buds sprout. This variation is challenging for farmers and land managers. Measuring and managing weeds typically concentrates on the activated part of the propagule bank, that is on young plants that have emerged from propagules. These are visible above ground as growing weed plants that compete with crops and produce new propagules. Many preventive and cultural means of weed control focus on getting propagules to germinate and then destroying the seedlings or sprouts from vegetative parts before they can harm the crop.

However, it is well known that the major part of the propagule bank stays inactive, ensuring a sufficient reservoir for future growth (Gallandt 2006; Bohan et al. 2011b). Historically, managing these inactive propagules in arable cropping was as important as controlling the active part of the propagule bank (Gerowitt 2016; Baessler and Klotz 2006). However, since the introduction of effective herbicides, the focus has moved to controlling germinated seeds and sprouted ramets because, to be effective, herbicides need at least a radicle (embryonic root) to be absorbed by plants. Attempts to reduce reliance on chemical control by herbicides need to both manage active propagules before they harm crops and also deal with inactive propagules which provide the reservoir for future weed infestation.

The population dynamics of weeds commonly involves active and inactive propagules which both need to be accounted for in weed management (Tørresen et al. 2017). Figure 1 illustrates these dynamics. The objective of this chapter is to review advances in understanding the importance of inactive weed propagules in both annual and perennial weeds, and what influences their fate in the soil. This focus provides new perspectives on the management of weed propagules which will help develop new strategies to manage plant propagules in arable systems. We will concentrate on sites under arable production rather than under perennial/permanent crops. Geographically, the chapter looks at temperate conditions, stretching on a gradient from Nordic to semi-arid areas, but excluding tropical and sub-tropical areas. In general, we have searched for the most recent studies though we have also used broader, more explanatory studies where these have provided a better understanding of a topic. The next section describes the current management of weed propagules in arable farming, in order to discriminate between practices focusing on inactive weed propagules from those trying to activate them in weed control. It also summarises current knowledge on what we know about weed propagule dissemination in arable production. Advances in managing weed propagules are then reviewed in the following sections, and, finally, the chapter highlights new avenues for research in weed propagule management.

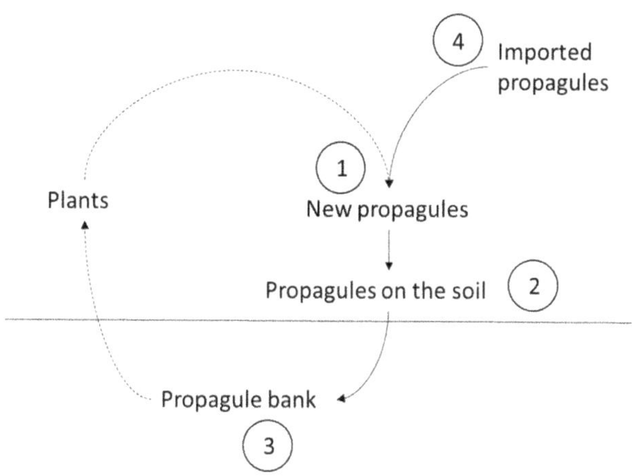

Figure 1 Inactive weed propagules and their position in population dynamics: new propagules either produced on-site (1) or imported into the field (4), before being incorporated into the soil (2) and in the soil propagule bank (3). Dashed lines - not the focus of this chapter.

2 Current and historical management of weed propagules: an overview

This section uses the concepts of active and inactive parts of the weed propagule reservoir to assess current weed management practices. We have identified three categories of weed management activities:

1. those related to cultivating the main crop,
2. those related to the period between the cultivation of the main crops, and
3. those related to movements of materials during arable farming.

2.1 Weed propagule management related to main crop management

Main crops cultivated in the field are crops rotated in time. Weed propagule management results from tilling the soil, preparing the seedbed, establishing, cultivating and harvesting the main crop. The multi-annual tillage system used for soil management affects both active and inactive propagules. The tillage system refers to the fundamental paradigm of how the soil is managed to establish the main crops. Common tillage strategies are based on ploughing, shallow conservation tillage or zero tillage, although multiple methods can be applied within a cropping system. Of those, only ploughing inverts the soil. The other two systems manage the soil without inversion and can be grouped as non-inversion tillage systems, which are preferred in conservation agriculture.

Early agronomists first developed ploughing to reduce weed infestation, and it remains a major component of arable weed management. Reducing or even avoiding ploughing in conservation agriculture or no-till systems usually leads to increased reliance on herbicides (Wiese and Steinmann 2020). The fact that most organic farmers still rely on ploughing reflects the continuing dependence on soil inversion for non-chemical weed control (Hofmeijer et al. 2021). However, ploughing today is increasingly considered harmful as it increases the risk of soil erosion; adversely affects soil structure, soil flora and microflora; and increases energy and labour costs (Nabel et al. 2021).

The tillage system strongly affects the fate of weed propagules in the soil because it influences the location of propagules and the environment they experience. They are affected by abiotic conditions resulting from burial depth, water and light availability. Propagules are also biologically connected with the activity of above-ground fauna, epigaeic fauna and sub-surface soil fauna and microbial communities. The activity of soil fauna results in predation of propagules, while soil microorganisms are responsible for the decay of propagules. Both predation and decay processes mainly affect inactive propagules and reduce the reservoir of propagules in the soil. All processes supporting soil biological activity can increase weed propagule losses by these means and contribute to successful weed management (see Section 5).

As mentioned, weed propagules are also dependent on abiotic conditions in the soil. These physical conditions will determine whether a healthy, germinated seedling emerges or not. A failure to emerge is called fatal germination (Martinková and Honěk 2013). Secondary dormancy can prevent seeds from fatal germination but not always; seedlings that do not reach the soil surface constitute losses to the soil seed bank reservoir (Tørresen et al. 2017). Though much less investigated and reported, healthy ramet sprouts also fail to emerge (Boström et al. 2013). However, as most ramets provide more resources than seeds in sustaining young plants as they make their way to the surface to become self-sustaining through photosynthesis, fatal sprouting is expected to be less frequent than fatal germination.

Inverting the soil affects inactive seeds and ramets by burying them. When weed propagule densities are very high, burial through ploughing may be beneficial. However, inverting the soil reduces the populations of many species of soil fauna and microbes (Krauss et al. 2020; Briones and Schmidt 2017) and buries the top soil, which is known to be more biologically active than deeper soil layers (Hao et al. 2021). The benefits gained from burying seeds thus may be offset by the decrease in soil biological processes, except for some specialised taxa such as earthworms, that are able to exploit the whole soil profile. In contrast, non-inversion tillage does not bury weed propagules but it has less effect on soil biological activity. However, for decay and predation to be effective against weed propagules in the long term, there needs to be a continuous maintenance of the soil microbiome and fauna through a conservation biocontrol approach.

Direct chemical or mechanical control in the main crop aims at reducing weeds which may cause yield losses by competing with crops for light, water and nutrients. These weed control measures in the main crops target active propagules because no seeds or ramets are produced. However, even pre-emergence herbicides do not affect inactive weed propagules. Herbicide take up only becomes possible with the emergence of the radicle (or embryonic root) of a seedling. The growing period offers no good opportunities to affect inactive propagules, but this changes at harvest. Combine harvesting collects seeds, as well as cutting and bailing above-ground biomass, including weed seeds and ramets, while up-rooting harvest equipment mainly collects ramets. Historically, farmers took the opportunity at harvest to take as many weed propagules away from their field as possible. While combine harvesters and other harvest machines commonly used in farming have not prioritised this activity in the past 50 years, given the reliance on herbicides for weed control in conventional systems, this situation is now changing (see Section 4).

2.2 Weed propagule management between main crops

In rotations of annual field crops, shorter or longer interim periods between the main crops allow for additional weed control. This is an opportunity to activate weed propagules by inducing germination, which then allows them to be removed as young plants. Stubble cultivation and the false seedbed technique, in particular, use this process, which activates propagules to emerge as seedlings or sprouts and then destroys the plants using tillage (which chops up the seedlings, leaves them vulnerable to desiccation through exposure on the surface or, alternatively, deprives them of light by burying them) or herbicides. Historically, the interim period between crops was also used to manually collect ramets, especially sprouted and non-sprouted rhizomes of *Elymus repens* (syn. *Agropyron repens, Elytrigia repens*) (Håkansson 1982, see Section 6.2). Competitive canopies of cover or catch crops, grown as subsidiary crops, can also suppress the growth of activated weed propagules in this period through competition for light, water and nutrients (Baraibar et al. 2017b). Shallow tillage to establish these subsidiary crops also eliminates propagules. Destroying plants by cultivation or suppressing them using subsidiary crops are competing approaches used during the interim period (Melander et al. 2016). Both have major effects on activated propagules but may also affect inactive propagules because both approaches increase microbial activity in the soil and thus support propagule decay. Repeated shallow disturbance aerates the top soil, promoting microbial activity, while subsidiary crops protect the soil and enrich it with root biomass.

2.3 Weed propagule management in movements across fields

The range of operations required in arable farming provides a third opportunity to manage weed propagules. Weed propagules are not only spread within fields. All activities where seeds, soil, fertilisers and machines move between fields can also transport weed propagules. Seeds and ramets are constantly transported short distances on farm equipment. The more contact there is between farms (e.g. due to sharing of equipment), the greater the transport of seeds and ramets. These mainly move inactive propagules, offering additional opportunities to manage them (see Section 7).

Historically, accidental propagule exchange via seeds and food or feed commodities has been very important in shaping weed vegetation worldwide (Mack 1991). In Europe, this movement of plant material started ca. 6000 years ago with the beginning of arable farming. Weeds developed alongside arable activities and some species even co-evolved with crops (Burrichter et al. 1993). Migration, trade and cooperation between farmers have significantly contributed to the species-rich arable flora in many European countries that are increasingly protected as part of the cultural heritage (Richner et al. 2015). Weed seeds historically accompanied the movement of crop seeds, while ramets accompanied materials such as rootstocks, root or tuber crops or onions, that is subterranean harvested products. The invention of seed cleaning machines in the nineteenth century, which separated weed from crop seeds, has reduced this movement at the local, regional and global levels.

3 Advances in managing inactive weed propagules

Advances in managing inactive weed propagules involve preventive, cultural, biological and physical control methods. Rather than basing this section on particular methods, we focus on the relevant target:

- The first target is collecting and physically destroying seeds and ramets associated with crops, preferably at harvest (Section 4).
- The second target is the elimination of seeds on the soil surface, which provide a food source for seed predators (i.e. they are mainly destroyed by biological means) (Section 5).
- The third target are weed propagules which are fully incorporated into the soil and can be biologically destroyed by soil microorganisms or by physical methods (Section 6).
- Finally, the process chains around arable farming offer additional possibilities to manage weed propagules (Section 7).

4 Managing weed propagules: collecting and destroying seeds in crops

As long as herbicides were an easy, cheap and efficient tool to destroy germinated seeds, collecting and destroying seeds before they were shed was little used as a control option in arable cropping. This changed fundamentally with the emergence of herbicide resistance. Australia remains the country facing the greatest problems with herbicide resistance, and it is not surprising that weed seed collection at harvest was first re-introduced on a significant scale there (Walsh et al. 2017). The practice has now spread successfully to many other parts of the world (Walsh et al. 2018).

When crops are maturing, top spraying of herbicides onto the crop stand is a selective weed control option (Pandey and Medd 1990). The herbicides affect the lemma of immature weed plants, hindering ripening (Medd et al. 1992). Top-spraying controls plants rather than viable but ungerminated weed seeds and is therefore not a control method for managing inactive weed propagules.

The preferential time for weed seed collecting is at harvest. Depending on the weed and crop species, varying quantities of seeds can potentially be collected depending on the extent to which they are shed before crop harvest. The success of weed collection has mainly been examined in cereal crops. Though levels vary according to factors such as species and weather conditions, all experiments found high numbers of weeds collected for relevant species, for example *Lolium* ssp. in Australia (85%, Walsh and Powles 2014) or *Avena fatua* in the US (75%, Shirtliffe and Entz 2005). In model studies in Scandinavia, Bitarafan and Andreasen (2020a) determined that between 16% and 53% of the seeds from the annual grass weeds *Alopecurus myosuroides* and *Apera spica venti* produced in winter wheat crops were collectable. Between 23.5% and 95.7% of the seeds produced by ten dicot weed species (*Anagallis arvensis* L., *Capsella bursa-pastoris* L. Medik, *Chenopodium album* L., *Geranium molle* L., *Persicaria maculosa* Gray, *Polygonum aviculare* L., *Silene noctiflora* L., *Sonchus arvensis* L., *Veronica persica* Poir and *Viola arvensis* Murray) in an oat crop were harvestable and thus considered good targets for seed collection at harvest. However, other species were classified as intermediate to poor targets (Bitarafan and Andreasen 2020b). Field studies in winter wheat and soybeans, where chaff was collected, indicate that a large portion of the harvestable seeds is indeed collected (Shergill et al. 2020). Despite these promising results, the strong dependence on the time of harvest and the technological accuracy required for chaff collection affect results in practice (Soni et al. 2019). Collecting seeds at harvest must consider the degree to which weed seed production and harvesting a crop coincide. Geddes and Davis (2021) have modelled this relationship in establishing the critical period for weed seed control.

There are two options after successfully collecting weed seeds at harvest:

- removing them as part of the chaff from the field, or
- destroying them during harvesting.

Moving seeds from fields requires a collecting bin. Typically the combine harvester pulls a separate chaff-collecting cart and blows the chaff into the collector. The resulting quantities of weed seeds and chaff collected are considerable and need to be deposited somewhere during harvesting (Unger and Glasner 2019). In Australia, the material dropped in windrows during harvesting is later burned on the field to kill the weed seeds (Walsh and Newman 2007). This is not allowed in Europe because of the risk of starting a wildfire. Composting the chaff is another option being explored, cereal and soybean are suitable as the process can destroy seed viability. The chaff collected has the potential to be used for other purposes such as producing bioenergy. Weiß and Glasner (2018) have, for example, explored processes like sorting and pelleting chaff to burn it for energy production.

The logistics of pulling a collecting cart and subsequent disposal of chaff can be complex and increase the costs of harvesting (Unger and Glasner 2019). This has led to developing ways to destroy the collected weed seeds instantly in the combine harvester, and milling or crushing weed seeds has now become an established commercial technology (Walsh et al. 2012). Crushing or milling weed seeds in the combine harvester has been studied in several crops, such as cereals and soybean. Three conditions influence success (Schwartz-Lazaro et al. 2017): the total amount of chaff, the feeding rate and the moisture content of the chaff.

Controlling these conditions is technically demanding. Testing the milling procedure found that seed size is not an important factor. Stationary milling was found to affect 12 species of weed seeds of different sizes equally. Seeds remaining intact after milling and buried in the soil for overwintering were significantly less viable than unprocessed seeds (Shergill et al. 2020). Beam et al. (2019) checked two weed species for their occurrence in the following crops after harvesting and milling seeds with the combine. While *Lolium perenne* in continuous winter wheat cultivation was reduced by 0–70%, *Amaranthus palmeri* in continuous soybean cultivation was reduced by 24%.

Milling or crushing are currently the preferred methods to kill weed seeds in the combine harvester. Jakobsen et al. (2019) investigated a thermal technique to destroy collected seeds, using hot exhaust gas from the engine of the combine harvester. In experiments, an exposure time of 4 and 6 seconds at 140°C destroyed the viability of all species investigated (*Alopecurus myosuroides, Centaurea cyanus, Geranium pusillum, Lapsana communis, Lolium perenne, Rumex crispus, Spergula arvensis* and *Tripleurospermum inodorum*). The conditions under which the exhaust gases can be used need to be evaluated with respect to legal safety limits. Treated like this, the dead weed seeds can stay on the field or be used as innovative energy sources and other applications (Glasner et al. 2019). Overall, the potential for further developing technologies to destroy weeds at the harvest stage is significant.

5 Managing weed propagules: predating weed seeds at the soil surface

The process of eating seeds is known as seed predation or granivory. Seeds can be eaten while they are on the mother plant (pre-dispersal predation) or after they have been dispersed (post-dispersal). Since most weed seed predation in agricultural systems is post-dispersal, this section focuses on the latter process. Post-dispersal seed predation happens mainly when seeds are on the soil surface or within the first centimetres of the soil since most (though not all) predators do not dig for buried seeds (see Paulsen et al. 2013). There are many organisms that eat seeds, either almost exclusively, like harvester ants, or as part of their diets, such as carabid beetles (Coleoptera), crickets, earthworms, isopods, granivorous mice and birds (Baraibar et al. 2009; Kulkarni et al. 2015; Westerman et al. 2003a, Fig. 2). The impact of weed seed predation by birds, isopods and earthworms is smaller compared to mammal and insect predation and has been less studied (but see Honek et al. 2009; Navntoft et al. 2009; Tschumi et al. 2018b). This section focuses on the predation of weed seeds since ramet predation is not very common.

Although much research has been done in the last 20 years, the impact of weed seed predation on weed population dynamics has not yet been fully assessed, mainly because the process seems to be very context-dependent (Davis et al. 2013). Reported weed seed predation rates range from 4 to 90% depending on the predator, weed species, location, year, landscape, crop and soil management practices, which makes generalization difficult. Predator species and populations differ greatly across different environments and latitudes (Peco et al. 2014). Ants, for example, seem to dominate in warmer

Figure 2 Weed seed predators: (a) Algerian mouse (*Mus spretus* Lataste) (photo Paula Westerman), (b) wood mouse (*Apodemus sylvaticus* L.) (photo Paula Westerman), (c) harvester ants (*Messor barbarus* L.) harvesting wild oats (*Avena sterilis*) (photo Barbara Baraibar) and (d) carabid bettle (*Harpalus refipes* DeGeer) (from https://www.wikidata.org/wiki/Q9038223).

areas (Baraibar et al. 2009; Chauhan et al. 2010; Evans and Gleeson 2016) while carabid beetles, crickets and granivorous mice are more abundant in temperate latitudes (Baraibar et al. 2012; Bohan et al. 2011a; Heggenstaller et al. 2006; Labruyere et al. 2016; Westerman et al. 2003a). Predation rates of these different organisms can vary substantially because of differences in seed consumption needs and quantities, the ability to cache seeds or not, and because they can respond differently to crop characteristics or to management practices such as tillage or pesticide use. All these factors may differ depending on when weed seed predation is measured and this can have profound impacts on reported predation rates. For example, rodents and some carabid predation rates are high when the ground is covered because it offers rodents security from their own predators and adequate moisture conditions to carabid beetles (Brown and Kotler 2004; Heggenstaller et al. 2006; Meiss et al. 2010; Shearin et al. 2008). In contrast, other predators like harvester ants do not necessarily need cover and can have high predation rates with and without ground cover (Baraibar et al. 2009). Predation rates measured at different crop stages may thus vary significantly, depending on factors such as timing and type of predator.

Another difficulty in estimating weed seed predation impact on plant population dynamics is the apparent response of plants in compensating for high seed predation rates. Pannwitt et al. (2021) showed that a reduction in seedlings of *Echinochloa crus-galli*, as a result of seed predation, was compensated by a density-dependent response that resulted in higher seedling survival and fecundity of the germinated seeds. Similarly, Swope and Parker (2010) found that granivory of *Centaurea solsticialis* by vertebrates, even if it ranged from 22 to 70%, had no effect on the number of emerged seedlings. Moreover, whilst the flowering density of *C. solsticialis* per area differed greatly, final seed production per plot did not. These results suggest that some species compensate for losses due to predation by using all flowers per unit area for seed production instead of reducing the use of some of them.

The previous discussion indicates that the degree to which field and crop management can be adjusted to favour weed seed predation and weed seed predators is site-specific and will depend on many factors such as location, crop rotation, the main weed species and the most abundant predators. However, there are some principles that can be used to guide management decisions to promote weed seed predation. These include:

- maximising weed seed exposure to predators and encounter rates, and
- providing suitable conditions for predators to locate and consume weed seeds.

These are discussed in the following sections.

5.1 Maximising weed seed exposure and encounter rates

One way to support weed seed predation is to increase the exposure of weed seeds on the soil surface after seed dispersal and maximise encounter rates. Seed encounter rates refer to the probability of a seed or a seed cache being located by predators, a necessary prerequisite for seed predation (Baraibar et al. 2011a; Holmes and Froud-Williams 2005). Even if seed exposure and encounter may seem similar, there are important differences between the two. Seeds may be exposed on the soil surface, but predators may not be able to locate them because of difficulty in accessing seeds, being distracted from weed seeds by other alternative prey or lack of synchronicity between seed exposure and predator demand for food. Models that study the dynamics of seeds after dispersal have concluded that factors related to seed availability are more important in determining overall seed losses due to predation than those related to seed demand (Westerman et al. 2006). This means that maximising seed encounter and exposure rates can increase the overall impact of seed predation on weed population dynamics.

As mentioned earlier, weed seed predators do not usually dig for buried seeds. Therefore, the most obvious way to increase exposure and encounter is to delay or eliminate any practice that can bury seeds after weed seed dispersal. Besides natural burial caused by soil cracks or rainfall events, tillage, harvest or application of manure can bury weeds seeds, thus making them less accessible to predators (Law and Gallagher 2018; Westerman et al. 2009). If weed seed dispersal is close to harvest, as is often the case, the harvest may be delayed to extend the period of weed seed availability before seeds are naturally incorporated into the seed bank (Westerman et al. 2006, 2009). Delaying harvest usually favours predators that need the crop canopy to safely locate and consume seeds such as carabid beetles, crickets and rodents (Fischer et al. 2021a; Heggenstaller et al. 2006), even if the drying out of the canopy may mitigate this effect. If crop harvest is delayed, however, caution must be taken to prevent crop losses due to predators if weed seed availability is low (Baraibar et al. 2011b). In the case of cover crops, termination can also be delayed in order to maximise weed seed exposure or to increase the time overlap between predators and exposed seeds (Gallandt et al. 2005).

Another way to maximise weed seed exposure and encounter rates is to eliminate or delay post-harvest tillage after weed seed dispersal. The soil can be left untilled to allow granivorous organisms to access the seeds for as long as possible (Baraibar et al. 2017a; Westerman et al. 2006). Longer exposure times increase the possibility of seed location and consumption (Pannwitt et al. 2017; Westerman et al. 2003b). Tillage can be performed later in the season when predator activity is lower and most weed seeds have been removed or are no longer accessible. Tillage implements vary in their effect on seed burial. Inversion tillage implements such as mouldboard ploughs tend to bury a larger proportion of seeds (Mohler et al. 2006; Scherner et al. 2016) thus making them unavailable to predators. Law and Gallagher (2018) reported that 57.5% of seed beads remained on the soil surface following the use of minimal tillage implements such as vertical coulters, rotary harrows or cultivators, while 98.2% of seed beads were recovered in a no-till control. Burial depth and seed exposure are a factor of seed size. Small seeds are usually more prone to naturally disappearing rapidly from the soil surface and moving through soil cracks into the soil seed bank but also disappear more rapidly after tillage (Bagavathiannan and Norsworthy 2013; Law and Gallagher 2018; Westerman et al. 2009). Weed seed predation is therefore likely to be lower for small seeds compared to larger ones, which means that small-seeded species may be favoured if weed seed predation is an important filter shaping weed communities (Honek et al. 2007).

Maximising weed seed encounter rates may be hampered, even if weed seed exposure is high if other competing food resources such as other seeds or insects are available at the same time as weed seeds (Baraibar et al. 2017a; Fischer et al. 2021b). Baraibar et al. (2017a) found that weed seed predation rates by harvester ants decreased after harvest in fields with a large amount of crop seeds on the soil surface left by the combine. For omnivorous carabids, Carbonne et al. (2020) showed that alternative prey, in this case, insects, decreased weed seed predation rates. Large amounts of alternative prey such as insects may also decrease weed seed predation rates through increasing intraguild predation (when one species both competes with and predates upon another species) (Schumacher et al. 2020; Tschumi et al. 2018a). In this case, omnivorous carabid beetles increased as a result of high insect populations. The high abundance of beetles then caused an increase in rodent predators consuming the beetles. As a result, overall seed predator numbers and weed seed predation rates decreased. High levels of alternative prey do not always elicit intraguild predation responses (Blubaugh et al. 2016) as sometimes high seed availability increases omnivorous carabid populations and decreases consumption of other target prey (Frank et al. 2011). While limiting the amount of insect prey in a field may be difficult, adjusting the harvesting equipment to limit crop seed return to the field when harvesting is a feasible option to increase weed seed encounter and weed seed predation rates.

5.2 Providing suitable conditions to increase seed predator populations

Increasing seed predator populations can be a way to increase weed seed predation although factors governing this process are still being investigated. Some research has found that beetle abundance is positively correlated to weed seed predation rates (González et al. 2020; Menalled et al. 2007; Trichard et al. 2013a), while other studies did not find this correlation (Fischer et al. 2021a; Fischer et al. 2011; Gallandt 2006; Saska et al. 2008). Sampling methods, weather and background seed availability have been suggested as reasons to explain this lack of correlation (Saska et al. 2008; Baraibar et al. 2017a). Nevertheless, increasing weed seed predator populations increases the chances that weed seeds are discovered and eaten.

The way to increase weed seed predator populations, however, can be very species- and genera-dependent. Granivorous mice and some insects populations favour ground cover (Heggenstaller et al. 2006; Ng et al. 2018), which can be provided by the crop (Heggenstaller et al. 2006), cover crops (Blubaugh et al. 2016; Gallandt et al. 2005 but see Lewis et al. 2020), field edges (Ng et al. 2018; Saska et al. 2007) or weeds (Bilenca et al. 2007). Other insects may not need the cover to thrive (Trichard et al. 2014; Baraibar et al. 2009) and may, instead, be more affected by tillage. Deep inversion tillage and rotary tillage can decrease weed seed predator populations, mainly of ground-dwelling arthropods (Shearin et al. 2007) by causing direct mortality and through significantly altering micro-habitat characteristics and food availability (Heggenstaller et al. 2006; Shearin et al. 2008). Minimum tillage implements such as vertical coulters, rotary harrows or cultivators do not seem to affect predator populations (Law and Gallagher 2018). However, some species, such as *Pterostichus melanarius* or *Harpalus rufipes*, may be sensitive to all tillage implements (Shearin et al. 2007). Tillage can also affect oviposition rates and larva populations of weed seed predators that spend part of their cycle underground, which is the case for some important granivorous carabids such as *H. rufipes* (Blubaugh and Kaplan 2015). In general, conservation tillage seems to promote weed seed predator populations and results in higher weed seed predation rates (Menalled et al. 2007; Petit et al. 2017; Trichard et al. 2014). However, the effect of tillage on seed predation can also depend on the landscape (i.e. the percentage of arable land) structure around the field (Fischer et al. 2011; Trichard et al. 2013b).

5.3 Final considerations

High seed availability needs to be combined with high predator activity to maximise losses to predation. If the synchronicity between these processes is low, increasing exposure may not result in the expected increase in weed seed predation rates. Studies that fail to take this into account can provide unrealistic expectations of the impact that weed seed predation has on overall population dynamics. It is therefore important to match rates of weed seed production with estimates of weed seed removal. Increasing the synchronicity between these two processes can be difficult but some management practices may help achieve it. Westerman et al. (2011) determined that late flowering crops such as sugar beet in The Netherlands can offer a better overlap between weed availability and seed predator activity than winter cereals, resulting in higher weed seed predation. Increasing exposure, as explained earlier, is another way to synchronise these two processes. Finally, there are a number of factors that may influence weed seed predation rates that are not (easily) manageable such as landscape structure (Fischer et al. 2011; Trichard et al. 2013b) or seed preference (Honek et al. 2007) but must also be taken into account when measuring removal rates.

6 Managing weed propagules below ground

Propagules below ground provide both the seed bank and the ramets resulting from the creeping root/rhizome system of arable perennials. Microbial decay, as a natural process below ground, reduces weed propagules and can be stimulated or disturbed by arable management. Physical interventions mainly destroy below-ground ramets from perennials.

6.1 Microbial decay in the soil

When buried, weed seeds are exposed to soil microorganisms, fungi and bacteria that can cause seed decay and death. Weed seed decay is related to soil microbial composition. Certain microorganisms, such as the saprophytes from the genera *Mucor*, *Rhizopus*, *Trichoderma*, *Cladosporium*, *Penicillium*, *Chaetomium* and *Aspergillus* (Wagner and Mitschunas 2008), as well as pathogenic species from *Phytium*, *Fusarium*, *Pyrenophora* or *Epicoccum* (references within Petit et al. 2018), have been described as responsible genera for weed seed decay. Although ramets from perennial weeds also experience decay, this section will focus on the decay of weed seeds because research on ramet decay of arable perennials is still very limited.

Weed seed decay by soil microorganisms seems to be a random and opportunistic saprophytic relation between a fungus and a seed (Wagner and Mitschunas 2008) or a host-pathogen interaction (Müller-Stöver et al. 2016; Petit et al. 2018), but the exact mechanisms by which a seed is killed by microorganisms are still being investigated (Petit et al. 2018). Research has shown that the ability of some microorganisms to attack seeds may be related to their physiological state. Franke et al. (2014) and Meyer et al. (2018) reported that *Fusarium tricinctum* attacked non-dormant *Bromus tectorum* L. seeds but disregarded dormant seeds. Chen et al. (2018) reported this effect by the same *Fusarium* species on two grass species in China (*Stipa bungeana* and *Lespedeza davurica*). These results suggest that a portion of so-called decayed seeds may instead be lost from the seed bank due to fatal germination caused by a fungal attack (so-called damping-off, Davis and Renner 2007). However, the extent to which this is a species-specific response, or is more widespread, is still unknown. Fungal species like *Pyrenophora semeniperda* or other Ascomycetes apparently do not require an immediate stimulus from the host seed in order to initiate growth and attack the seed and have been reported to have a mere saprophytic relationship with seeds (Chee-Sanford 2008; Meyer et al. 2018). Seed decay may be related to a suite of complex interactions among microorganisms and seeds within the soil matrix that we are just beginning to understand. For example, seeds of some species, like velvetleaf (*Abutilon theophrasti*), decay in high proportions when exposed to microorganisms in laboratory conditions but are relatively long-lived seeds in field conditions, thus suggesting that there are many factors at play that regulate the decay or persistence of weed seeds in the soil (Chee-Sanford et al. 2006).

The use of microorganisms to manage the weed seed bank has been explored for some time (Chee-Sanford et al. 2006; Kremer 1993; Kremer and Li 2003; Wagner and Mitschunas 2008) although only a few specific management options have been widely applied. One of the main challenges in using soil microbial communities for weed seed management is the limited ability to identify and characterise microbial populations and their functions in the soil. Methodological advances such as DNA sequencing, metagenomics (Kao-Kniffin et al. 2013) and metatranscriptomics (which allows the study of expressed functions) have the potential to help identify and characterise new useful microbial agents and help translate discoveries more rapidly into practice (Müller-Stöver et al. 2016).

Managing the inactive part of the weed seed bank through seed decay can follow two approaches:

- identifying and selecting microorganisms to target specific weed species and inoculating them into the soil, or
- conserving and promoting native seed-decaying microorganisms.

There are still only a few studies that use the first approach to target specific weed species in the seed bank and they have been mainly performed in pots or Petri dishes with limited applicability in field conditions (Ehlert et al. 2014; Franke et al. 2014; Fuerst et al. 2018; Müller-Stöver et al. 2009). These studies try to identify the microbial species responsible for weed seed decay, isolate them and apply them to the soil to stimulate selective decay of weeds (but not crops), acting as a bioherbicide. Other studies, mainly looking at invasive species in rangelands, target weed seeds while they are still on the plant (Medd and Campbell 2005). This approach requires a high specificity of the microbial agent to the target weed seed and still has limited applicability in cropping systems. Moreover, since the attack on the seeds can still be ongoing after the seed has been shed, the efficacy of those microorganisms also depends on the soil conditions of the location where the seed is dispersed.

Knowledge gaps in the use of microorganisms include:

- the identification of appropriate species to use,
- the development of techniques to inoculate the specific agent into the soil,
- ensuring pathogen specificity so a microorganism affects target weeds but not the crop,
- understanding how an inoculated microorganism interacts with the resident microbial community, and
- how well a microorganism persists in the soil.

These gaps require further research.

The other approach is to use the soil microbiome to stimulate weed seed decay. This approach has focused on measuring decay rates from resident microbial populations and relating them to crop and soil management practices, sometimes without identifying the responsible microorganism(s) (Chee-Sanford et al. 2006; Müller-Stöver et al. 2016; Petit et al. 2018). Management practices studied include those that alter the physical or chemical structure of the soil and which, therefore, may also change microbial communities. Increasing soil organic matter through diverse crop rotations and the incorporation of cover crops has been related to increased soil biological activity (Kremer and Li 2003) and to consequent increases in weed seed decay rates from generalists pathogenic fungi (Anderson 2015; Davis et al. 2006; Kremer and Li 2003; Mohler et al. 2012). However, this result has not held in all circumstances (Frost et al. 2019a; Gómez et al. 2014; Mohler et al. 2018; Nikolić et al. 2020; Ullrich et al. 2011), suggesting that other factors besides organic matter influence biological activity related to seed decay. Differential microbial populations or different levels of weed species susceptibility to seed decay may explain seemingly contradictory results (Gómez et al. 2014).

Different fertilization amendments can also change microbial communities (Jangid et al. 2008; Lupatini et al. 2017). Given their composition, it has been suggested that organic amendments such as compost or manure can increase weed seed decay rates more than synthetic fertilisers but differences in the nature of organic amendments can also alter those rates (De Cauwer et al. 2011; Fennimore and Jackson 2003). Finally, different tillage practices can also change soil conditions and potentially alter the microbial composition. However, weed seed decay rates do not necessarily differ in no-till compared to tilled systems or in full-tillage compared to strip-tillage systems (Frost et al. 2019b; Gallandt et al. 2004). Besides microbial composition, tillage largely influences light, oxygen, temperature and moisture conditions experienced by seeds and its effects may therefore be more related to changes in seed germination than to decay.

So far, weed seed decay is still a poorly understood and used tool to manage weed seed banks. Current advances in characterising soil microbial populations and understanding their functions may help advance the use of this management technique.

6.2 Physical and chemical methods for collecting and destroying weed propagules in the soil

Vegetative subterranean ramets are propagules of perennials. Seeds of perennials are part of the seedbank in arable fields. Since seeds are discussed elsewhere, this section focuses on ramets.

The vegetative organs of arable perennials in the soil grow mainly in three different ways:

- creeping rhizomes,
- creeping roots, and
- tap roots.

It is also important to note that the unfragmented rootstock itself is important for survival and re-infestation of weed propagules in soil. Rhizomes of *Elymus repens* and *Sorghum halepense*, as well tubers formed from rhizomes of *Cyperus esculentus* creep shallow. *Cynodon dactylon* produces stolons, that are morphological stems creeping in the top soil. Morphologically, rhizomes and stolons are subterranean stems running just below or at the surface that determines that they have an apical dominance. The four species belonging to the Poaceae family of plants, *Equisetum arvense* (a member of the Equisetopsidae family of fern plants), produce rhizomes running exceptionally deep in the soil (up to 160 cm). *Cirsium arvense, Sonchus arvense* and *Convolvulus arvense* have creeping adventitious roots. These are horizontally-running roots to promote clonal spatial growth and distribution (Klimešová and Martínková 2004).

Tap roots, for example *Rumex* ssp., *Taraxacum officinalis* or *Tussilago farfara*, are best adapted to survive in grassland systems, in which the above-ground biomass is regularly cut or grazed but the soil is not tilled. Control in grasslands is based on detecting patches and destroying the tap root as deep as possible. These species also occur in patches in arable fields and may increase if farms move to zero tillage. However, tap root species are not as well adapted to arable conditions as creeping root species. This section, therefore, focuses on creeping perennials.

Since the subterranean propagule-producing creeping system of perennial weeds involves adventitious roots or rhizomes, farmers have used mechanical means to collect propagule fragments (Fig. 3). Harrowing and hoeing whilst the main crop is growing often has a limited effect against perennials (Melander et al. 2016). The interim period between main crops offers more opportunities to reduce ramets.

There are two types of mechanical treatments that target creeping (rather than sprouted) ramets:

- one aims to lift them up close to the surface, collect and dry them out, and
- the other cuts ramets in short non-viable pieces.

Both approaches require that machines are pulled through the soil to a depth where the target ramets are located. Different machines are already in

Figure 3 Reaching propagules below ground, (a) rhizomes of *Elymus repens* (photo Jukka Salonen), (b) adventitious roots of *Cirsium arvense* (photo Marian Weigel), (c) shallow cutting equipment (photo Jukka Salonnen), (d) deep cutting equipment (photo Marian Weigel).

commercial use, particularly in Northern Europe. The Danish Kvik-up cultivator (www.kvikagro.com) and the Finish Kvik-finn (www.lyckegard.com) run at a depth of 10–15 cm. Cutting machines, for example like that developed by Kverneland (www.kvernelandgroup.com), aim at going deeper (up to 30 cm) and cutting either vertically or horizontally through the soil (Ringselle et al. 2020). Vertical cutting produces fragments, which are smaller the more often the cutter is used and is mainly used for weed species with shallow root systems. Horizontally cutting acts by chopping up creeping roots and fragments from deep roots, and is therefore suitable for weed species with deep root systems. Both approaches avoid inverting the soil between the main crops. However, the intensity of soil disturbance varies. Vertical cutting requires several passes pulling the coulters vertically through the ground. Horizontal cutting requires power to pull the blades to the targeted depth but keeps the soil surface almost undisturbed.

In practice, it is hard to test whether only sprouted ramets or non-sprouted and inactive ramets are affected by mechanical treatments. Since digging up ramets is impractical in field experiments, the number of emerging sprouts is used to assess the efficacy of mechanical treatments. We, therefore, refer to reported results and make a qualified estimate of whether non-sprouted ramets are affected.

The shallow rhizome system of *Elymus repens* (2.5–7.5 cm; Ringselle et al. 2020) can be treated mechanically with equipment moving through the upper soil layer, pulling and collecting rhizomes close to the surface and leaving them out to dry. Ringselle et al. (2020) refer to this as the desiccation strategy and suggest that dry rhizomes biomass can be collected and destroyed. Ringselle et al. (2020 and references therein) have provided an extended report on scientific results and practical experience using these mechanical techniques in Denmark, Canada, Sweden, Finland and Norway. Used on different soils and in autumn as well as in spring, the pulling effect on *E. repens* rhizomes is evident, although efficacy varies. Sprouting seems not to be necessary, enabling machines to pull out inactive rhizomes.

Pulling out does not work with adventitious roots running deeper in the soil (below 15 cm) while cutting works on rhizomes as well as on creeping roots. In rhizomes, apical dominance is disrupted by cutting, increasing the risk of more sprouting when fragments do not die after cutting. Repeated vertically cutting reduced the rhizome biomass of *E. repens* by up to 63% (Ringselle et al. 2018). Although not tested specifically, we assume that vertical fine cutting destroys both sprouted and non-sprouted ramets. Horizontal cutting is assumed to work on *Cirsium arvense* (Ringselle et al. 2020; Favreleier et al. 2020). Experimental field studies are running in Northern Europe to test efficacy against creeping perennials (Zhang et al. 2020).

Although arable soils carry thousands of weed seeds per m², there have been no advances in ways to destroy these seeds in soils mechanically. Intact, inactive seeds are tiny and hard to detect in the soil and to crush. Indeed, they are extremely well adapted to stand the various mechanical soil treatments.

Physical treatments using heat affect inactive seeds and ramets in the soil. Heating up soil can be achieved by adding hot water or steam (Melander and Kristensen 2011). If the climate allows, a more energy-efficient way is to use solar energy to heat up the soil, known as solarisation. Solarisation periodically uses a plastic cover for the soil to increase the temperature (Candido et al. 2011). Both steaming and solarisation can produce soil temperatures of > 40°C, high enough to destroy intact seeds and ramets (Birthisel and Gallandt 2019). Survival seems to depend on the thermo-resistance of weed seeds, which is further discussed in Section 7. Heating not only affects weed propagules but also soil microorganisms and fauna (van Loenen et al. 2003). Undesirable ecological side effects and high energy costs in temperate climates limit the use of the heating methods to speciality crops and high-value cropping systems, for example perennial crops like strawberries (Fennimore and Goodhue 2016). In the case of solarisation, research is investigating replacing the plastic cover with a biodegradable substrate but there is still a need to prove its effects on weed propagules (Di Mola et al. 2021).

When taking out the infested soil is possible, for example in pot or bed cultivation, other methods to heat up the soil can be used. Hess et al. (2018) tested the heating of soil taken from nature conservation areas infested with invasive species in a microwave batch system. The authors reported a significant reduction of three species of seeds and ramets. However, applications on complete arable fields were not the focus of the paper.

Chemical management of ungerminated/unsprouted propagules in the soil is limited. 'Soil active' herbicides target propagules in the soil, but they only affect active germinated seeds. Soil fumigants like methylbromide, which might penetrate an intact seed coat, are banned for field use in many parts of the world (Samtani et al. 2011). Ito and Ito (2021) recently reported chemical control of creeping perennial fragments in pot studies using soil-injected herbicides. They found chemical control reduced bud sprouting of seven species, including species which have creeping roots or rhizomes. However, they did not investigate whether chemicals were absorbed only by sprouted propagules or also by inactive propagules. Leon and van der Laat (2018) studied the effect of fermentation residues like vinasse solutions on seeds in the soil. In these experiments, germination was suppressed though intact seeds were not killed. The authors reported a twofold increase in the number of dormant seeds for *Amaranthus palmeri*, *Senna obtusifolia* and *Digitaria ciliaris* when imbibed or imbibed and germinated in vinasse solutions.

7 Managing weed propagules and seeds in modern circular process chains

In the past, cleaning harvested crops was the best way to manage inactive weed propagules. Although modern value chains for arable farming products are highly organised and regulated according to international and national rules, global trade remains a source for spreading weed seeds. As an example, the global grain trade still faces problems with the introduction of species (Wilson et al. 2016) or biotypes resistant to herbicides (Shimono et al. 2015).

The move to close material, energy and nutrient cycles on a regional and local scale increases the amount of plant biomass moving around spatially. Traditionally the biomass from a harvested arable crop was used mainly to feed livestock. Fodder crops and maize are cultivated for this purpose. The harvested biomass needs to undergo a preservation process to ensure a year-round supply. Ensiling or drying the biomass achieves this but weed seeds are known for their ability to survive these preservation processes and digestion by ruminant livestock (Westerman and Gerowitt 2013 and references therein). Livestock residues containing weed seeds are an additional source of weed propagules when they are applied as manure on arable fields (Aper et al. 2014).

A new circular process chain is evolving with the use of plant biomass as a renewable source for energy production in biogas plants. This involves moving biomass from fields into biogas plants and then moving residues back to the field. Although the products may be different, the crops and many of the processes are similar. The fermentation process in biogas plants, for example is an anaerobic process similar to ruminant digestion (Westerman and Gerowitt 2013). There is still a lack of knowledge on how far weed propagules survive a biogas process chain. The residue of biogas production is a valuable organic fertiliser which is used on many other farms. This process carries a much higher risk of unwanted spread of weed seeds than individual on-farm use of livestock manure.

Weeds commonly accompany maize biomass harvested for bioenergy use. A field survey assessing the number of ripe weed seeds contaminating the crop biomass indicated a huge range of seed numbers produced in maize crops (Westerman and Gerowitt 2012). *Chenopodium album* and *Echinochloa crus-galli* were most likely to contaminate the crop since, at harvest, they produced the highest number of ripe seeds above cutting height (15–30 cm). Unlike combined crops like cereals, where it is possible to select weed from grain seeds (see Section 4), chopping the total biomass at harvest means it is not possible to select weed seeds. As a result, weed seeds enter the biogas process chain.

Further steps in this chain are ensiling and fermenting biomass. Silage types differ in terms of ensiled substrate and ensiling conditions, which influence seed death rates (Müller and Hahn 2020). Biogas fermentation operates at mesophilic (30–37°C) or thermophilic (50–57°C) temperatures. A continuous flow-through or a batch system feeds the biomass into the fermenter. Different retention times of the biomass are required for optimal biogas production: 17–45 days in mesophilic biogas plants and ca. 15 days in thermophilic plants. Reactors can also be constructed as mixtures of batch and continuous flow (e.g. continuous in-flow but batch-type discharge) (Deublein and Steinhauser 2008). This means no two biogas installations are identical. The effective retention time and how the biomass is moved (batch or continuous type) lead to varying temperature gradients in the reactors (Deublein and Steinhauser 2008). During the anaerobic biogas process, seeds are affected by pH value (6.8–8), by the operating temperature and by microorganisms and chemicals used in the process like enzymes and acids. High temperatures and a long exposure time are the main factors in inactivating weed seeds. The time until at least half of viable seeds are inactivated varies from hours to weeks, depending on the species (Westerman and Gerowitt 2013). Although, weed species clearly differ in their ability to survive silage and anaerobic digestion, overall, the full process chain only allows a few weed seed species to survive (Simard and Lambert-Beaudet 2016; Piltz et al. 2017).

While responses on fermentation differ widely among weed species, most seeds are inactivated by ensiling, – except those with hard-seedness (Hahn et al. 2021). Hard-seedness means that seed coats or testas are water-impermeable. Anaerobic digestion in biogas plants also seems an efficient way to kill most weed seeds. However, differences found between populations of a single weed species, between the type of biogas plants (batch or continuous reactors) and even between two subsequent runs of a single reactor highlight the variation in chances of seed survival (Westerman et al. 2012a,b). The longer the substrate is kept under full reaction conditions, the more completely it will be degraded. In general, thermophilic are more efficient than mesophilic reactors. In mesophilic reactors, the average retention time should last a minimum of 10 days. Experiments by Johansen et al. (2013) suggest grass weeds and seeds of perennial species do not survive these conditions. Among the best surviving species are *Abutilon theophrasti*, *Malva neglecta* and *C. album* (Westerman et al. 2012a). Because they live in the soil, ramets of perennials normally do not enter the harvested biomass but, if they do, we can assume they are not able to survive process conditions. Though most species only survive in small numbers, the large quantities of seeds that enter the biogas process chain increase the chance of weed seed spread. In Northern Europe, the combination of high seed production and rates of survival in commercial biogas reactors makes it likely that weed seeds will survive the biogas chain, especially *C. album*.

Biomass from flower mixtures sown to enrich the biodiversity of arable landscapes is becoming an additional substrate for biogas plants (Cossel and Lewandowski 2016). Seeds from wildflower plants are similar to weed seeds and may, therefore, survive and spread (Hahn et al. 2018). However, the 'hard seedness' trait (also called physical dormancy) identified by Westerman and Gerowitt (2013) facilitates survival in various anaerobic processes, from which survival in the biogas fermentation process can be deduced too. This trait describes the ability of seeds to resist absorbing water due to water-impermeable coats. The trait is common not only in the Fabaceae and Malvaceae plant families but also occurs in other families. Wildflower mixtures cropped for biomass use should avoid species known for 'hard seedness'. It is important to distinguish this from another trait: 'seed coat thickness'. The term is often included in trait databases and characterises the physical resistance of seeds to any external attack (Schutte et al. 2014), while hard seedness is connected to water-impermeable layers in the seed coat. Seeds that do not absorb water are less sensitive to heat stress than those that can absorb water to a greater or lesser degree (Westerman and Gerowitt 2013 and references therein). This thermo-resistance of species characterised by seed hardness explains their resistance when steamed, heated in wet soils, passed through the intestinal tract of animals or passed through an anaerobic biomass fermenter. It will be important for research to address whether these two traits are linked.

8 New avenues for research

Weed control requires a systematic combination of approaches through an integrated weed management (IWM) programme. Most weed control measures, including herbicides (regardless of whether they are synthetic or bio-based), hand weeding, mechanical and physical weed control, target the management of weed seedlings, sprouts or other above-ground green plant materials. Measures like false and stale seedbeds aim to activate propagules. Weed suppression from cover crops, competitive crop cultivars, increasing crop seeding density, decreasing row distance or ridge cultivation aim at either hindering emergence or reducing growth and development of plants, due to competition from crop plants for light, water and nutrients.

However, fewer tools are available to manage weeds that are inactive (non-germinating seeds or ramets). As we previously described, the tillage system (inversion or non-inversion) is of fundamental importance for the management of inactive propagules. In Section 1, we described four targets in managing inactive weed propagules in arable farming systems (Fig. 1). These interventions offer additional control measures at harvest, on the soil surface, in the soil and in the process chain. Within the framework of IWM, the need to reduce pesticide use and support organic farming, it makes sense to use these options and to integrate them into an on-farm weed management strategy as much as possible. Research needs to address predicting how weeds respond to such changes in the weed management system.

8.1 Systemically integrating the management of inactive weed propagules

Weeds adapt to management practices. Herbicide resistance illustrates this ability of weeds to adapt but there are many more examples. There is therefore no reason to think that weeds will not also adapt to soil propagule management strategies. For example, many weed species have evolved the ability to retain seeds until crop harvest because it results in an increase in seed dispersal within the field through the action of combine harvesters and other machinery (Maity et al. 2021). In Australia selection has occurred within some weed species towards shorter times from emergence to flowering and with reduced plant height as a result of the continuous use of harvest weed seed control techniques (Ashworth et al. 2016; Sun et al. 2021). In a similar way to above-ground weed management, agroecological weed control (MacLaren et al. 2020) also needs to successfully manage below-ground weed populations sustainably over the long term.

Agroecological weed management aims at decreasing the abundance and density of problematic weed species to ensure the economic benefits of crop production (not only yield, but the balance between costs and benefits) while encouraging more diverse, and less competitive weed communities. Weed diversity sustains ecosystem processes like weed seed predation, farmland bird populations and other ecosystem services like pollination or pest management (Norris and Kogan 2005; Marshall et al. 2003). However, above-ground weed management aiming at restoring weed diversity requires a diverse propagule bank. Successful management of inactive propagules in weed populations could result in a decreased weed propagule bank. Research on managing weed propagules should therefore consider how to preserve their diversity and functionality at the agroecosystem level while limiting the negative impacts of weeds on crops. The paradigm 'low weed density - but high weed diversity' is still difficult to accept as well as to achieve by farmers, but certainly a new perspective in weed management.

8.2 The role of dormancy in propagules

Dormancy describes the ability of propagules not to germinate or sprout. Seeds in particular are either quiescent when the external conditions do not allow for germination or dormant when external conditions (temperature, water and light) are appropriate but still do not result in germination. The latter occurs due to water-impermeable layers (physical dormancy) or due to a specific chemical or environmental cue (Dalling et al. 2020). In terms of dormancy, the weed seed bank in the soil can be categorised as active and inactive parts (Tørresen et al. 2017 and references therein). Ott et al. (2019) apply the same concept to below-ground perennial buds, referring to them as the active bud bank and the dormant bud bank. Despite decades of research, the role of seed traits on persistence and dormancy is still poorly understood (Dalling et al. 2020). Dormancy in weed seeds is also far from being fully understood, and this is even more true for vegetative below-ground buds of creeping perennials (Ott et al. 2019).

Dormancy is beneficial in all plant ecosystems but is of vital importance in arable farming. Dormancy is an 'insurance' for the survival of the plant population as the behaviour of the individual propagule is highly stochastic. In arable systems dormancy ensures unpredictability. Elaborated dormancy in weed propagules might be 'the answer of nature to all the weed management attempts in farming'. Successfully managing inactive propagules would dramatically reduce the 'insurance value' of dormancy. While dormancy regulates if and how many individual weeds can be targeted through management of germinated seeds and sprouted buds, the differences between dormant/non-dormant propagules have little an effect on collecting them at harvest (4), predating them on the soil surface (5), destroying them in the soil (6) or controlling them in process chains (7). Managing inactive propagules would alter the interplay of arable management and dormancy in weed propagules. It is very likely that weeds will adapt their dormancy pattern to this situation since weeds adapt to any repeated management measures (see Section 8.1). Systematic experiments and simulations are two ways to better understand dormancy patterns in weed propagules if inactive seeds are well controlled. Tracing dormancy patterns experimentally requires long-term approaches and sophisticated measurements of 'dormancy' (Davis et al. 2016). Modelling will require appropriate data on seed behaviour (Gardarin and Colbach 2015).

8.3 Seed versus site limitation in arable systems

Taking inactive propagules out of the population cycle is the basis for destroying them in the various approaches described in Section 4–7. For annual species, the success of these approaches in weed management implies that arable weed populations are seed limited. The concept of limitations in plant recruitment originates from basic plant ecology and describes the impact of seed additions or removal on plant populations (Harper 1977). In populations that are seed limited, a relationship exists between the number of seeds added and the number of seedlings in vegetation. In contrast, microsite or establishment limitations assumes that changes in the number of seeds will have little or no effect on plant density. When seeds are added at a range of increasing densities, the resulting seed-plant curve will level off at densities above which full microsite limitation is attained as the plant population is at its carrying capacity (Münzbergová and Herben 2005 and references therein). The concept of seed versus microsite limitation is based on the ecology of natural, undisturbed or semi-natural vegetation like forests and grasslands (Clark et al. 2007; Poulsen et al. 2007; Münzbergová and Herben 2005). Two reviews (Turnbull et al. 2000; Maron and Crone 2006) concluded that species are more likely to experience seed limitation in disturbed or early successional habitats and microsite limitation in late-successional habitats.

Application of the limitation concept to arable conditions is not frequent. Boyd and van Acker (2004) and Selig et al. (2021) conducted seed addition experiments with weeds on arable sites and confirmed both types of limitations working under these conditions. Münzbergová and Herben (2005) found seed-microsite limitation gradients working on a local scale like an arable field. Unlike natural and semi-natural habitats, in arable fields, both limitations are management-driven to a high degree, either by the removal of seeds or microsites. Further research to understand, when and why which limitation dominates is urgently required. Answering these questions may have important implications for quantifying the effect of managed seed losses on weed populations. Consequences for weed management could be far-reaching. When a population size is fully microsite limited, more seeds do not matter, because they will not recruit plants anyway. In this situation, it is more rational to limit microsites as much as possible (through preventive arable management methods) rather than seeds. However, taking off seeds on-site may provoke the population to become seed limited. Then microsite limitation becomes less important. However, density-dependent seed production as observed in experiments in grasslands for *Centaurea solsticialis* (Swope and Parker 2010) and maize cropping for *Echinochloa crus-galli* (Pannwitt et al. 2021) may foil this tactic because more seeds per plant are produced in low weed density situations (see Section 5). This reaction would push the population towards and beyond the carrying capacity to avoid any seed limitation.

What limits perennials weeds in their population size at the local field scale could follow similar processes. Ott et al. (2019) use the term 'bud limitation' for all below-ground buds in plant populations including rhizomatous and adventitious root buds. The authors state that the history of disturbance management on arable fields selects for quick regeneration, that is population growth, of arable creeping perennials. Together with their capability to regrow from fragmented ramets, this requires specific research with a lateral clonal spread under arable conditions.

9 References

Anderson, R. L. (2015). Integrating a complex rotation with no-till improves weed management in organic farming: a review, *Agronomy for Sustainable Development* 35(3), 967-974.

Aper, J., de Cauwer, B., de Roo, S., Lourenço, M., Fievez, V., Bulcke, R. and Reheul, D. (2014). Seed germination and viability of herbicide resistant and susceptible chenopodium album populations after ensiling, digestion by cattle and manure storage, *Weed Research* 54(2), 169-177.

Ashworth, M. B., Walsh, M. J., Flower, K. C., Vila-Aiub, M. M. and Powles, S. B. (2016). Directional selection for flowering time leads to adaptive evolution in *Raphanus raphanistrum* (wild radish), *Evolutionary Applications* 9(4), 619-629.

Baessler, C. and Klotz, S. (2006). Effects of changes in agricultural land-use on landscape structure and arable weed vegetation over the last 50 years, *Agriculture, Ecosystems and Environment* 115(1-4), 43-50.

Bagavathiannan, M. V. and Norsworthy, J. K. (2013). Postdispersal loss of important arable weed seeds in the midsouthern United States, *Weed Science* 61(4), 570 -579.

Baraibar, B., Canadell, C., Torra, J., Royo-Esnal, A. and Recasens, J. (2017a). Weed seed fate during summer fallow: the importance of seed predation and seed burial, *Weed Science* 65(4), 515-524.

Baraibar, B., Hunter, M. C., Schipanski, M. E., Hamilton, A. and Mortensen, D. A. (2017b). Weed suppression in cover crop monocultures and mixtures, *Weed Science* 66(1), 121-133.

Baraibar, B., Carrión, E., Recasens, J. and Westerman, P. R. (2011a). Unravelling the process of weed seed predation: developing options for better weed control, *Biological Control* 56(1), 85-90.

Baraibar, B., Ledesma, R., Royo-Esnal, A. and Westerman, P. R. (2011b). Assessing yield losses caused by the harvester ant Messor barbarus (L.) in winter cereals, *Crop Protection* 30(9), 1144-1148.

Baraibar, B., Daedlow, D., de Mol, F. and Gerowitt, B. (2012). Density dependence of weed seed predation by invertebrates and vertebrates in winter wheat, *Weed Research* 52(1), 79-87.

Baraibar, B., Westerman, P. R., Carrión, E. and Recasens, J. (2009). Effects of tillage and irrigation in cereal fields on weed seed removal by seed predators, *Journal of Applied Ecology* 46(2), 380-387.

Beam, S. C., Mirsky, S., Cahoon, C., Haak, D. and Flessner, M. (2019). Harvest weed seed control of Italian ryegrass [Lolium perenne L. ssp. multiflorum (Lam.) Husnot], common ragweed (Ambrosia artemisiifolia L.), and Palmer amaranth (Amaranthus palmeri S. Watson), *Weed Technology* 33(4), 627-632.

Benvenuti, S. (2007). Weed seed movement and dispersal strategies in the agricultural environment, *Weed Biology and Management* 7(3), 141-157.

Bilenca, D. N., González-Fischer, C. M., Teta, P. and Zamero, M. (2007). Agricultural intensification and small mammal assemblages in agroecosystems of the Rolling Pampas, central Argentina, *Agriculture, Ecosystems and Environment* 121(4), 371-375.

Birthisel, S. K. and Gallandt, E. R. (2019). Trials evaluating solarization and Tarping for improved stale seedbed preparation in the Northeast USA, *Organic Farming* 5(1), 52-65.

Bitarafan, Z. and Andreasen, C. (2020a). Seed production and retention at maturity of blackgrass (Alopecurus myosuroides) and silky windgrass (Apera spica-venti) at wheat harvest, *Weed Science* 68(2), 151-156.

Bitarafan, Z. and Andreasen, C. (2020b). Seed retention of ten common weed species at oat harvest reveals the potential for harvest weed seed control, *Weed Research* 60(5), 343-352.

Blubaugh, C. K., Hagler, J. R., Machtley, S. A. and Kaplan, I. (2016). Cover crops increase foraging activity of omnivorous predators in seed patches and facilitate weed biological control, *Agriculture, Ecosystems and Environment* 231, 264-270.

Blubaugh, C. K. and Kaplan, I. (2015). Tillage compromises weed seed predator activity across developmental stages, *Biological Control* 81, 76-82.

Bohan, D. A., Boursault, A., Brooks, D. R. and Petit, S. (2011a). National-scale regulation of the weed seedbank by carabid predators, *Journal of Applied Ecology* 48(4), 888-898.

Bohan, D. A., Powers, S. J., Champion, G., Haughton, A. J., Hawes, C., Squire, G., Cussans, J. and Mertens, S. K. (2011b). Modelling rotations: can crop sequences explain arable weed seedbank abundance?, *Weed Research* 51(4), 422-432.

Boström, U., Andersson, L., Forkman, J., Hakman, I., Liew, J. and Magnuski, E. (2013). Seasonal variation in sprouting capacity from intact rhizome systems of three perennial weeds, *Weed Research* 53(5), 387-398.

Boyd, N. and van Acker, R. (2004). Seed and microsite limitations to emergence of four annual weed species, *Weed Science* 52(4), 571-577.

Briones, M. J. I. and Schmidt, O. (2017). Conventional tillage decreases the abundance and biomass of earthworms and alters their community structure in a global meta-analysis, *Global Change Biology* 23(10), 4396-4419.

Brown, J. S. and Kotler, B. P. (2004). Hazardous duty pay and the foraging cost of predation, *Ecology Letters* 7(10), 999-1014.

Burrichter, E., Hüppe, J. and Pott, R. (1993). Agrarwirtschaftlich bedingte Vegetationsbereicherung und -verarmung in historischer Sicht, *Phytocoenologia* 23(1-4), 427-447.

Candido, V., D'Addabbo, T., Miccolis, V. and Castronuovo, D. (2011). Effect of different solarizing materials on weed suppression and lettuce response, *Phytoparasitica* 40(2), 185-194.

Carbonne, B., Petit, S., Neidel, V., Foffova, H., Daouti, E., Frei, B., Skuhrovec, J., Řezáč, M., Saska, P., Wallinger, C., Traugott, M. and Bohan, D. A. (2020). The resilience of weed seedbank regulation by carabid beetles, at continental scales, to alternative prey, *Scientific Reports* 10(1), 19315.

Cauwer, B., de D'hose, T., Cougnon, M., Leroy, B., Bulcke, R. and Reheul, D. (2011) Impact of the quality of organic amendments on size and composition of the weed seed bank, *Weed Research* 51(3), 250-260.

Chauhan, B. S., Migo, T., Westerman, P. R. and Johnson, D. E. (2010). Post-dispersal predation of weed seeds in rice fields, *Weed Research* 50(6), 553-560.

Chee-Sanford, J. C. (2008). Weed seeds as nutritional resources for soil Ascomycota and characterization of specific associations between plant and fungal species, *Biology and Fertility of Soils* 44(5), 763-771.

Chee-Sanford, J. C., Williams, M. M., Davis, A. S. and Sims, G. K. (2006). Do microorganisms influence seed-bank dynamics?, *Weed Science* 54(3), 575-587.

Chen, T., Nan, Z., Zhang, X., Hou, F., Christensen, M. and Baskin, C. (2018). Does dormancy protect seeds against attack by the pathogenic fungus Fusarium

tricinctum in a semiarid grassland of Northwest China?, *Plant and Soil* 422(1-2), 155-168.
Clark, C. J., Poulsen, J. R., Levey, D. J. and Osenberg, C. W. (2007). Are plant populations seed limited? A critique and meta-analysis of seed addition experiments, *The American Naturalist* 170(1), 128-142.
Dalling, J. W., Davis, A. S., Arnold, A. E., Sarmiento, C. and Zalamea, P. C. (2020). Extending plant defense theory to seeds, *Annual Review of Ecology, Evolution, and Systematics* 51(1), 123-141.
Davis, A. S., Anderson, K. I., Hallett, S. G. and Renner, K. A. (2006). Weed seed mortality in soils with contrasting agricultural management histories, *Weed Science* 54(2), 291-297.
Davis, A. S., Fu, X., Schutte, B. J., Berhow, M. A. and Dalling, J. W. (2016). Interspecific variation in persistence of buried weed seeds follows trade-offs among physiological, chemical, and physical seed defenses, *Ecology and Evolution* 6(19), 6836-6845.
Davis, A. S. and Renner, K. A. (2007). Influence of seed depth and pathogens on fatal germination of Velvetleaf (Abutilon theophrasti) and giant foxtail (*Setaria faberi*), *Weed Science* 55(1), 30-35.
Davis, A. S., Taylor, E. C., Haramoto, E. R. and Renner, K. A. (2013). Annual postdispersal weed seed predation in contrasting field environments, *Weed Science* 61(2), 296-302.
Deublein, D. and Steinhauser, A. (2008). *Biogas from Waste and Renewable Resources*, Weinheim, Germany, Wiley-VCH Verlag GmbH & Co. KGaA.
Di Mola, I., Ventorino, V., Cozzolino, E., Ottaiano, L., Romano, I., Duri, L. G., Pepe, O. and Mori, M. (2021). Biodegradable mulching vs traditional polyethylene film for sustainable solarization: chemical properties and microbial community response to soil management, *Applied Soil Ecology* 163, 103921.
Ehlert, K. A., Mangold, J. M. and Engel, R. E. (2014). Integrating the herbicide imazapic and the fungal pathogen Pyrenophora semeniperda to control Bromus tectorum, *Weed Research* 54(4), 418-424.
Evans, T. A. and Gleeson, P. V. (2016). Direct measurement of ant predation of weed seeds in wheat cropping, *Journal of Applied Ecology* 53(4), 1177-1185.
Favrelière, E., Ronceux, A., Pernel, J. and Meynard, J.-M. (2020). Nonchemical control of a perennial weed, Cirsium arvense, in arable cropping systems: a review, *Agronomy for Sustainable Development* 40(4). DOI: 10.1007/s13593-020-00635-2.
Fennimore, S. A. and Goodhue, R. E. (2016). Soil disinfestation with steam: a review of economics, engineering, and soil pest control in California strawberry, *International Journal of Fruit Science* 16(sup1), 71-83.
Fennimore, S. A. and Jackson, L. E. (2003). Organic amendment and tillage effects on vegetable field weed emergence and Seedbanks 1, *Weed Technology* 17(1), 42-50.
Fischer, C., Riesch, F., Tscharntke, T. and Batáry, P. (2021). Large carabids enhance weed seed removal in organic fields and in large-scale, but not small-scale agriculture, *Landscape Ecology* 36(2), 427-438.
Fischer, C., Thies, C. and Tscharntke, T. (2011). Mixed effects of landscape complexity and farming practice on weed seed removal, *Perspectives in Plant Ecology, Evolution and Systematics* 13(4), 297-303.
Frank, S. D., Shrewsbury, P. M. and Denno, R. F. (2011). Plant versus prey resources: influence on omnivore behavior and herbivore suppression, *Biological Control* 57(3), 229-235.

Franke, J., Geary, B. and Meyer, S. E. (2014). Identification of the infection route of a Fusarium seed pathogen into nondormant *Bromus tectorum* seeds, *Phytopathology* 104(12), 1306–1313.

Frost, M. D., Haramoto, E. R., Renner, K. A. and Brainard, D. C. (2019). Tillage and cover crop effects on weed seed persistence: do light exposure and fungal pathogens play a role?, *Weed Science* 67(1), 103–113.

Fuerst, E. P., James, M. S., Pollard, A. T. and Okubara, P. A. (2018). Defense enzyme responses in dormant wild oat and wheat caryopses challenged with a seed decay pathogen, *Frontiers in Plant Science* 8, 2259.

Gallandt, E. R. (2006). How can we target the weed seedbank?, *Weed Science* 54(3), 588–596.

Gallandt, E. R., Fuerst, E. P. and Kennedy, A. C. (2004). Effect of tillage, fungicide seed treatment, and soil fumigation on seed bank dynamics of wild oat (Avena fatua), *Weed Science* 52(4), 597–604.

Gallandt, E. R., Molloy, T., Lynch, R. P. and Drummond, F. A. (2005). Effect of cover-cropping systems on invertebrate seed predation, *Weed Science* 53(1), 69–76.

Gardarin, A. and Colbach, N. (2015). How much of seed dormancy in weeds can be related to seed traits?, *Weed Research* 55(1), 14–25.

Geddes, C. M. and Davis, A. S. (2021). The critical period for weed seed control: a proposed framework to limit weed seed return, *Weed Research* 61(4), 282–287.

Gerowitt, B. (2016). Zum Nutzen der Artenvielfalt bei Ackerunkräutern für das Unkrautmanagement, *Julius-Kühn-Archiv* 452, 13–18.

Glasner, C., Vieregge, C., Robert, J., Fenselau, J., Bitarafan, Z. and Andreasen, C. (2019). Evaluation of new harvesting methods to reduce weeds on arable fields and collect a new feedstock, *Energies* 12(9), 1688.

Gómez, R., Liebman, M. and Munkvold, G. (2014). Weed seed decay in conventional and diversified cropping systems, *Weed Research* 54(1), 13–25.

González, E., Seidl, M., Kadlec, T., Ferrante, M. and Knapp, M. (2020). Distribution of ecosystem services within oilseed rape fields: effects of field defects on pest and weed seed predation rates, *Agriculture, Ecosystems & Environment* 295, 106894.

Gonzalez-Andujar, J. L., Chantre, G. R., Morvillo, C., Blanco, A. M. and Forcella, F. (2016). Predicting field weed emergence with empirical models and soft computing techniques, *Weed Research* 56(6), 415–423.

Hahn, J., de Mol, F. and Müller, J. (2021). Ensiling reduces seed viability: implications for weed management, *Frontiers in Agronomy* 3, 708851. DOI: 10.3389/fagro.2021.708851.

Hahn, J., Parzych, D., Schulz, J., Westermann, P. R. and Gerowitt, B. (2018). Wildflower seeds in the biogas reactor: screening the risk of survival of different species, *Julius-Kühn-Archiv* 458, 41–46.

Håkansson, S. (1982). Multiplication, growth and persistence of perennial weeds. In: *Biology and Ecology of Weeds. Springer, Dordrecht*, pp. 123–135. DOI: 10.1007/978-94-017-0916-3.

Hao, J., Chai, Y. N., Lopes, L. D., Ordóñez, R. A., Wright, E. E., Archontoulis, S. and Schachtman, D. P. (2021). The effects of soil depth on the structure of microbial communities in agricultural soils in Iowa, USA, *Applied and Environmental Microbiology* 87, e02673-20. DOI: 10.1128/AEM.02673-20.

Haring, S. C., and Flessner, M. L. (2018). Improving soil seed bank management. *Pest Management Science* 74 (11), 2412–2418. DOI: 10.1002/ps.5068.

Harper, J. L. (1977). *Population Biology of Plants*, HarperCollins Publishers Ltd. 857 pp.

Hatcher P. E. (2017). Perennial Weeds. In: Hatcher, P. E. and Froud-Williams, R. J. (Eds.) *Weed research: Expanding horizons*, Hoboken, NJ, John Wiley & Sons Ltd, 389–412.

Heggenstaller, A. H., Menalled, F. D., Liebman, M. and Westerman, P. R. (2006). Seasonal patterns in post-dispersal seed predation of Abutilon theophrasti and Setaria faberi in three cropping systems, *Journal of Applied Ecology* 43(5), 999–1010.

Hess, M. C. M., de Wilde, M. , Yavercovski, N., Willm, L., Mesléard, F. and Buisson, E. (2018). Microwave soil heating reduces seedling emergence of a wide range of species including invasives, *Restoration Ecology* 26, S160–S169.

Hofmeijer, M. A. J., Melander, B., Salonen, J., Lundkvist, A., Zarina, L. and Gerowitt, B. (2021). Crop diversification affects weed communities and densities in organic spring cereal fields in northern Europe, *Agriculture, Ecosystems and Environment* 308, 107251.

Holmes, R. J. and Froud-Williams, R. J. (2005). Post-dispersal weed seed predation by avian and non-avian predators, *Agriculture, Ecosystems and Environment* 105(1–2), 23–27.

Honek, A., Martinkova, Z., Saska, P. and Koprdova, S. (2009). Role of post-dispersal seed and seedling predation in establishment of dandelion (Taraxacum agg.) plants, *Agriculture, Ecosystems and Environment* 134(1–2), 126–135.

Honek, A., Martinkova, Z., Saska, P. and Pekar, S. (2007). Size and taxonomic constraints determine the seed preferences of Carabidae (Coleoptera), *Basic and Applied Ecology* 8(4), 343–353.

Ito, M. and Ito, K. (2021). Establishing targeted control of creeping perennial weeds with soil-active chemical injections: assessment of subterranean bud responses in contact, *Weed Biology and Management* 21(1), 28–33.

Jakobsen, K., Jensen, J. A., Bitarafan, Z. and Andreasen, C. (2019). Killing weed seeds with exhaust gas from a combine harvester, *Agronomy* 9(9), 544.

Jangid, K., Williams, M. A., Franzluebbers, A. J., Sanderlin, J. S., Reeves, J. H., Jenkins, M. B., Endale, D. M., Coleman, D. C. and Whitman, W. B. (2008). Relative impacts of land-use, management intensity and fertilization upon soil microbial community structure in agricultural systems, *Soil Biology and Biochemistry* 40(11), 2843–2853.

Johansen, A., Nielsen, H. B., Hansen, C. M., Andreasen, C., Carlsgart, J., Hauggard-Nielsen, H. and Roepstorff, A. (2013). Survival of weed seeds and animal parasites as affected by anaerobic digestion at meso- and thermophilic conditions, *Waste Management (New York, N.Y.)* 33(4), 807–812.

Kao-Kniffin, J., Carver, S. M. and DiTommaso, A. (2013). Advancing weed management strategies using metagenomic techniques, *Weed Science* 61(2), 171–184.

Klimešová, J. and Martínková, J. (2004). Intermediate growth forms as a model for the study of plant clonality functioning: an example with root sprouters, *Evolutionary Ecology* 18(5–6), 669–681.

Krauss, M., Berner, A., Perrochet, F., Frei, R., Niggli, U. and Mäder, P. (2020). Enhanced soil quality with reduced tillage and solid manures in organic farming – a synthesis of 15 years, *Scientific Reports* 10(1), 4403. DOI: 10.1038/s41598-020-61320-8.

Kremer, R. J. (1993). Management of weed seed banks with microorganisms, *Ecological Applications* 3(1), 42–52.

Kremer, R. J. and Li, J. (2003). Developing weed-suppressive soils through improved soil quality management, *Soil and Tillage Research* 72(2), 193–202.

Kulkarni, S. S., Dosdall, L. M. and Willenborg, C. J. (2015). The role of ground beetles (Coleoptera: Carabidae) in weed seed consumption: a review, *Weed Science* 63(2), 355–376.

Labruyere, S., Bohan, D. A., Biju-Duval, L., Ricci, B. and Petit, S. (2016). Local, neighbour and landscape effects on the abundance of weed seed-eating carabids in arable fields: a nationwide analysis, *Basic and Applied Ecology* 17(3), 230–239.

Law, J. J. and Gallagher, R. S. (2018). Seed distribution and invertebrate seed predation in No-Till and minimum-till maize systems, *Agronomy Journal* 110(6), 2488–2495.

Leon, R. G. and van der Laat, R. (2018). Herbicidal and seed dormancy induction activity of fermentation residual vinasse, *Weed Science* 66(3), 317–323.

Lewis, D. G., Cutulle, M. A., Schmidt-Jeffris, R. A. and Blubaugh, C. K. (2020). Better together? Combining cover crop mulches, organic herbicides, and weed seed biological control in reduced-tillage systems, *Environmental Entomology* 49(6), 1327–1334.

Lupatini, M., Korthals, G. W., de Hollander, M. , Janssens, T. K. S. and Kuramae, E. E. (2017). Soil microbiome is more heterogeneous in organic than in conventional farming system, *Frontiers in Microbiology* 7, 2064.

Mack, R. N. (1991). The commercial seed trade: an early disperser of weeds in the United States, *Economic Botany* 45(2), 257–273.

MacLaren, C., Storkey, J., Menegat, A., Metcalfe, H. and Dehnen-Schmutz, K. (2020). An ecological future for weed science to sustain crop production and the environment: a review, *Agronomy for Sustainable Development* 40(4), 24.

Maity, A., Lamichaney, A., Joshi, D. C., Bajwa, A., Subramanian, N., Walsh, M., and Bagavathiannan, M. (2021). Seed shattering: a trait of evolutionary importance in plants. *Frontiers in Plant Science* 12 (June), 1–17. DOI: 10.3389/fpls.2021.657773.

Maron, J. L. and Crone, E. (2006). Herbivory: effects on plant abundance, distribution and growth, *Proceedings of the Royal Society of Biological Sciences* 273(1601), 2575–2584.

Marshall, E. J. P., Brown, V. K., Boatman, N. D., Lutman, P. J. W., Squire, G. R. and Ward, L. K. (2003). The role of weeds in supporting biological diversity within crop fields, *Weed Research* 43(2), 77–89.

Martinková, Z. and Honěk, A. (2013). Fatal germination in barnyardgrass (*Echinochloa crus-galli*), *Plant Protection Science* 49(4), 193–197.

Medd, R. W. and Campbell, M. A. (2005). Grass seed infection following inundation with Pyrenophora semeniperda, *Biocontrol Science and Technology* 15(1), 21–36.

Medd, R. W., McMillan, M. G. and Cook, A. S. (1992). Spray-topping of wild oats (Avena spp.) in wheat with selective herbicides, *Plant Protection Quarterly* 7, 62–65.

Meiss, H., Le Lagadec, L., Munier-Jolain, N., Waldhardt, R. and Petit, S. (2010). Weed seed predation increases with vegetation cover in perennial forage crops, *Agriculture, Ecosystems and Environment* 138(1–2), 10–16.

Melander, B., and Kristensen, J. K. (2011). Soil steaming effects on weed seedling emergence under the influence of soil type, soil moisture, soil structure and heat duration. *Annals of Applied Biology* 158 (2), 194-203. DOI: 10.1111/J.1744-7348.2010.00453.X.

Melander, B., Rasmussen, I. A. and Olesen, J. E. (2016). Incompatibility between fertility building measures and the management of perennial weeds in organic cropping systems, *Agriculture, Ecosystems and Environment* 220, 184–192.

Menalled, F. D., Smith, R. G., Dauer, J. T. and Fox, T. B. (2007). Impact of agricultural management on carabid communities and weed seed predation, *Agriculture, Ecosystems and Environment* 118(1–4), 49–54.

Meyer, S. E., Beckstead, J. and Allen, P. S. (2018). Niche specialization in Bromus tectorum seed bank pathogens, *Seed Science Research* 28(3), 215–221.

Mohler, C., Frisch, J. and McCulloch, C. (2006). Vertical movement of weed seed surrogates by tillage implements and natural processes, *Soil and Tillage Research* 86(1), 110–122.

Mohler, C. L., Dykeman, C., Nelson, E. B. and Ditommaso, A. (2012). Reduction in weed seedling emergence by pathogens following the incorporation of green crop residue, *Weed Research* 52(5), 467–477.

Mohler, C. L., Taylor, A. G., DiTommaso, A., Hahn, R. R. and Bellinder, R. R. (2018). Effects of incorporated rye and hairy vetch cover crop residue on the persistence of weed seeds in the soil, *Weed Science* 66(3), 379–385.

Müller, J. and Hahn, J. (2020). Ensilability of biomass From effloresced flower strips as co-substrate in bioenergy production, *Frontiers in Bioengineering and Biotechnology* 8, 14. DOI: 10.3389/fbioe.2020.00014.

Müller-Stöver, D., Kohlschmid, E. and Sauerborn, J. (2009). A novel strain of Fusarium oxysporum from Germany and its potential for biocontrol of Orobanche ramosa, *Weed Research* 49(2), 175–182.

Müller-Stöver, D., Nybroe, O., Baraibar, B., Loddo, D., Eizenberg, H., French, K., Sønderskov, M., Neve, P., Peltzer, D. A., Maczey, N. and Christensen, S. (2016). Contribution of the seed microbiome to weed management, *Weed Research* 56(5), 335–339.

Münzbergová, Z. and Herben, T. (2005). Seed, dispersal, microsite, habitat and recruitment limitation: identification of terms and concepts in studies of limitations, *Oecologia* 145(1), 1–8.

Nabel, M., Selig, C., Gundlach, J., von der Decken, H. and Klein, M. (2021). Biodiversity in agricultural used soils: threats and options for its conservation in Germany and Europe, *Soil Organisms* 93(1), 1–11.

Navntoft, S., Wratten, S. D., Kristensen, K. and Esbjerg, P. (2009). Weed seed predation in organic and conventional fields, *Biological Control* 49(1), 11–16.

Ng, K., McIntyre, S., Macfadyen, S., Barton, P. S., Driscoll, D. A. and Lindenmayer, D. B. (2018). Dynamic effects of ground-layer plant communities on beetles in a fragmented farming landscape, *Biodiversity and Conservation* 27(9), 2131-2153.

Nikolić, N., Squartini, A., Concheri, G., Stevanato, P., Zanin, G. and Masin, R. (2020). Weed seed decay in no-till field and planted riparian buffer zone, *Plants (Basel, Switzerland)* 9(3), 293.

Norris, R. F. and Kogan, M. (2005). Ecology of interactions between weeds and arthropods, *Annual Review of Entomology* 50, 479–503.

Ott, J. P., Klimešová, J. and Hartnett, D. C. (2019). The ecology and significance of below-ground bud banks in plants, *Annals of Botany* 123(7), 1099–1118.

Pandey, S. and Medd, R. W. (1990). Integration of seed and plant kill tactics for control of wild oats: an economic evaluation, *Agricultural Systems* 34(1), 65–76.

Pannwitt, H., Westerman, P. R., de Mol, F. and Gerowitt, B. (2021). Demographic processes allow *Echinochloa crus-galli* to compensate seed losses by seed predation, *Agronomy* 11(3), 565.

Pannwitt, H., Westerman, P. R., de Mol, F., Selig, C. and Gerowitt, B. (2017). Biological control of weed patches by seed predators; responses to seed density and exposure time, *Biological Control* 108, 1–8.

Paulsen, T. R., Colville, L., Kranner, I., Daws, M. I., Högstedt, G., Vandvik, V. and Thompson, K. (2013). Physical dormancy in seeds: a game of hide and seek? *New Phytologist*, 198 (2), 496-503. DOI: 10.1111/nph.12191.

Peco, B., Laffan, S. W. and Moles, A. T. (2014). Global patterns in post-dispersal seed removal by invertebrates and vertebrates, *PLoS ONE* 9(3), e91256.

Petit, S., Cordeau, S., Chauvel, B., Bohan, D., Guillemin, J.-P. and Steinberg, C. (2018). Biodiversity-based options for arable weed management: a review, *Agronomy for Sustainable Development* 38(5), 1-21.

Petit, S., Trichard, A., Biju-Duval, L., McLaughlin, Ó.-B. and Bohan, D. A. (2017). Interactions between conservation agricultural practice and landscape composition promote weed seed predation by invertebrates, *Agriculture, Ecosystems and Environment* 240, 45-53.

Piltz, J. W., Stanton, R. A. and Wu, H. (2017). Effect of ensiling and in sacco digestion on the viability of seeds of selected weed species, *Weed Research* 57(6), 382-389.

Poulsen, J. R., Osenberg, C. W., Clark, C. J., Levey, D. J. and Bolker, B. M. (2007). Plants as reef fish: fitting the functional form of seedling recruitment, *The American Naturalist* 170(2), 167-183.

Richner, N., Holderegger, R., Linder, H. P. and Walter, T. (2015). Reviewing change in the arable flora of Europe: a meta-analysis, *Weed Research* 55(1), 1-13.

Ringselle, B., Bertholtz, E., Magnuski, E., Brandsæter, L. O., Mangerud, K. and Bergkvist, G. (2018). Rhizome fragmentation by vertical disks reduces Elymus repens growth and benefits Italian ryegrass-white clover crops, *Frontiers in Plant Science* 8, 2243.

Ringselle, B., de Cauwer, B., Salonen, J. and Soukup, J. (2020). A review of non-chemical management of couch grass (*Elymus repens*), *Agronomy* 10(8), 1178. DOI: 10.3390/agronomy10081178.

Samtani, J. B., Ajwa, H. A., Weber, J. B., Browne, G. T., Klose, S., Hunzie, J. and Fennimore, S. A. (2011). Evaluation of non-fumigant alternatives to methyl bromide for weed control and crop yield in California strawberries (*Fragaria ananassa L.*), *Crop Protection* 30(1), 45-51.

Saska, P., van der Werf, W., Vries, E. and Westerman, P. R. (2008) Spatial and temporal patterns of carabid activity-density in cereals do not explain levels of predation on weed seeds, *Bulletin of Entomological Research* 98(2), 169-181.

Saska, P., Vodde, M., Heijerman, T., Westerman, P. and van der Werf, W. (2007). The significance of a grassy field boundary for the spatial distribution of carabids within two cereal fields, *Agriculture, Ecosystems and Environment* 122(4), 427-434.

Scherner, A., Melander, B. and Kudsk, P. (2016). Vertical distribution and composition of weed seeds within the plough layer after eleven years of contrasting crop rotation and tillage schemes, *Soil and Tillage Research* 161, 135-142.

Schumacher, M., Dieterich, M. and Gerhards, R. (2020). Effects of weed biodiversity on the ecosystem service of weed seed predation along a farming intensity gradient, *Global Ecology and Conservation* 24, e01316.

Schutte, B. J., Davis, A. S., Peinado, S. A. and Ashigh, J. (2014). Seed-coat thickness data clarify seed size-seed-bank persistence trade-offs in Abutilon theophrasti (Malvaceae), *Seed Science Research* 24(2), 119-131.

Schwartz-Lazaro, L. M., Norsworthy, J. K., Walsh, M. J. and Bagavathiannan, M. V. (2017). Efficacy of the integrated Harrington seed destructor on weeds of soybean and rice production systems in the Southern United States, *Crop Science* 57(5), 2812-2818.

Selig, C., de Mol, F., Westerman, P. R. and Gerowitt, B. (2021). Quantifying seed and establishment limitation to seedling recruitment of arable weeds–an example of *Echinochloa crus-galli*, *Weed Science*, 70 (1), 87–94. DOI: 10.1017/wsc.2021.68.

Shearin, A. F., Reberg-Horton, S. C. and Gallandt, E. R. (2007). Direct effects of tillage on the activity density of ground beetle (Coleoptera: Carabidae) weed seed predators, *Environmental Entomology* 36(5), 1140–1146.

Shearin, A. F., Reberg-Horton, S. C. and Gallandt, E. R. (2008). Cover crop effects on the activity-density of the weed seed predator Harpalus rufipes (Coleoptera: Carabidae), *Weed Science* 56(3), 442–450.

Shergill, L. S., Bejleri, K., Davis, A. and Mirsky, S. B. (2020). Fate of weed seeds after impact mill processing in midwestern and mid-atlantic United States, *Weed Science* 68(1), 92–97.

Shimono, Y., Shimono, A., Oguma, H., Konuma, A. and Tominaga, T. (2015). Establishment of Lolium species resistant to acetolactate synthase-inhibiting herbicide in and around grain-importation ports in Japan, *Weed Research* 55(1), 101–111.

Shirtliffe, S. J. and Entz, M. H. (2005). Chaff collection reduces seed dispersal of wild oat (Avena fatua) by a combine harvester, *Weed Science* 53(4), 465–470.

Simard, M.-J. and Lambert-Beaudet, C. (2016). Weed seed survival in experimental mini-silos of corn and alfalfa, *Canadian Journal of Plant Science* 96(3), 448–454.

Soni, N., Nissen, S. J., Westra, P., Norsworthy, J. K., Walsh, M. J. and Gaines, T. A. (2019). Seed retention of winter annual grass weeds at winter wheat harvest maturity shows potential for harvest weed seed control, *Weed Technology* 34(2), 266–271.

Sun, C., Ashworth, M. B., Flower, K., Vila-Aiub, M. M., Rocha, R. L. and Beckie, H. J. (2021). The adaptive value of flowering time in wild radish (Raphanus raphanistrum), *Weed Science* 69(2), 203–209.

Swope, S. M. and Parker, I. M. (2010). Widespread seed limitation affects plant density but not population trajectory in the invasive plant Centaurea solstitialis, *Oecologia* 164(1), 117–128.

Tørresen, K. S., Karlsson, L. M. and Gonzalez-Andujar, J. L. (2017). Seed biology and population dynamics. In: Hatcher, P. E. and Froud-Williams, R. J. (Eds), *Weed Research: Expanding Horizons*, Hoboken, NJ, John Wiley & Sons Ltd, pp. 85–113.

Trichard, A., Alignier, A., Biju-Duval, L. and Petit, S. (2013). The relative effects of local management and landscape context on weed seed predation and carabid functional groups, *Basic and Applied Ecology* 14(3), 235–245.

Trichard, A., Ricci, B., Ducourtieux, C. and Petit, S. (2014). The spatio-temporal distribution of weed seed predation differs between conservation agriculture and conventional tillage, *Agriculture, Ecosystems and Environment* 188, 40–47.

Tschumi, M., Ekroos, J., Hjort, C., Smith, H. G. and Birkhofer, K. (2018a). Predation-mediated ecosystem services and disservices in agricultural landscapes, *Ecological Applications* 28(8), 2109–2118.

Tschumi, M., Ekroos, J., Hjort, C., Smith, H. G. and Birkhofer, K. (2018b). Rodents, not birds, dominate predation-related ecosystem services and disservices in vertebrate communities of agricultural landscapes, *Oecologia* 188(3), 863–873.

Turnbull, L. A., Crawley, M. J. and Rees, M. (2000). Are plant populations seed-limited? A review of seed sowing experiments, *Oikos* 88(2), 225–238.

Ullrich, S. D., Buyer, J. S., Cavigelli, M. A., Seidel, R. and Teasdale, J. R. (2011). Weed seed persistence and microbial abundance in long-term organic and conventional cropping systems, *Weed Science* 59(2), 202–209.

Unger, J. S. and Glasner, C. (2019). Cost analysis of chaff harvesting concepts in Germany, *Agronomy* 9(10), 579.

van Loenen, M. C. A., Turbett, Y., Mullins, C. E., Feilden, N. E. H., Wilson, M. J., Leifert, C. and Seel, W. E. (2003). Low temperature-short duration steaming of soil kills soil-borne pathogens, nematode pests and weeds, *European Journal of Plant Pathology* 109(9), 993–1002.

von Cossel, M. and Lewandowski, I. (2016). Perennial wild plant mixtures for biomass production: impact of species composition dynamics on yield performance over a five-year cultivation period in southwest Germany, *European Journal of Agronomy* 79, 74–89.

Wagner, M. and Mitschunas, N. (2008). Fungal effects on seed bank persistence and potential applications in weed biocontrol: a review, *Basic and Applied Ecology* 9(3), 191–203.

Walsh, M. and Newman, P. (2007). Burning narrow windrows for weed seed destruction, *Field Crops Research* 104(1–3), 24–30. DOI: 10.1016/J.FCR.2007.05.012.

Walsh, M., Ouzman, J., Newman, P., Powles, S. and Llewellyn, R. (2017). High levels of adoption indicate that harvest weed seed control is now an established weed control practice in Australian cropping, *Weed Technology* 31(3), 341 –347.

Walsh, M. J., Broster, J. C., Schwartz-Lazaro, L. M., Norsworthy, J. K., Davis, A. S., Tidemann, B. D., Beckie, H. J., Lyon, D. J., Soni, N., Neve, P. and Bagavathiannan, M. V. (2018). Opportunities and challenges for harvest weed seed control in global cropping systems, *Pest Management Science* 74(10), 2235–2245.

Walsh, M. J., Harrington, R. B. and Powles, S. B. (2012). Harrington seed destructor: a new nonchemical weed control tool for global grain crops, *Crop Science* 52(3), 1343–1347.

Walsh, M. J. and Powles, S. B. (2014). High seed retention at maturity of annual weeds infesting crop fields highlights the potential for harvest weed seed control, *Weed Technology* 28(3), 486–493.

Weiß, B. D. and Glasner, C. (2018). Evaluation of the process steps of pretreatment, pellet production and combustion for an energetic utilization of wheat chaff, *Frontiers in Environmental Science* 6, 36. DOI: 10.3389/fenvs.2018.00036.

Westerman, P., Luijendijk, C. D., Wevers, J. D. A., and Van Der Werf, W. (2011). Weed seed predation in a phenologically late crop. *Weed Research* 51 (2), 157–164. DOI: 10.1111/J.1365-3180.2010.00834.X.

Westerman, P. R., Dixon, P. M. and Liebman, M. (2009). Burial rates of surrogate seeds in arable fields, *Weed Research* 49(2), 142–152.

Westerman, P. R. and Gerowitt, B. (2012). The probability of maize biomass contamination with weed seeds, *Journal of Plant Diseases and Protection* 119(2), 68–73.

Westerman, P. R. and Gerowitt, B. (2013). Weed seed survival during anaerobic digestion in biogas plants, *The Botanical Review* 79(3), 281–316.

Westerman, P. R., Heiermann, M., Pottberg, U., Rodemann, B. and Gerowitt, B. (2012a). Weed seed survival during mesophilic anaerobic digestion in biogas plants, *Weed Research* 52(4), 307–316.

Westerman, P. R., Hildebrandt, F. and Gerowitt, B. (2012b). Weed seed survival following ensiling and mesophilic anaerobic digestion in batch reactors, *Weed Research* 52(3), 286–295.

Westerman, P. R., Hofman, A., Vet, L. E. M. and van der Werf, W. (2003a). Relative importance of vertebrates and invertebrates in epigeaic weed seed predation in organic cereal fields, *Agriculture, Ecosystems and Environment* 95(2–3), 417–425.
Westerman, P. R., Wes, J. S., Kropff, M. J. and van der Werf, W. (2003b). Annual losses of weed seeds due to predation in organic cereal fields, *Journal of Applied Ecology* 40(5), 824–836.
Westerman, P. R., Liebman, M., Heggenstaller, A. H. and Forcella, F. (2006). Integrating measurements of seed availability and removal to estimate weed seed losses due to predation, *Weed Science* 54(3), 566–574.
Wiese, A. and Steinmann, H.-H. (2020). Yield effects associated with glyphosate use in non-GMO arable farming: a review, *Crop Protection* 134, 105148.
Wilson, C. E., Castro, K. L., Thurston, G. B. and Sissons, A. (2016). Pathway risk analysis of weed seeds in imported grain: a Canadian perspective, *NeoBiota* 30, 49. DOI: 10.3897/neobiota.30.7502.
Zhang, H., Andert, S., Brandsæter, L. O., Rasmussen, J., Robin, M.-H., Salonen, J., Tørresen, K. S., Valantin-Morison, M. and Gerowitt, B. (2020). Future management of arable perennials - an introduction to the project AC/DC-weeds, *Julius-Kühn-Archiv* 464, 280–285.

Chapter 4

Advances in understanding allelopathic interactions between weeds and crops

Çağla Görkem Eroğlu[1] and Aurélie Gfeller[1], Agroscope, Plant Production Systems, Herbology in Field Crops, Switzerland; Anna Elizabeth Louw-Gaume, Agroscope, Corporate Strategy, Switzerland; and Judith Wirth, Agroscope, Plant Production Systems, Herbology in Field Crops, Switzerland

1 Introduction

Farmers and agricultural scientists are facing a major challenge to ensure food security for the rapidly growing world population through sustainable crop production practices. Both the challenges of demographic pressure and environmental ecology require innovative and smart solutions to counteract further negative consequences in the future, despite the current difficulties in reconciling these goals (Gaffney et al., 2019).

Weed management is the most representative example of this apparent discrepancy, since the use of chemical herbicides is a cheap and practical solution that has guaranteed farmers weed-free plots for the past 50–70 years. However, the continued use of herbicides is a debatable issue, given their detrimental impacts on the environment and the potential consequences on animal and human health. Today, social and political pressures are mounting to either withdraw them from the market or restrict their use (Barzman and Dachbrodt-Saaydeh, 2011). For example, the Swiss population voted in June 2021 in two popular initiatives aiming at limiting and even banning the use of synthetic pesticides in agriculture: the initiative for clean drinking water and the initiative for a synthetic pesticide-free Switzerland (Schmidt et al., 2019). The

[1] Authors have contributed equally.

http://dx.doi.org/10.19103/AS.2021.0098.04

alternative means of weed control, as part of integrated weed management (IWM) strategies, is a key objective. The phenomenon of allelopathy is a promising avenue in this regard.

The term 'allelopathy' can be defined as the inhibitory or stimulatory effect of one plant on another via the production of chemical compounds (called allelochemicals) and their release into the environment (Rice, 1984). In general, there are two plant partners: the donor plant, which produces and releases the allelochemicals, and the receiver plant, which is the plant 'responding' to the released compounds. It appears that all plants release compounds into the environment, but the responses of the receiver plants to the release of allelochemicals are difficult to characterise, especially while categorising them in terms of 'negative' or 'positive' to fit the common definition of allelopathy. Some plants release compounds into the environment with variations caused by various parameters such as diffusion distance, quantity, chemical composition, and organ localisation. Indeed, the adaptation to neighbouring plants requires a high level of plasticity in wild plants, and to define the contribution of allelopathy in terms of the expression of plastic trait responses is challenging (Callaway et al., 2003; Uesugi et al., 2019).

The precise definition of allelopathy has been the subject of controversy with many opinions depending on a scientific background, an idea that holds true for the authors of this paper (plant molecular biologist, plant ecophysiologist, plant biotechnologist and agronomist). The challenge is indeed to reconcile different scientific approaches in a multidisciplinary future model on how allelopathy could contribute to IWM with effective and long-lasting solutions for a farm that should operate as a profitable economic entity. From the perspective of a biologist interested in chemical ecology, there is a need to understand plant-plant interactions, especially between crops and weeds, to promote interactions that are neither positive nor negative but may result in environmental adaptation such as niche differentiation. In addition, it is essential to identify new allelochemicals that show significant efficacies on weeds as alternatives for chemical herbicides. Moreover, from an agronomical perspective, it is important to focus on the underlying principles of the allelopathic crop–weed interactions that could contribute to future weed control in the field. It is also important to remind the reader to consider all factors carefully when selecting crops with the allelopathic potential to manage the likely high expectations from farmers and even agricultural advisors for high-yielding weed-suppressive crops.

In line with these priorities, the chapter covers two topics. In Section 2, we discuss allelopathy in crop–weed interactions while the second part focuses on the practical aspects of allelopathy with reference to IWM. Section 3 includes a detailed discussion on the research findings on buckwheat (*Fagopyrum esculentum*), as our research group has studied its potential allelopathic properties during the past ten years.

2 Understanding allelopathy in crop-weed interactions

2.1 Allelochemical classes and plant defence

Considering allelochemicals from an evolutionary viewpoint shows that allelopathic compounds have high structural diversity with a wide degree of multi-functionality. Most allelopathic compounds are secondary metabolites, and as by-products of primary metabolism, they are not directly involved in plant development. The production of secondary metabolites requires an expenditure of energy and resources with functions in signal transduction and defence that contribute to the adaptation of plants to their environment (Bourgaud et al., 2001; Wink, 2003).

Wink (2003) stated that the production of allelochemicals by plants should be understood as the optimisation of plant resources to control a wide range of potential enemies. For example, gramine is an alkaloid produced by barley (*Hordeum vulgare*), and its efficacy and toxicity have been demonstrated on fungi (Wippich and Wink, 1985; Matsuo et al., 2001), bacteria (Sepulveda and Corcuera, 1990), mammals (Gallagher et al., 1964; Goelz et al., 1980), insects (Corcuera, 1984) and plants (Liu and Lovett, 1993; Kremer and Ben-Hammouda, 2009).

The example of rice (*Oryza sativa*) illustrates the complexity of allelochemical induction and the possible functions of allelochemicals (Fig. 1). The diterpenoids momilactone A and B were first identified in rice husk and were subsequently found to be secreted from the roots of various rive cultivars (Kato-Noguchi and Ino, 2003; Kato-Noguchi, 2008; Kong et al., 2004; Kato et al., 1973). Momilactone synthesis is induced by various external stimuli such as the phytohormone jasmonic acid (JA) (Yoshida et al., 2017), UV light (Kato-Noguchi et al., 2007a), root exudates (Zhang et al., 2018a), drought and salinity (Xuan et al., 2016), soil microorganisms (Xie et al., 2017) and elicitors from insects and fungi (Wari et al., 2019; Schmelz et al., 2014), suggesting that they might be critical compounds in stress tolerance. The induction of momilactones is associated with plant responses in two categories: physiology and defence. They probably protect plant leaves against UV light (not demonstrated) and preserve seed dormancy in rice husks (Kato et al., 1973). Moreover, they are implicated in the growth inhibition of neighbouring plants (Kato-Noguchi and Peters, 2013). Momilactone A is accumulated at higher concentrations upon fungal infections. Various mutant rice lines that over-accumulate momilactone A showed increased resistance to pathogenic fungi *Magnaporthe grisea*, *Rhizoctonia solani*, *Blumeria graminearum* and *Fusarium oxysporium* and pathogenic microbes *Xanthomonas oryzae* (Sawada et al., 2004; Mori

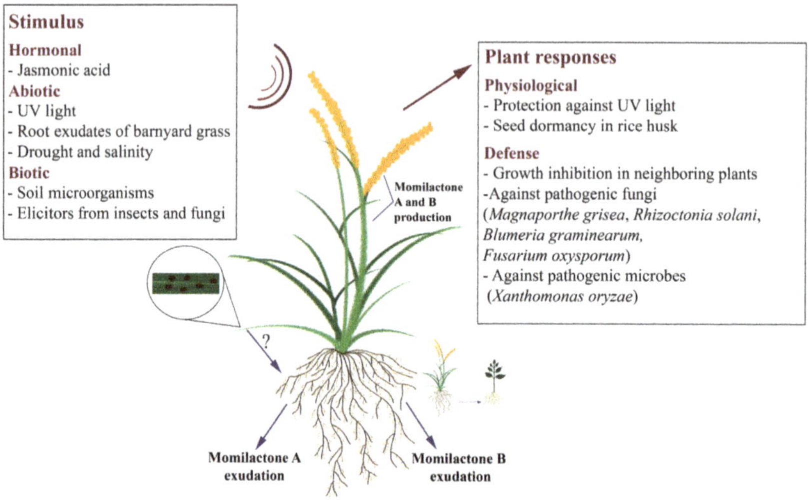

Figure 1 The induction of momilactone production and subsequent plant responses in rice.

et al., 2007; Hasegawa et al., 2010; Kurusu et al., 2010; Gu et al., 2019) While momilactone B has a higher allelopathic activity, momilactone A has a higher activity against fungal pathogens (Kato-Noguchi and Peters, 2013).

In terms of the structural diversity of allelochemicals, many chemical classes can be identified, from alkaloids (Liu and Lovett, 1993) to terpenoids (Kato-Noguchi and Peters, 2013), phenolics (Li et al., 2010), quinones (Dayan et al., 2010) and flavonoids (Weston and Mathesius, 2013; Huang et al., 2015). A wide range of modes-of-action belong to these chemical classes, and the authors would like to refer to other references for an overview of the exhaustive list of mechanisms and targets of the different allelochemicals (Reigosa et al., 2006; Dayan and Duke, 2014).

2.2 Production of allelochemicals

Plants are sessile organisms coping with changing environmental conditions that affect their growth and survival, but with the ability to integrate signals and adapt to changes in resource supply. Plants collect information from their belowground and aboveground environments with regard to nutrient availability and light and can detect chemical cues such as volatile compounds, leachates and root exudates (Wang et al., 2021).

The production of allelochemicals in living plants is an inducible process, except for the release of allelochemicals during residue degradation (decaying plant material). It is influenced by various biotic factors such as the neighbouring plants (Hazrati et al., 2020; Hazrati et al., 2021) and their microbial underground partners, and abiotic factors from the environment such as temperature (Hess et al., 1992) and light (Dayan, 2006), while the developmental stage (Liu and Lovett, 1993) of the donor plant is also a factor of interference.

The inducible production of allelopathic compounds in a neighbouring plant is a topic that has received considerable scientific interest during the past years (Section 2.4). Molecular communication between plants is an essential component to study plant-plant interactions. Specific messenger molecules that are a part of signal transduction contribute to an integrated response at the plant level in the neighbouring plant (van Dam and Bouwmeester, 2016).

When two plants grow next to each other, the primary mechanism for the recognition of plant neighbours is through changes in light quality. For example, changes in red to far-red light ratios and the blue light caused by the neighbours can induce changes in stem and/or petiole growth as well as redirect leaf growth (Smith et al., 1990). It was also suggested to potentially affect the production of secondary molecules with allelopathic potential (Kegge et al., 2015). In sorghum (*Sorghum bicolor*), changing the wavelength of light caused variations in the levels of sorgoleone synthesised (Dayan, 2006), whereas exposure to low-intensity light can increase the level of hordenine production in barley (Lovett et al., 1994).

Plants can also produce a blend of unique volatile organic compounds (VOCs) that contribute to communication via air (aboveground compartment) and can trigger a response in receiver plants at the level of growth, reproduction and defence, with the overall result of improved resilience (Novoplansky, 2009). When barley plants were exposed to VOCs emitted by another barley cultivar, more biomass was allocated to the root (Ninkovic, 2003). In tobacco (*Nicotiana tabacum*), the perception of the volatile phytohormone ethylene is necessary to promote shade avoidance (Pierik et al., 2003).

One important route for allelopathic communication between plants is the root exudation of a wide variety of chemical compounds, including VOCs, into the rhizosphere (belowground soil compartment). In petri dish experiments, root VOCs from the bitou bush (*Chrysanthemoides monilifera* spp. *Rotundata*) negatively affected seed germination and seedling growth of different native plants from Australia (Ens et al., 2009; Jassbi et al., 2010), but the role of root VOCs as mediators of plant-plant interactions under field conditions still remains to be further investigated (Delory et al., 2016).

2.3 Rhizosphere model for belowground microbial interactions in allelopathy

In general, rhizosphere research is still very much an unknown science, which is partly due to the complexity of studying the hidden and heterogeneous soil environment (Shelef et al., 2019). Many questions are still unanswered in terms of the mechanistic understanding and functionality of allelochemicals in the rhizosphere, such as the perception of belowground root exudates by neighbouring plants, signal transmission pathways, the effects of other rhizosphere microorganisms in signal transmission and the plants' response to various stimuli. The communication between the roots and the rhizosphere community is based on chemical compounds (van Dam and Bouwmeester, 2016; Wang et al., 2021) which are protected from degradation by oxygen and light, making belowground chemical signals more stable and possibly more reliable than those above ground (Karlovsky, 2008). However, plants produce and secrete root exudates consisting of secondary metabolites which can signal to and interfere with the other soil organisms (Venturi and Keel, 2016). Root exudates provide nutrients for the microbial community, and there is a known relationship between root exudation and enhanced microbial activity and diversity in the rhizosphere. Since root exudates are rich in organic carbon, they serve as substrate and attract microorganisms, thereby altering the chemical composition of the rhizosphere (Karlovsky, 2008; Bakker et al., 2013).

Allelopathy is a sophisticated process with various factors to consider in terms of understanding the activity of allelochemicals in the soil. Upon release into the environment, rhizospheric microorganisms affect the allelopathic interactions of root-exuded compounds through degradation mechanisms that could either improve the allelopathic interactions, by resulting in the accumulation of phytotoxic products, or render them inactive. For instance, the benzoxazinoid (BX) allelochemicals found in crops such as wheat (*Titricum sp.*), rye (*Secale cereale*) and maize (*Zea mays*) are subjected to microbial degradation in the soil. Bacterial enzymes convert DIBOA (2,4-dihydroxy-1,4-benzoxazin-3-one) to BOA (2-benzoxazolinone) and DIMBOA (2,4-dihyd roxy-7-methoxy-1,4-benzoxazin-3-one) to MBOA (6-methoxybenzoxazolin-2-one) through heterocyclic ring contraction, making the compounds more stable, which allows them to remain in the soil for a longer period of time. However, they are less active than their precursors (Macías et al., 2005a; Macías et al., 2005b; Schütz et al., 2019). DIBOA indirectly promotes plant fitness by attracting *Pseudomonas putida* upon pathogen attack, and this might lead to systemic defence priming in maize plants (Neal et al., 2012; Schandry and Becker, 2020).

BOA and MBOA further degrade to the aminophenoxazines APO (2-Aminophenoxazin3-one) and AMPO (2-amino-7-methoxyphenoxazin-3-one) and can also degrade to their *N*-acetyl derivatives AAPO (2-acetamido-phenoxazin-3-one) and AAMPO (2-acetamido-7-methoxyphenoxazin-3-one) through the action of non-pathogenic organisms. While AMPO was shown to have no phytotoxic effects, APO has higher phytotoxicity than BOA and DIBOA (Macías et al., 2005a). Additionally, AZOB (2,2`-oxo-l,l`-azobenzene), another derivative of BOA, has a higher inhibiting effect on barnyard grass and garden cress than its precursor BOA (Inderjit, 2005; Chase et al., 1991; Nair et al., 1990). Soil microorganisms degrade rice flavone glycosides and the resulting products have adverse effects on microorganisms and fungi (Macías et al., 2019). Sorghum root hairs exude the weed-inhibiting allelochemical sorgoleone throughout the crop's growing season, but it undergoes mineralisation, a process that involves complete microbial degradation into inorganic compounds (Gimsing et al., 2009). The phenolic compounds such as *p*-coumaric, ferulic, *p*-hydroxybenzoic and *trans*-cinnamic acids are degraded by microorganisms that utilise the root exudates as an energy source hence influences the dynamics of plant–plant interactions.

The previous studies suggest that various compounds from different plant species either degrade at slower rates or in negligible amounts in sterile soil (Gimsing et al., 2009; Macías et al., 2005b), suggesting not only the action of root exudates but also that the microorganisms in the soil might be involved in these complex underground interactions.

In the next paragraph, a rhizosphere model shows in more detail the two-way communication involving root exudates between plant partners for various crop–weed interactions.

2.4 Allelochemical interactions in wheat, rice, buckwheat and sorghum

In this section, we illustrate how four different pairs of heterospecific neighbouring plant species perceive each other via root exudates: wheat/different species, rice/barnyard grass (*Echinochloa crus-galli*), buckwheat/redroot pigweed (*Amaranthus retroflexus*) and sorghum/velvetleaf (*Abutilon theophrasti*).

The possibility of wheat allelopathy was suggested a very long time ago (Schreiner and Reed, 1907) and it was reported that the roots of wheat (and other crop plants) could exude compounds that inhibit their own seedlings. It could be shown that the production of DIMBOA was induced by root exudates released from neighbouring plants (Li et al., 2016; Zhang et al., 2016) (Fig. 2). By using a mesh to avoid direct root contact between wheat and the different weed species, the allelopathic effect changed, suggesting that root contact plays a role in wheat allelopathy and might be restricted to a given weed species (Zhang et al., 2016). In addition, Kong et al. (2018) reported that wheat could respond to at least 100 plant species by producing DIMBOA and as loliolide and JA were present in root exudates from different species, it was suggested that these molecules are involved in the belowground signaling events. Wheat plants can detect, early in their development, conspecific (of their own species) and heterospecific (from different species) neighbours by these

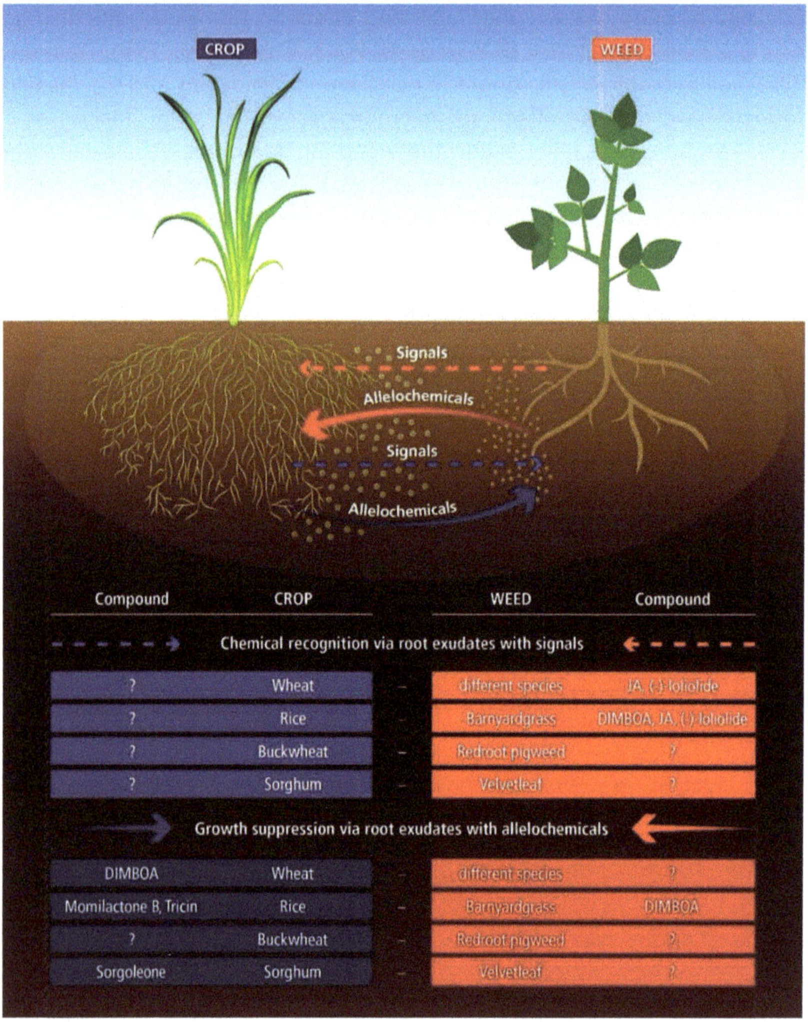

Figure 2 Examples for plant-plant interactions that are mediated by root exudates. Crops (wheat, rice, buckwheat and sorghum) can recognise chemical signals in weed (different species, barnyard grass, redroot pigweed and velvet leaf) root exudates (red broken arrow), but no information is available on crop recognition by weeds (blue broken arrow). Furthermore, crops can exudate allelochemicals to suppress weed growth (blue full arrow) and weeds might also release allelochemicals into the rhizosphere (red full arrow). The question mark (?) indicates that no compounds have been identified yet. The small dots around the roots represent soil microorganisms that might modify allelochemicals and their signals (Section 2.3).

ubiquitous signaling chemicals and subsequently increase the production of the allelochemical DIMBOA. However, in wheat, DIMBOA levels appear not to correlate well with weed inhibition, suggesting that weed suppression is caused by multiple factors.

Dilday et al. (1998) first reported the possibility of weed-inhibitory effects in the rice rhizosphere, and today a wide variety of rice allelochemicals such as momilactone A and B, phenolic acids, phenylalkanoic acids, hydroxamic acids, fatty acids, terpenes and indoles are known (Kato-Noguchi, 2008; Kato-Noguchi, 2011b), as mentioned previously in Section 2.1. It is easier to characterise rice-weed interactions since rice is grown in paddy soils, a liquid environment that helps with the collection of root exudates. Rice has the further benefit of including both allelopathic and non-allelopathic varieties, helping with the design of experimental set-ups to discriminate between allelopathic effects (see Section 3.3.2). The best-characterised example of induction strategies and plant response in rice is the rice–barnyard grass interaction. Allelopathic rice varieties can detect barnyard grass and will increase their production of allelochemicals (Kong et al., 2006; Zhao et al., 2005; Kato-Noguchi, 2011a) (Fig. 2). Barnyard grass root exudates can induce the production of rice allelochemicals (Kong et al., 2006; Yang and Kong, 2017), suggesting that signaling chemicals are present in the barnyard grass root exudates. It has been shown that DIMBOA is a signaling chemical emitted into the soil by barnyard grass (Guo et al., 2017), and the neighbouring rice will subsequently induce its own allelopathic response by increasing the secretion of the allelochemical momilactone B (Zhang et al., 2018; Kato-Noguchi, 2011a). Studies of the production of the rice allelochemicals momilactone B and tricin in the presence of different biotypes of barnyard grass have also confirmed the hypothesis that allelopathic rice detects the presence of barnyard grass through the presence of loliolide and JA in barnyard grass as signaling compounds (Li et al., 2019).

Our research efforts showed that buckwheat modifies its root exudation when co-cultivated with redroot pigweed, with a growth-repressive effect on redroot pigweed seedlings. Some of the unidentified compounds in the root exudates were only present when the two species were co-cultivated, suggesting some level of recognition between species, while the induction in buckwheat appears to be mediated by the presence of redroot pigweed (Fig. 2) (Gfeller et al., 2018b). However, the authors have not investigated the identity of the signaling compound(s) and the mechanism of the growth repression in redroot pigweed up to now. In another study where buckwheat was grown in culture solution for ten days with lettuce, a dose-response suppressive activity on root and hypocotyl elongation of lettuce seedlings was found (Kato-Noguchi et al., 2007b). Tin et al. (2009) identified caprolactam (azepan-2-one) as a candidate allelopathic molecule responsible for this type of elongation inhibition. Water extracts from buckwheat-grown soil showed significant repressive activity on root elongation of barnyard grass and common purslane (*Portulaca oleracea*), whereas hairy galinsoga (*Galinsoga quadriradiata*), livid amaranth (*Amaranthus blitum*) and lettuce (*Lactuca sativa*) did not respond (Kalinova et al., 2005; Tominaga and Uezu, 1995). Kalinova et al. (2007) showed that soil from a buckwheat stand had significant suppressive activity against lettuce radicle elongation after three days of growth. Methanol and boiling water extracts of the same soil revealed the presence of several phytotoxic molecules that include a gallic acid derivative, palmitic acid methyl ester, vanillic acid, rutin and a 4-hydroxyacetophenone derivative, but it was not clear whether these compounds originated from root exudates, leachates or the necrotic parts of buckwheat. The authors addressed this result further by analysing the agar medium on which buckwheat was grown for 12 days, and the identified compounds included a quercetin derivative, palmitic acid, squalene, epicatechin, vitexin and very interestingly, the same gallic acid derivative that was originally present in the soil extract.

The main allelochemical of sorghum is sorgoleone, which is specific for the Sorghum genus and is synthesised by the tips of root hairs (Weston et al., 2013). Environmental factors influence sorgoleone production (Hess et al., 1992) and plant hormones like auxin can also stimulate sorgoleone synthesis (Uddin et al., 2010). A more indirect way is also through methyl-jasmonate and JA that act as plant hormones responsible for root growth and hair formation (Uddin et al., 2013). Moreover, it was suggested that sorghum seedlings can secrete sorgoleone after germination and can also respond to the presence of the neighbouring plant (velvetleaf) by releasing more sorgoleone (Dayan, 2006) (Fig. 2).

2.5 Experimental methodology and allelopathic trait selection

Most studies on allelochemicals are being conducted under laboratory conditions. To identify/quantify chemical compounds in exudates/leachates, the ideal situation is to extract these compounds under optimal and sterile conditions with minimal interference for further chemical characterisation such as mass spectrometric analyses.

The most common approaches to study allelopathy have been compared by Zhang et al. (2021). The basic idea is to identify an allelochemical or a cocktail of allelochemicals under laboratory conditions in a first step, and secondly, to design protocols for chemical detection and activity characterisation, such as persistence, under field conditions. Since a major area of agronomic interest is the discovery of weed-suppressive crops/cultivars as part of an IWM strategy, it is a prerequisite to confirm a high level of persistence and activity for a potential allelochemical under field conditions. In this regard, testing a plant extract from donor plants on different receiver plant species under controlled laboratory conditions is the general approach in order to reach a conclusion on growth-suppressive effects. However, the demonstration of the efficacy of allelopathy is hard to confirm in the field, which is a complex ecosystem with a multitude of interactions.

One aspect to consider in the interpretation of research intended to characterise allelopathic potential is that effects due to resource competition are an integral part of allelopathy. It is almost impossible to completely separate competition from allelopathy, either under laboratory or under field conditions. Enhanced weed suppression can result from competitive advantages of the crop, such as plant height, leaf shape, leaf angle, absorption of water and nutrients, and/or growth-repressive allelochemicals that are released into the environment. Specific competitive traits like plant height or leaf angle can be determined quite easily, but monitoring water and nutrient uptake is more difficult. It is further helpful to consider allelopathy not as a measurable plant trait *per se* but a concept to illustrate the mechanisms implicated in the regulation, production, release and action of chemical compounds that affect the surrounding environment of a plant. The validation of suitable and measurable traits that reflect allelopathic potential under field conditions requires careful investigations. The traits studied for the receiver plant (ideally a weed) include features such as weed occurrence, size, biomass, seed set, different physiological traits such as chlorophyll fluorescence while for the allelopathic crop neighbour, the trait characteristics include allelochemical content in the crop and other phenotypic traits (Weidenhamer et al., 2014). In rice, the specific leaf area is correlated with rice allelopathic potential (Gaofeng et al., 2018). A fact worth mentioning is that for the farmer, the question of whether allelopathy is implicated in weed suppression is of secondary importance, as long as the desired result is achieved. However, to integrate allelopathy successfully into the weed control programmes of the future, it is essential to understand the underlying mechanisms.

2.6 Swiss case study: buckwheat

Our research group began to study the weed-suppressive effect of buckwheat almost a decade ago, after field observations in 2009 and 2010 in Switzerland showed that buckwheat field stands were basically weed free. This prompted us to follow up with a series of studies to understand the growth-suppressive properties of buckwheat on various other plants.

Our first approach tested the effects of the soil in which buckwheat had been grown (in the field and in pots) on lettuce and redroot pigweed growth, in petri dish experiments, but no growth-suppressive effects were found. In parallel, we tested the effect of water extracts from the same soil samples on lettuce growth, and this study did not identify growth-suppressive effects either. We draw the following three conclusions: Firstly, there are either no allelopathic molecules in the soil solution (not soluble in water), or they are rapidly degraded; secondly, the growth-inhibiting effect is due to a long-term and constant exposure of small quantities of allelochemicals and; thirdly, the root must be in direct contact to mediate allelopathic effects (Gfeller and Wirth, 2015).

Further efforts were focused on dissecting the effects of resource competition from allelopathy as suggested in the literature (Falquet et al., 2015). In field trials with two shading levels, redroot pigweed biomass was similar, demonstrating that light interception by buckwheat was not the primary mechanism responsible for redroot pigweed growth suppression (Gfeller et al., 2018b).

We also developed a method to separate resource competition for water, nutrients and light from allelopathic root interactions in pot trials (Fig. 3). In these investigations, water and nutrient supply were kept constant and in sufficient amounts while the effect of shading was evaluated by the presence or absence of vertical nets, and impenetrable plastic barriers separated the rhizospheres of the weed (redroot pigweed) and the crop (buckwheat) to prevent the roots of the different plants from interacting. We found that, in the absence of shading, redroot pigweed growth was repressed by at least 65% by direct root interactions of a potentially allelopathic nature (Falquet et al., 2014). In the next step, the experimental setup was improved to study the effects of chemical diffusion. The roots of buckwheat and redroot pigweed were separated with a permeable mesh. The growth of redroot pigweed was evaluated in the presence and absence of buckwheat. The results showed that buckwheat suppressed the growth of redroot pigweed by 41% (Gfeller et al., 2018a) and 68% (Gfeller et al., 2018b) when roots were directly interacting, whereas buckwheat suppressed redroot pigweed growth by 53% (Gfeller et al., 2018a) and 46% (Gfeller et al., 2018a) without physical root interactions, probably through the diffusion of allelopathic compounds. The originality of

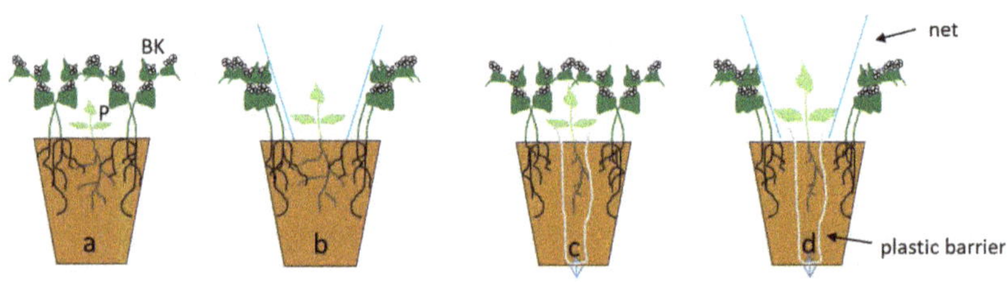

Figure 3 Experimental design of pot trials with buckwheat (BK) and redroot pigweed (P) plants. BK plants were grown on the outer sides of the pots while P was sown in the center of the pots in different conditions: without any barrier separating them (a), nets separating the aerial parts to prevent shading (b), impermeable plastic bags separating the roots of P from BK to prevent root interactions (c), both nets separating the aerial parts and plastic bags separating the roots (d). Adapted from Falquet et al. (2014).

our approach was to test the buckwheat-redroot pigweed interactions with plants and not with seedlings; and the pot trial was performed for 28 days while we followed redroot pigweed growth for 55 days in the field.

In parallel, studies to cultivate buckwheat in glass sand (Fig. 4) were performed to obtain 'clean' root exudates after extraction with methanol and further analysis by high-resolution mass spectrometry, a step that is inevitable for chemical characterisation.

We found that the BK-P root exudates inhibited redroot pigweed root growth by 49% (Fig. 5a). Moreover, the characterisation of root exudates by UHPLC-HRMS and principal component analysis (PCA) showed that BK and BK-P had different metabolic profiles (Fig. 5b). We concluded that buckwheat

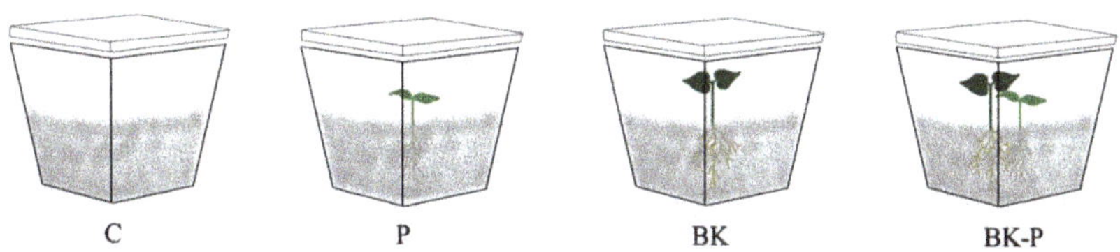

Figure 4 Experimental setup of buckwheat and redroot pigweed glass sand cultures for root exudate collection. From right to left: plastic culture box filled with glass sand without any plants for control (C), containing redroot pigweed only (P), buckwheat only (BK) and buckwheat and redroot pigweed growing together (BK-P). Adapted from Gfeller et al. (2018a).

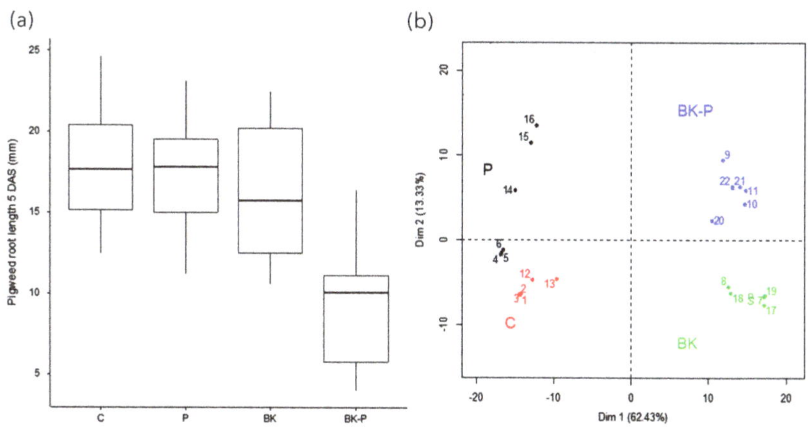

Figure 5 Results from experiments with sand cultures of buckwheat and redroot pigweed. (a) Redroot pigweed root length five days after sowing (DAS) when treated with different root exudates obtained from glass sand cultures; Tukey's HSD, P-value <.05, n = 9. (b) Principal component analysis (PCA) score plots of dimensions (Dim) 1 and 2. PCA on markers obtained from the different root exudates from sand culture and separated by UHPLC-HRMS. The root exudates were obtained from 11-day-old sand cultures of boxes shown above, n = 6. Adapted from Gfeller et al. (2018a).

changes its root exudation in the presence of redroot pigweed, which indicates heterospecific recognition (Gfeller et al., 2018a).

However, we could not isolate the compounds responsible for the growth-repressive effects and it was difficult to separate the outcome from the potential effects attributable to competition for resources (Gfeller et al., 2018a). To select interesting allelochemical candidates in future experiments, we will concentrate on the differences between allelochemical production by a buckwheat plant grown with redroot pigweed and one grown without the weed plant. No recent studies on buckwheat-weed interactions are available and the latest works have focused on residue degradation of buckwheat roots (Szwed et al., 2020; Szwed et al., 2019), a topic that is further addressed in the following section.

3 Allelopathy: a future component of IWM

The concept presented so far in this chapter has focused on the growth-suppressive properties of some plants (a crop or a weed plant) on a neighbouring plant. The next logical step will be to apply this knowledge advantageously at the farmer's level, and/or in an agricultural setting for improved efficiency in terms of weed control, such as in crop rotations or as a complementary management tool. Different possibilities exist to exploit growth-suppressive effects, such as the development of new (bio)herbicides based on allelochemicals and the use of allelopathic crops.

Since the 1950s, farmers have been using chemical herbicides (with high efficacy of over 95%), which are cheap, easy-to-use and guarantee weed-free fields, and therefore the natural compounds with lower efficiency were not likely to be used by farmers. However, pressure is rising from various actors along the food chain for more sustainable agriculture, more biodiversity and pesticide-free food/products. In the future, IWM will be based on the combination of different weed management tools.

3.1 Development of new herbicides based on allelochemical templates

Allelochemical compounds constitute an incredible reservoir of new molecules whose modes of action have been shaped by evolutionary processes, with the advantage that they differ from the modes of action known for traditional synthetic herbicides (De Souza Barros et al., 2020). The idea of an unexplored reserve of future new modes of action is strengthened by an analysis of registration data for plant protection products in the United States. For the period 1997–2010, about 30% of insecticides and fungicides registered were for natural products or derivatives of natural products, while the proportion for herbicides was only 8% (Cantrell et al., 2012). This avenue of detecting new herbicides is also particularly interesting in view of the alarming trend where weeds are developing resistance to traditional herbicides. The other advantages might include benefits with regard to improved soil biodegradability of molecules of natural origin compared to synthetic compounds (Dayan et al., 2009). The current research focuses on strategies where partial weed control is no longer a knock-out criterion in the search for new herbicidal compounds (Ciriminna et al., 2019; Duke et al., 2014).

3.1.1 New chemical herbicides

To develop a new chemical herbicide based on the structure of a natural molecule, it is essential to establish a structure–activity relationship for the given molecule of interest to design more effective molecular analogues (Dayan and Duke, 2014). This possibility is particularly interesting if the starting molecule acts according to a novel mode of action but potentially involves the loss of the superior biodegradability attributed to natural compounds, depending on the chemical modifications made.

An excellent example of herbicides developed from an allelochemical compound is the triketones (e.g. the maize herbicide mesotrione) based on the molecule leptospermone, which is produced by both the bottlebrush plant (*Callistemon citrinus*) and the Manuka tree (*Leptospermum scoparium*) (Lee et al., 1997; Dayan et al., 2011). Triketones are bleaching herbicides that inhibit hydroxyphenylpyruvate dioxygenase (HPPD), an enzyme that plays a crucial role in plastoquinone and tocopherol biosynthesis in plants (Beaudegnies et al., 2009). Triketones are the latest herbicide site-of-action introduced on the market (Dayan and Duke, 2020). Many molecular target sites of natural phytotoxins used for the development of new herbicides are known (Dayan and Duke, 2014), but so far, commercial herbicides with a new herbicide site-of-action have not been developed yet (Dayan and Duke, 2020). An allelochemical-based benzothiazine derivate, originating from the rice allelochemical tricin, was developed and applied to paddy fields and resulted in effective weed control of the dominant weeds (Zhao et al., 2019).

3.1.2 New bioherbicides

In addition to the focus on new allelochemicals with growth-suppressive characteristics based on precursors with structural similarities, another approach is to identify plant extracts with allelochemical properties. Bioherbicides are defined as products of natural origin for weed control and include phytotoxic plant-based secondary metabolites (Cordeau et al., 2016). However, most of the current bioherbicides on the market are based on fungal or bacterial microorganisms and only very few contain natural plant extracts. One example of the latter is a product that contains the active ingredient pelargonic acid and other saturated fatty acids (Cordeau et al., 2016). However, pelargonic acid is not an allelochemical, but an acid that occurs naturally in different vegetables and fruits and partially controls broadleaf and grass weeds (Ciriminna et al., 2019). An example of a bioherbicide based on an allelochemical is sorgaab, a water extract of the green parts of mature sorghum plants that inhibits weeds in wheat (Cheema and Khaliq, 2000; Cheema et al., 2008; Głąb et al., 2017). Moreover, in 2005, a patent (https://patents.google.com/patent/KR200600837 74A/en) was registered in South Korea for the rice momilactones A and B, and the respective products are under commercial development (Zhao et al., 2018).

3.2 Allelopathic crops

The use of allelopathic crops in IWM relies on the cultivation of plants with high allelopathic potential of economic interest (Wu et al., 1999). From a biological point of view, allelopathy can be an active and plant-regulated process related to the chemical response of a living organism to its environment or a passive process related to the presence of a plant decaying in the environment of another plant. In agricultural systems, within the same field, both processes may happen simultaneously, mediated by the same allelopathic crop.

3.2.1 Release of allelochemicals

The release of allelochemical compounds into the environment includes processes such as the leaching of aerial parts, volatilisation, decomposition of plant residues and root exudation (Fig. 6). Although not all of these

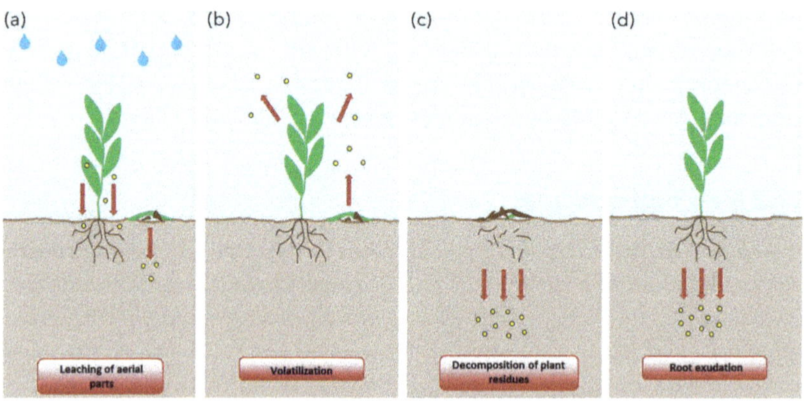

Figure 6 Representation of the different kinds of allelochemical release. Leaching of aerial parts (a), volatilization (b), decomposition of plant residues (c) and root exudation (d).

mechanisms are achievable targets in terms of IWM, we will discuss all the possible mechanisms in the following sections.

Organic and inorganic metabolites may be released from plants by rain, dew and mist in a process referred to as *leaching* (Tukey, 1966) (Fig. 6a), and a classical example is the fern species, *Pteridium aquilinum*, whose fronds release phytotoxic compounds into the environment after rainfall (García-Jorgensen et al., 2020). The water that runs off the trunks and foliage of various eucalyptus species has also been reported to be particularly phytotoxic (May and Ash, 1990; Song et al., 2019). However, to our knowledge, no allelopathic leachates from arable crops for weed control have been reported.

Volatilisation (Fig. 6b) appears to be the preferable route for the environmental spread of monoterpenes produced by the two Mediterranean plants, sage (*Salvia leucophylla* Greene) (Muller and Muller, 1964) and pine (*Pinus halepensis*) (Santonja et al., 2019), a process favoured by the climate of the Mediterranean region (Reigosa et al., 2006). The colloidal matter present in the soil causes their fixation by adsorption from where they can exert their toxic effect on the surrounding plants (Muller and del Moral, 1966). Just like the case of leachates, allelopathic volatiles have not been utilised for weed control in arable crops. A recent review presented an understanding of the role of plant volatiles as mediators of plant interaction (Ninkovic et al., 2021).

The use of plant residues can be an effective tool for weed management, since decaying plant residues can negatively affect plant growth and performance through various release mechanisms (Fig. 6c) (Zhang et al., 2021). Examples include the monocotyledons rye (Barnes and Putnam, 1983; Flood and Entz, 2018) rice (Chou and Lin, 1976) and dicotyledons such as sunflower (*Helianthus annuus*) (Leather, 1983; Alsaadawi et al., 2012), hairy vetch (*Vicia villosa*) (Teasdale and Mohler, 1993; Campiglia et al., 2010), buckwheat (Szwed et al., 2020) and red clover (*Trifolium pratense*) (Ohno et al., 2000; Marcinkevičienė et al., 2013).

In Sections 2.3 and 2.4, the role of root exudates (Fig. 6d) have been described for crop-weed interactions in wheat, rice, buckwheat and sorghum, but allelochemicals have also been identified in root exudates of various other plant species and agricultural crops (Wang et al., 2021). Furthermore, the BXs exudated by rye to the rhizosphere are taken up by the neighbour hairy vetch and subsequently detected in hairy vetch shoots (Hazrati et al., 2020).

3.2.2 Current agricultural and farming practices based on the principles of allelopathy

In agriculture, the three farmer practices that implement the principles of allelopathy in the field include intercropping, the use of cover crops and the use of plant residues.

Intercropping involves growing two or more compatible crops simultaneously on the same field with the intention of improving yield by enhancing resource utilisation; and secondly, it could be a practice where one crop (the intercrop) is used as soil cover to control weeds in the field without having a negative effect on the main crop. One example of how to implement this practice was seen in the tall interrow crops like cotton, maize or soybean. Intercropping of cotton with sorghum and sunflower strongly suppressed weeds and had a positive effect on cotton yield (Kandhro et al., 2014); sunn hemp (*Crotalaria juncea*) with its high contents of phenolics and terpenoids also suppressed weeds in cotton (Blaise et al., 2020). Sorghum intercropping in maize controlled purple nutsedge (*Cyperus rotundus*) (Mahmood et al., 2013a); and the forage legume silver leaf desmodium (*Desmodium uncinatum*) used as an intercrop reduced *Striga hermonthica* infestation in maize (*Z. mays*) (Hooper et al., 2010). In soybean cultivation, buckwheat grown in the interrow provided good weed control (Biszczak et al., 2020). Moreover, fenugreek (*Trigonella foenum-graecum*) was reported to produce flavonol glycosides which showed allelopathic activity (Omezzine et al., 2014) and provided excellent weed control when grown as an intercrop in coriander (*Coriandrum sativum*) (Pouryousef et al., 2015).

In the case of cover crops (CC), the idea is to plant the CC in between two main crops to cover the soil and not to harvest the CC. In this way, the CC provides multiple ecosystem services. Weeds can be suppressed by 70% to 95% through direct competition for resources, by allelochemicals in living field stands (Gerhards and Schappert, 2020; Blanco-Canqui et al., 2015) and by CC residue degradation. The threshold value to apply in this case is that above 3 t/ha of CC biomass, weed suppression occurs for all CC (Gebhard et al., 2013; Gfeller et al., 2018b), but below this threshold, only some CCs like Brassicaceae and black oat (*Avena strigosa*) successfully suppressed weeds, which might be due to growth-suppressive root exudates from the CC (Gfeller et al., 2018b). In field trials, it is almost impossible to separate the competitive and allelopathic effects of CCs on weed growth. Our experiments (also discussed in Section 2.7) showed that light interception by the CC was not the primary mechanism responsible for redroot pigweed growth suppression (Gfeller et al., 2018b).

Kunz et al. (2016) also investigated the question of how to separate the effects due to resource competition from allelochemical effects in field studies. It was reported that the aboveground dry biomass and canopy cover of mustard (*Sinapis alba*), fodder radish (*Raphanus sativus* var. niger) and spring vetch (*Vicia sativa*) did not correlate with the density of the predominant weeds goosefoot (*Chenopodium album*), chamomile (*Matricaria chamomilla*) and chickweed (*Stellaria media*). Although CC suppressed weeds by 60%, the competition for the light seemed to play a minor role in total weed-suppressive ability. By correlating the results of two experiments at two different scales, a field trial that evaluated weed density in different CC systems and a Petri dish assay on the germination capacity of several weeds treated with aqueous extracts of the CCs grown in the field trial, the authors concluded that 50% of the variation in weed density could be explained by allelopathy (Kunz et al., 2016).

The physical and biochemical characteristics of plant residues may alter weed germination and growth, but the specific mechanisms involved are difficult to study. Several field studies have reported the weed-suppressive effects of crop residues or mulch and the allelopathic effects were confirmed under controlled conditions in experiments but not in field settings. Studies that focused on the effects of plant residues from different crops or cover crops on the germination and growth of several weeds under field and laboratory conditions have shown that the weed species appear to have varying sensitivity towards allelopathic cover crop residues (Sturm et al., 2018). Biochemical effects on weed suppression in the field across various treatments and locations indicating the importance of studying environmental factors in well-designed set-ups (Swanton et al., 2015). Although it is known that certain plant residues release allelochemicals that inhibit seed germination and growth (Jabran et al., 2015; Kelton et al., 2012; Nichols et al., 2015), a very interesting finding is that phytotoxic crop residue effects are stronger on small-seeded weeds than on large-seeded crops (Kruidhof et al., 2010; Petersen et al., 2001). This result is promising, considering future IWM strategies.

The allelopathic effects of crop residues will be described for different crops in the next section. The *Brassica* species produce a large number of allelochemicals including glucosinolates, brassinosteroids and isothiocyanates with weed-suppressive potential in several cropping systems (Rehman et al., 2019). Even if some Brassica species showed inhibition of wheat germination and seedling growth (Bialy et al., 1990), it could be a sustainable tool for IWM. Some successful examples are: brassica residue incorporation at 6 t/ha in mung bean reduced weed dry weight and density by 61% and 52%, respectively (Ullah et al., 2020); and turnip (*Brassica rapa*) mulches released inhibitory isothiocyanates that were part of the observed weed suppression in the field (Petersen et al., 2001).

Buckwheat residues in the soil can suppress various weeds, and the two most important classes of compounds identified are flavonoids and phenolic acids (Falquet et al., 2015). Based on the hypothesis that phytotoxic compounds from buckwheat tissues are released during plant decomposition, several studies focused on assessing the inhibitive effect of buckwheat extracts under laboratory conditions. Leaf extracts showed the greatest inhibition (followed by shoot and inflorescence extracts) on the root elongation of lettuce and several weeds (Golisz et al., 2007; Hayashi, 1998; Ohsawa and Nakatani, 2005). In pot bioassays, the incorporation of root residues showed no inhibitory effect on the growth of Powell's amaranth (*Amaranthus powellii*) (Kumar et al., 2009). In contrast, Szwed et al. (2019) stated that buckwheat root residues had a much stronger allelopathic effect on several weed species than residues of the aerial buckwheat parts. No studies were found examining the effects of buckwheat residues incorporated in field trials.

Field studies on the incorporation of sunflower (*Helianthus annuus*) residues significantly inhibited weed growth (Alsaadawi et al., 2012; Leather, 1983). Furthermore, a genotype-dependent effect was reported for eight different sunflower genotypes while chemical analysis revealed higher concentrations of phenolic compounds in the most suppressive genotypes, compared with the least-suppressive genotypes.

Monocotyledonous crop residues like rice, black oat, maize, wheat and rye appear to release similar allelochemicals (BX and phenolic acids) and also with similar effects on weed growth. Cereal rye (*Secale cereale*) is one of the most studied allelopathic crops (Jabran et al., 2015). It produces a persistent ground cover, its mulch decomposes slowly and it efficiently controls summer annual weeds (Mirsky et al., 2013). In addition to BX exuded by rye roots (Belz and Hurle, 2005), phenolics appear to be a rye decomposition product (Otte et al., 2020). Interestingly, the timing of allelochemical release plays an important role in potentially maximising allelopathic effects, as shown for coumaric and vanillic acids (phenolic compounds), which were exuded at higher rates during the first week after field termination of rye (Otte et al., 2020). However, the phenolic acid concentrations measured in rye were three-fold lower than the toxicity thresholds previously reported for coumaric, vanillic and ferulic acids in horticultural and field crops (Otte et al., 2020; Chou and Patrick, 1976). Furthermore, it appears that the allelopathic effects of wheat, particularly on ryegrass *(Lolium perenne L.)* and field forget-me-not (*Myosotis arvensis L.)* cannot be attributed to the synergistic effects of otherwise weakly active allelopathic compounds (Jia et al., 2006). Finally, there is no conclusion on whether tillage influences phenolic acid release from rye. Otte et al. (2020) observed no big differences between tillage and no-tillage systems, whereas Kruidhof et al. (2014) observed a maximised allelopathic effect of tillage two weeks after rye termination.

Rice straw may produce and release allelochemicals into the paddy, which suppress the growth of plants germinating later (Chung et al., 2001; Inderjit et al., 2004), but no difference exists between the allelopathic and non-allelopathic varieties. Kong et al. (2006) observed that rice residues released growth-repressive momilactone B and lignin-related phenolic acids into the soil during decomposition.

To conclude this section, the authors would like to re-emphasize the issue regarding the design and selection of new allelochemical molecules for the purpose of growth inhibition in crops of interest (Section 3.1.1). It is important to characterise the biodegradability of allelochemicals and to investigate their environmental persistence and their potential impact on present and future crops. Residue management, for example, could not only have an immediate negative effect on the development of weeds but also a prolonged negative effect on the succeeding crop.

3.3 Breeding for allelopathic traits in crops

3.3.1 Genetic variation in plants with allelopathic potential

It is evident from previous discussions (Section 2.3 on allelopathic and non-allelopathic rice varieties) that potential exists amongst several crop varieties such as alfalfa, oat, wheat and rapeseed (Zubair et al., 2017; Fernández-Aparicio and Rubiales, 2019; Shamaya et al., 2018; Raman et al., 2018). The genetic control of allelopathic properties was studied intensively in rice (Subrahmaniam et al., 2018; Jensen et al., 2001). The focus of the research efforts was to understand the biochemical pathways of the momilactone allelochemicals, and specifically momilactone B (Kato-Noguchi and Ino, 2003; Kato-Noguchi and Peters, 2013; Shimura et al., 2007). Moreover, it was possible to identify the underlying genetic basis for the production of momilactone B and the inactivation of two selected genes (copalyl diphosphate synthase 4 and kaurene synthase-like 4) that decreased the allelopathic potential of three mutants compared to the wild type (Xu et al., 2012). Rice germplasm was also screened for allelopathic genetic potential (Pheng et al., 2009), but a drawback is that studies to understand natural variation are not designed to accommodate the possibility that allelopathy is also an inducible process mediated by the presence of weeds. The exogenous application of the signaling hormones methyl jasmonate and methyl salicylate lead to differential induction of the allelopathic potential in two rice cultivars, and interestingly, the cultivar with the higher allelopathic potential responded more strongly than the cultivar with the lower potential (Mahmood et al., 2013b), showing that plant hormones can also affect genetic potential.

Gramine biosynthesis starts from tryptophan and then it is further converted into two intermediates (Gross et al., 1974). In the barley cultivar Proctor, the first known stable intermediate from tryptophan could not be identified, a finding that might explain the lack of allelopathy in this cultivar (Hanson et al., 1983). Indeed, the amount of gramine produced by barley varies greatly depending on the cultivar (Hanson et al., 1983; Liu and Lovett, 1993) and this was used as a selection criterion in a study of 127 landraces and cultivars covering the gene pool from Nordic countries (collected over 100 years) in a screen for allelopathic activity against ryegrass. Interestingly, the level of gramine was lower in the new cultivars (Bertholdsson, 2004). Similar results were observed in a study with Tunisian barley landraces and modern accessions (Bouhaouel et al., 2018; Bouhaouel et al., 2020). Both research groups suggested that the old landraces were more allelopathic and that modern selection methodologies favour other traits and might even counter-select for allelopathic traits. A recent study screened 18 accessions of barley from the Middle East, one accession from Tibet and a modern cultivar for their gramine and hordenine content in different plant parts, and this study also demonstrated the impact of domestication on the production and distribution of the two allelochemicals in barley (Maver et al., 2020). Similar conclusions were drawn Tibugari et al. (2019) for sorgoleone in 353 different African sorghum accessions where new sorghum accessions had very little sorgoleone compared to some landraces and wild sorghum.

3.3.2 Breeding programmes and allelopathic rice varieties

Breeding for improved weed suppressiveness is a function of weed-competitive ability that is the outcome of the interaction of several traits, e.g. plant height, leaf area (Dimaano et al., 2017) and allelopathic activity. The identification of quantitative trait loci (QTLs) for allelopathic functions represents a strategy to enhance allelopathic activity in crops by using marker-assisted selection (Schulz et al., 2013). However, not many allelopathic crop cultivars have been developed, although breeding programmes exist in crops like wheat and rice (Bertholdsson et al., 2012).

Huagan-3 was the first allelopathic rice cultivar to be developed based on the identification of a QTL linked to an increased allelopathic effect of rice (Kong et al., 2011). It was released by the administration of Guangdong province (China) in 2009 and by the Ministry of Agriculture of China in 2015. Another three allelopathic rice cultivars (N-liangyou-201, Hualiangyou-78 and Huagan-2205) were released by the local administrations of Guangdong province, Guangxi province and Anhui province between 2017 and 2019. Currently, farmers cultivate these allelopathic rice cultivars on at least 50 000 ha in the provinces of Anhui, Guandong, Guangxi and Hainan in South China (Kong CH, 2020, pers. comm.).

Breeding for rice allelopathy in Asia is an enduring process, and one such breeding programme was stopped after several years of research and development due to problems with autotoxicity and other problems associated with weed susceptibility, since Asian rice fields are infested with mixtures of weed species. Another problem is that rice breeders mainly focus on high-yielding cultivars and breeding for allelopathic cultivars may not be their priority. Moreover, hand weeding is still affordable in some Asian countries, which means that investments in crop allelopathy for use in IWM are not attractive (Bhagirath SC, pers. comm.).

One discrepancy for modern-day breeders is to breed crops that are both high-yielding and have strong allelopathic potential. This dilemma exists also for disease-resistant crops as they are lower-yielding, and the quality of the harvested product is also considered to be lower (Brown and Rant, 2013). In organic farming, the use of disease-resistant plants is also associated with yield loss, but compensation comes from higher product prices.

An interesting possibility could be to focus on studies to increase the allelopathic potential of cover crops, since they do not have the constraint of high quality and high yield. However, the authors are not aware of any breeding programmes with this objective.

4 Conclusion

The performance of crops in agricultural systems with low herbicide input is dependent on a detailed understanding of weed biology, weed population dynamics and crop-weed interactions. Furthermore, each agrosystem is specific, dynamic and influenced by numerous crop-weed interactions with varying biotic and abiotic environmental factors. In this complex environment, only a few examples of crop allelopathy have been documented (Fig. 2). In the first part of this chapter, we discussed the challenges of studying allelopathy in terms of experimental methodology and allelopathic trait selection. The biggest challenge remains to identify and characterise compounds involved in rhizosphere interactions and signaling, which further affect the root growth and performance of neighbouring plants.

We also presented a case study based on ten years of research and expertise in understanding the allelopathic interactions of buckwheat. During the past years, we studied the belowground interactions between buckwheat and redroot pigweed, but have expanded the research efforts recently to other *cover crops* and *weeds* based on observations of growth suppression in other plant-plant interactions in both the field and under controlled conditions (Gfeller et al., 2018a,b). In particular, we are interested in the root growth effects between plants in different crop-weed set-ups (monocot/monocot, monocot/dicot and dicot/dicot).

Some current research objectives that require scientific expertise across various disciplines are:

- Patterns of root exudation as a consequence of direct crop-weed root interactions,
- Selection of potential candidate chemical compounds in root exudates induced by plant neighbours,
- Root morphological traits induced by neighbouring plants and candidate allelochemicals,
- Potential gene candidates in weed model plants (transcriptional level), and
- Understanding the soil behaviour of candidate chemical compounds.

Once it becomes possible to characterise the promising allelochemicals in the cover crop which are induced by the presence of the neighbouring weed plants, further applications in IWM can be developed, and it will be possible to meet consumer expectations for more sustainable food production.

5 Where to look for further information

Recently, the journal Plant, Cell and Environment published a special Issue on plant-plant interactions covering the communication among plants and their mechanisms in an ecological context including light and volatile signaling, and underground communication networks.

Plant, Cell and Environment, Special Issue: Plant-Plant interactions, Volume44, Issue 8, August 2021.

We would like to endorse the special issue in Trends in Plant Science: Unraveling the Secrets of the Rhizosphere focusing on the interactions of plants with rhizosphere microorganisms. It also covers the mechanisms behind the belowground interactions and gives insights on root-root interactions and methodological aspects to study belowground interactions.

Trends in Plant Science, Special Issue: Unravelling the Secrets of the Rhizosphere, Volume 21, Issue 3, 169-278, March 2016.

6 References

Alsaadawi, I. S., Sarbout, A. K. and Al-Shamma, L. M. 2012. Differential allelopathic potential of sunflower (*Helianthus annuus* L.) genotypes on weeds and wheat (*Triticum aestivum* L.) crop. *Archives of Agronomy and Soil Science* 58(10), 1139-1148.

Bakker, P. A., Berendsen, R. L., Doornbos, R. F., Wintermans, P. C. and Pieterse, C. M. 2013. The rhizosphere revisited: root microbiomics. *Frontiers in Plant Science* 4, 165 .

Barnes, J. P. and Putnam, A. R. 1983. Rye residues contribute weed suppression in no-tillage cropping systems. *Journal of Chemical Ecology* 9(8), 1045-1057.

Barzman, M. and Dachbrodt-Saaydeh, S. 2011. Comparative analysis of pesticide action plans in five European countries. *Pest Management Science* 67(12), 1481-1485.

Beaudegnies, R., Edmunds, A. J., Fraser, T. E., Hall, R. G., Hawkes, T. R., Mitchell, G., Schaetzer, J., Wendeborn, S. and Wibley, J. 2009. Herbicidal 4-hydroxyphenylpyruvate dioxygenase inhibitors–a review of the triketone chemistry story from a Syngenta perspective. *Bioorganic and Medicinal Chemistry* 17(12), 4134-4152.

Belz, R. G. and Hurle, K. 2005. Differential exudation of two benzoxazinoidsone of the determining factors for seedling allelopathy of *Triticeae* species. *Journal of Agricultural and Food Chemistry* 53(2), 250-261.

Bertholdsson, N. O. 2004. Variation in allelopathic activity over 100 years of barley selection and breeding. *Weed Research* 44(2), 78-86.

Bertholdsson, N. O., Andersson, S. C. and Merker, A. 2012. Allelopathic potential of *Triticum* spp., *Secale* spp. and *Triticosecale* spp. and use of chromosome substitutions and translocations to improve weed suppression ability in winter wheat. *Plant Breeding* 131(1), 75–80.

Bialy, Z., Oleszek, W., Lewis, J. and Fenwick, G. R. 1990. Allelopathic potential of glucosinolates (mustard oil glycosides) and their degradation products against wheat. *Plant and Soil* 129(2), 277–281.

Biszczak, W., Różyło, K. and Krakska, P. 2020. Yielding parameters, nutritional value of soybean seed and weed infestation in relay-strip intercropping system with buckwheat. *Acta Agriculturae Scandinavica, Section B–Soil and Plant Science* 70(8), 640–647.

Blaise, D., Manikandan, A., Verma, P., Nalayini, P., Chakraborty, M. and Kranthi, K. R. 2020. Allelopathic intercrops and its mulch as an integrated weed management strategy for rainfed Bt-transgenic cotton hybrids. *Crop Protection* 135, 105214.

Blanco-Canqui, H., Shaver, T. M., Lindquist, J. L., Shapiro, C. A., Elmore, R. W., Francis, C. A. and Hergert, G. W. 2015. Cover crops and ecosystem services: insights from studies in temperate soils. *Agronomy Journal* 107(6), 2449–2474.

Bouhaouel, I., Gfeller, A., Boudabbous, K., Fauconnier, M. L., Amara, H. S. and Du Jardin, P. 2018. Physiological and biochemical parameters: new tools to screen barley root exudate allelopathic potential (*Hordeum vulgare* L. subsp. *vulgare*). *Acta Physiologiae Plantarum* 40(2), 1–14.

Bouhaouel, I., Gfeller, A., Boudabous, K., Fauconnier, M.-L., Ayed, O. S., Amara, H. S. and Du Jardin, P. 2020. Effects of physico-chemical and biological properties of soil on the allelopathic activity of barley (*Hordeum vulgare* L. subsp. *vulgare*) root exudates against Bromus diandrus Roth. and Stelleria media L. weeds. *Allelopathy Journal* 49(1), 17–34.

Bourgaud, F., Gravot, A., Milesi, S. and Gontier, E. 2001. Production of plant secondary metabolites: a historical perspective. *Plant Science* 161(5), 839–851.

Brown, J. K. M. and Rant, J. C. 2013. Fitness costs and trade-offs of disease resistance and their consequences for breeding arable crops. *Plant Pathology* 62, 83–95.

Callaway, R. M., Pennings, S. C. and Richards, C. L. 2003. Phenotypic plasticity and interactions amongst plants. *Ecology* 84 (5), 1115–1128.

Campiglia, E., Caporali, F., Radicetti, E. and Mancinelli, R. 2010. Hairy vetch (Vicia villosa Roth.) cover crop residue management for improving weed control and yield in no-tillage tomato (*Lycopersicon esculentum Mill.*) production. *European Journal of Agronomy* 33(2), 94–102.

Cantrell, C. L., Dayan, F. E. and Duke, S. O. 2012. Natural products as sources for new pesticides. *Journal of Natural Products* 75(6), 1231–1242.

Chase, W. R., Nair, M. G., Putnam, A. R. and Mishra, S. K. 1991. 2,2′-oxo-1,1′-azobenzene: microbial transformation of rye (Secale cereale L.) allelochemical in field soils by *Acinetobacter calcoaceticus*: III. *Journal of Chemical Ecology* 17(8), 1575–1584.

Cheema, Z. A. and Khaliq, A. 2000. Use of sorghum allelopathic properties to control weeds in irrigated wheat in a semi arid region of Punjab. *Agriculture, Ecosystems and Environment* 79(2–3), 105–112.

Cheema, Z. A., Khaliq, A. and Farooq, M. 2008. Sorghum allelopathy for weed management in wheat. In: Zeng, R. S., Mallik, A. U. and Luo, S. M. (Eds) *Allelopathy in Sustainable Agriculture and Forestry*. Springer, New York, NY, 255–270.

Chou, C.-H. and Lin, H.-J. 1976. Autointoxication mechanism of Oryza sativa I. Phytotoxic effects of decomposing rice residues in soil. *Journal of Chemical Ecology* 2(3), 353–367.

Chou, C.-H. and Patrick, Z. A. 1976. Identification and phytotoxic activity of compounds produced during decomposition of corn and rye residues in soil. *Journal of Chemical Ecology* 2(3), 369-387.
Chung, I. M., Ahn, J. K. and Yun, S. J. 2001. Identification of allelopathic compounds from rice (*Oryza sativa L.*) straw and their biological activity. *Canadian Journal of Plant Science* 81(4), 815-819.
Ciriminna, R., Fidalgo, A., Ilharco, L. M. and Pagliaro, M. 2019. Herbicides based on pelargonic acid: herbicides of the bioeconomy. *Biofuels, Bioproducts and Biorefining* 13(6), 1476-1482.
Corcuera, L. J. 1984. Effects of indole alkaloids from Gramineae on aphids. *Phytochemistry* 23(3), 539-541.
Cordeau, S., Triolet, M., Wayman, S., Steinberg, C. and Guillemin, J. P. 2016. Bioherbicides: dead in the water? A review of the existing products for integrated weed management. *Crop Protection* 87, 44-49.
Dayan, F. E. 2006. Factors modulating the levels of the allelochemical sorgoleone in Sorghum bicolor. *Planta* 224(2), 339-346.
Dayan, F. E., Cantrell, C. L. and Duke, S. O. 2009. Natural products in crop protection. *Bioorganic and Medicinal Chemistry* 17(12), 4022-4034.
Dayan, F. E. and Duke, S. O. 2014. Natural compounds as next-generation herbicides. *Plant Physiology* 166(3), 1090-1105.
Dayan, F. E. and Duke, S. O. 2020. Discovery for new herbicide sites of action by quantification of plant primary metabolite and enzyme pools. *Engineering* 6(5), 509-514.
Dayan, F. E., Howell, J. L., Marais, J. P., Ferreira, D. and Koivunen, M. 2011. Manuka oil, A natural herbicide with preemergence activity. *Weed Science* 59(4), 464-469.
Dayan, F. E., Rimando, A. M., Pan, Z. Q., Baerson, S. R., Gimsing, A. L. and Duke, S. O. 2010. Sorgoleone. *Phytochemistry* 71(10), 1032-1039.
De Souza Barros, V. M., Pedrosa, J. L. F., Gonçalves, D. R., Medeiros, F. C. L. D., Carvalho, G. R., Gonçalves, A. H. and Teixeira, P. V. V. Q. 2020. Herbicides of biological origin: a review. *Journal of Horticultural Science and Biotechnology*, 96(3), 288-296
Delory, B. M., Delaplace, P., Fauconnier, M.-L. and Du Jardin, P. 2016. Root-emitted volatile organic compounds: can they mediate belowground plant-plant interactions? *Plant and Soil* 402(1-2), 1-26.
Dilday, R., Yan, W., Moldenhauer, K. and Gravois, K. 1998. Allelopathic activity in rice for controlling major aquatic weeds. *Allelopathy in Rice*, 7-26.
Dimaano, N. G. B., Ali, J., Cruz, P. C. S., Baltazar, A. M., Diaz, M. G. Q., Acero, B. L. Jr. and Li, Z. 2017. Performance of newly developed weed-competitive rice cultivars under lowland and upland weedy conditions. *Weed Science* 65(6), 798-817.
Duke, S. O., Owens, D. K. and Dayan, F. E. 2014. The growing need for biochemical bioherbicides. In: Aaron D. G., Joel R. C., Stephen O. D. and James, N. S. (Eds) *Biopesticides: State of the Art and Future Opportunities*. American Chemical Society, 31-43.
Ens, E. J., Bremner, J. B., French, K. and Korth, J. 2009. Identification of volatile compounds released by roots of an invasive plant, bitou bush (*Chysanthemoides monilifera* spp. *rotundata*), and their inhibition of native seedling growth. *Biological Invasions* 11(2), 275-287.
Falquet, B., Gfeller, A., Pourcelot, M., Tschuy, F. and Wirth, J. 2015. Weed suppression by common buckwheat: a review. *Environment Control in Biology* 53(1), 1-6.

Falquet, B., Roux, D., Henriet, L., Tschuy, F. and Wirth, J. 2014. Simple method to separate resource competition from allelopathic root interactions. *Allelopathy Journal* 34, 227–240.

Fernández-Aparicio, M. and Rubiales, D. 2019. Selecting wheat and oat germplasm with allelopathic action against parasitic weeds. In: *Spanish Symposium on Physiology and Breeding of Cereals*.

Flood, H. E. and Entz, M. H. 2018. Effects of a fall rye cover crop on weeds and productivity of *Phaseolus* beans. *Canadian Journal of Plant Science* 99(1), 22–33.

Gaffney, J., Bing, J., Byrne, P. F., Cassman, K. G., Ciampitti, I., Delmer, D., Habben, J., Lafitte, H. R., Lidstrom, U. E., Porter, D. O., Sawyer, J. E., Schussler, J., Setter, T., Sharp, R. E., Vyn, T. J. and Warner, D. 2019. Science-based intensive agriculture: sustainability, food security, and the role of technology. *Global Food Security* 23, 236–244.

Gallagher, C. H., Koch, J. H., Moore, R. M. and Steel, J. D. 1964. Toxicity of Phalaris tuberosa for sheep. *Nature* 204, 542–545.

Gaofeng, X., Shicai, S., Fudou, Z., Yun, Z., Hisashi, K.-N. and David, R. C. 2018. Relationship between allelopathic effects and functional traits of different allelopathic potential rice accessions at different growth stages. *Rice Science* 25(1), 32–41.

García-Jorgensen, D. B., Hansen, H. C. B., Abrahamsen, P. and Diamantopoulos, E. 2020. A novel model concept for modelling the leaching of natural toxins: results for the case of ptaquiloside. *Environmental Science. Processes and Impacts* 22(8), 1768–1779.

Gebhard, C. A., Büchi, L., Liebisch, F., Sinaj, S., Ramseier, H. and Charles, R. 2013. Screening de légumineuses pour couverts végétaux: azote et adventices. *Recherche Agronomique Suisse* 4(9), 384–393.

Gerhards, R. and Schappert, A. 2020. Advancing cover cropping in temperate integrated weed management. *Pest Management Science* 76(1), 42–46.

Gfeller, A., Glauser, G., Etter, C., Signarbieux, C. and Wirth, J. 2018a. *Fagopyrum esculentum* alters its root exudation after *Amaranthus retroflexus* recognition and suppresses weed growth. *Frontiers in Plant Science* 9, 50.

Gfeller, A., Herrera, J. M., Tschuy, F. and Wirth, J. 2018b. Explanations for Amaranthus retroflexus growth suppression by cover crops. *Crop Protection* 104, 11–20.

Gfeller, A. and Wirth, J. 2015. Is growing buckwheat allelopathic? In: Price, J. E. (Ed) *New Developments in Allelopathy Research*. Nova Science Publishers, 15–29; *Julius-Kühn-Archiv* 452, p. 431. Hauppauge, NY, 11788 USA, ISBN: 978-1-63483-390-5.

Gimsing, A. L., Baelum, J., Dayan, F. E., Locke, M. A., Sejero, L. H. and Jacobsen, C. S. 2009. Mineralization of the allelochemical sorgoleone in soil. *Chemosphere* 76(8), 1041-1047.

Głąb, L., Sowiński, J., Bough, R. and Dayan, F. E. 2017. Allelopathic potential of sorghum (*Sorghum bicolor* (L.) *Moench*) in weed control: a comprehensive review. *Advances in Agronomy*, 145, 43–95.

Goelz, M. F. B., Rothenbacher, H., Wiggins, J. P., Kendall, W. A. and Hershberger, T. V. 1980. Some hematological and histopathological effects of the alkaloids gramine and hordenine on meadow voles (*Microtus pennsylvanicus*). *Toxicology* 18(2), 125–131.

Golisz, A., Lata, B., Gawronski, S. W. and Fujii, Y. 2007. Specific and total activities of the allelochemicals identified in buckwheat. *Weed Biology and Management* 7(3), 164–171.

Gross, D., Nemeckova, A. and Schütte, H. R. 1974. *Zur Biosynthese Des Gramins*. Biochem. Physiol. Pflanzen, 166, pp.281–7.

Gu, C. Z., Xia, X. M., Lv, J., Tan, J. W., Baerson, S. R., Pan, Z. Q., Song, Y. Y. and Zeng, R. S. 2019. Diterpenoids with herbicidal and antifungal activities from hulls of rice (Oryza sativa). *Fitoterapia* 136, 104183.

Guo, L., Qiu, J., Ye, C., Jin, G., Mao, L., Zhang, H., Yang, X., Peng, Q., Wang, Y., Jia, L., Lin, Z., Li, G., Fu, F., Liu, C., Chen, L., Shen, E., Wang, W., Chu, Q., Wu, D., Wu, S., Xia, C., Zhang, Y., Zhou, X., Wang, L., Wu, L., Song, W., Wang, Y., Shu, Q., Aoki, D., Yumoto, E., Yokota, T., Miyamoto, K., Okada, K., Kim, D. S., Cai, D., Zhang, C., Lou, Y., Qian, Q., Yamaguchi, H., Yamane, H., Kong, C. H., Timko, M. P., Bai, L. and Fan, L. 2017. *Echinochloa crus-galli* genome analysis provides insight into its adaptation and invasiveness as a weed. *Nature Communications* 8(1), 1-10.

Hanson, A. D., Ditz, K. M., Singletary, G. W. and Leland, T. J. 1983. Gramine accumulation in leaves of barley grown under high-temperature stress. *Plant Physiology* 71(4), 896-904.

Hasegawa, M., Mitsuhara, I., Seo, S., Imai, T., Koga, J., Okada, K., Yamane, H. and Ohashi, Y. 2010. Phytoalexin accumulation in the interaction between rice and the blast fungus. *Molecular Plant-Microbe Interactions: MPMI* 23(8), 1000-1011.

Hayashi, H. 1998. Effects of leaf and root exudates of common buckwheat on the early growth of weeds. In: Campbell, C. and Przybylski, R. (Eds) *VII International Symposium on Buckwheat Advances in Buckwheat Research 1998 Winnipeg*, Manitoba, Canada.

Hazrati, H., Fomsgaard, I. S. and Kudsk, P. 2020. Root-exuded benzoxazinoids: uptake and translocation in neighboring plants. *Journal of Agricultural and Food Chemistry* 68(39), 10609-10617.

Hazrati, H., Fomsgaard, I. S. and Kudsk, P. 2021. Targeted metabolomics unveil alteration in accumulation and root exudation of flavonoids as a response to interspecific competition. *Journal of Plant Interactions* 16(1), 53-63.

Hess, D. E., Ejeta, G. and Butler, L. G. 1992. Selecting sorghum genotypes expressing a quantitative biosynthetic trait that confers resistance to Striga. *Phytochemistry* 31(2), 493-497.

Hooper, A. M., Tsanuo, M. K., Chamberlain, K., Tittcomb, K., Scholes, J., Hassanali, A., Khan, Z. R. and Pickett, J. A. 2010. Isoschaftoside, a C-glycosylflavonoid from *Desmodium uncinatum* root exudate, is an allelochemical against the development of Striga. *Phytochemistry* 71(8-9), 904-908.

Huang, H., Xiao, X., Ghadouani, A., Wu, J., Nie, Z., Peng, C., Xu, X. and Shi, J. 2015. Effects of natural flavonoids on photosynthetic activity and cell integrity in *Microcystis aeruginosa*. *Toxins* 7(1), 66-80.

Inderjit 2005. Soil microorganisms: an important determinant of allelopathic activity. *Plant and Soil* 274(1-2), 227-236.

Inderjit, Rawat, D. S. and Foy, C. L. 2004. Multifaceted approach to determine rice straw phytotoxicity. *Canadian Journal of Botany* 82(2), 168-176.

Jabran, K., Mahajan, G., Sardana, V. and Chauhan, B. S. 2015. Allelopathy for weed control in agricultural systems. *Crop Protection* 72, 57-65.

Jassbi, A. R., Zamanizadehnajari, S. and Baldwin, I. T. 2010. Phytotoxic volatiles in the roots and shoots of Artemisia tridentata as detected by headspace solid-phase microextraction and gas chromatographic-mass spectrometry analysis. *Journal of Chemical Ecology* 36(12), 1398-1407.

Jensen, L. B., Courtois, B., Shen, L., Li, Z., Olofsdotter, M. and Mauleon, R. P. 2001. Locating genes controlling allelopathic effects against barnyardgrass in upland rice. *Agronomy Journal* 93(1), 21-26.

Jia, C., Kudsk, P. and Mathiassen, S. K. 2006. Joint action of benzoxazinone derivatives and phenolic acids. *Journal of Agricultural and Food Chemistry* 54(4), 1049-1057.

Kalinova, J., Triska, J. and Vrchotova, N. 2005. Biological activity of phenolic compounds present in buckwheat plants. *Allelopathy Journal* 16(1), 123-129.

Kalinova, J., Vrchotova, N. and Triska, J. 2007. Exudation of allelopathic substances in buckwheat (Fagopyrum esculentum Moench). *Journal of Agricultural and Food Chemistry* 55(16), 6453-6459.

Kandhro, M. N., Tunio, S., Rajpar, I. and Chachar, Q. 2014. Allelopathic impact of sorghum and sunflower intercropping on weed management and yield enhancement in cotton. *Sarhad Journal of Agriculture* 30(3).

Karlovsky, P. 2008. Secondary metabolites in soil ecology. *Secondary Metabolites in Soil Ecology*. Springer, Berlin, Heidelberg, 1-19.

Kato-Noguchi, H. 2008. Allelochemicals released from rice plants. *Japanese Journal of Plant Science* 2(1), 18-25.

Kato-Noguchi, H. 2009. Stress-induced allelopathic activity and momilactone B in rice. *Plant Growth Regulation* 59(2), 153-158.

Kato-Noguchi, H. 2011a. The chemical cross talk between rice and barnyardgrass. *Plant Signaling and Behavior* 6(8), 1207-1209.

Kato-Noguchi, H. 2011b. Convergent or parallel molecular evolution of momilactone A and B: potent allelochemicals, momilactones have been found only in rice and the moss *Hypnum plumaeforme*. *Journal of Plant Physiology* 168(13), 1511-1516.

Kato-Noguchi, H. and Ino, T. 2003. Rice seedlings release momilactone B into the environment. *Phytochemistry* 63(5), 551-554.

Kato-Noguchi, H., Ino, T. and Ota, K. 2008. Secretion of momilactone A from rice roots to the rhizosphere. *Journal of Plant Physiology* 165(7), 691-696.

Kato-Noguchi, H., Kujime, H. and Ino, T. 2007a. UV-induced momilactone B accumulation in rice rhizosphere. *Journal of Plant Physiology* 164(11), 1548-1551.

Kato-Noguchi, H., Sugimoto, H. and Yamada, M. 2007b. Buckwheat seedlings may inhibit other plant growth by allelopathic substances. *Environment Control in Biology* 45(1), 27-32.

Kato-Noguchi, H. and Peters, R. J. 2013. The role of momilactones in rice allelopathy. *Journal of Chemical Ecology* 39(2), 175-185.

Kato, T., Kabuto, C., Sasaki, N., Tsunagawa, M., Aizawa, H., Fujita, K., Kato, Y., Kitahara, Y. and Takahashi, N. 1973. Momilactones, growth inhibitors from rice, *Oryza sativa L. Tetrahedron Letters* 14(39), 3861-3864.

Kegge, W., Ninkovic, V., Glinwood, R., Welschen, R. A., Voesenek, L. A. and Pierik, R. 2015. Red: far-red light conditions affect the emission of volatile organic compounds from barley (*Hordeum vulgare*), leading to altered biomass allocation in neighbouring plants. *Annals of Botany* 115(6), 961-970.

Kelton, J., Price, A. J. and Mosjidis, J. 2012. Allelopathic weed suppression through the use of cover crops. *Weed Control*. 2, 978-953.

Kong, C., Liang, W., Xu, X., Hu, F., Wang, P. and Jiang, Y. 2004. Release and activity of allelochemicals from allelopathic rice seedlings. *Journal of Agricultural and Food Chemistry* 52(10), 2861-2865.

Kong, C. H., Chen, X. H., Hu, F. and Zhang, S. Z. 2011. Breeding of commercially acceptable allelopathic rice cultivars in China. *Pest Management Science* 67(9), 1100-1106.

Kong, C. H., Li, H. B., Hu, F., Xu, X. H. and Wang, P. 2006. Allelochemicals released by rice roots and residues in soil. *Plant and Soil* 288(1-2), 47-56.

Kong, C. H., Zhang, S. Z., Li, Y. H., Xia, Z. C., Yang, X. F., Meiners, S. J. and Wang, P. 2018. Plant neighbor detection and allelochemical response are driven by root-secreted signaling chemicals. *Nature Communications* 9(1), 3867.

Kremer, R. J. and Ben-Hammouda, M. 2009. Allelopathic plants. 19. Barley (*Hordeum vulgare L*). Allelopathy Journal 24(2), 225–242.

Kruidhof, H. M., Gallandt, E. R., Haramoto, E. R. and Bastiaans, L. 2010. Selective weed suppression by cover crop residues: effects of seed mass and timing of species' sensitivity. *Weed Research* 51(2), 177–186.

Kruidhof, H. M., Van Dam, N. M., Ritz, C., Lotz, L. A. P., Kropff, M. J. and Bastiaans, L. 2014. Mechanical wounding under field conditions: a potential tool to increase the allelopathic inhibitory effect of cover crops on weeds? *European Journal of Agronomy* 52, 229–236.

Kumar, V., Brainard, D. C. and Bellinder, R. R. 2009. Suppression of Powell Amaranth (*Amaranthus powellii*) by buckwheat residues: role of allelopathy. *Weed Science* 57(1), 66–73.

Kunz, C., Sturm, D. J., Varnholt, D., Walker, F. and Gerhards, R. 2016. Allelopathic effects and weed suppressive ability of cover crops. *Plant, Soil and Environment* 62(2), 60–66.

Kurusu, T., Hamada, J., Nokajima, H., Kitagawa, Y., Kiyoduka, M., Takahashi, A., Hanamata, S., Ohno, R., Hayashi, T. and Okada, K. 2010. Regulation of microbe-associated molecular pattern-induced hypersensitive cell death, phytoalexin production, and defense gene expression by calcineurin B-like protein-interacting protein kinases, OsCIPK14/15, in rice cultured cells. *Plant Physiology* 153(2), 678–692.

Leather, G. R. 1983. Sunflowers (*Helianthus annuus*) are allelopathic to weeds. *Weed Science* 31(1), 37–42.

Lee, D. L., Prisbylla, M. P., Cromartie, T. H., Dagarin, D. P., Howard, S. W., Provan, W. M., Ellis, M. K., Fraser, T. and Mutter, L. C. 1997. The discovery and structural requirements of inhibitors of p-hydroxyphenylpyruvate dioxygenase. *Weed Science*, 45(5), 601–609.

Li, L.-L., Zhao, H.-H. and Kong, C.-H. 2019. (-)-Loliolide, the most ubiquitous lactone, is involved in barnyardgrass-induced rice allelopathy. *Journal of Experimental Botany* 71(4), 1540–1550.

Li, Y. H., Xia, Z. C. and Kong, C. H. 2016. Allelobiosis in the interference of allelopathic wheat with weeds. *Pest Management Science* 72(11), 2146–2153.

Li, Z. H., Wang, Q., Ruan, X., Pan, C. D. and Jiang, D. A. 2010. Phenolics and plant allelopathy. *Molecules* 15(12), 8933–8952.

Liu, D. L. and Lovett, J. V. 1993. Biologically active secondary metabolites of barley. II. Phytotoxicity of barley allelochemicals. *Journal of Chemical Ecology* 19(10), 2231–2244.

Lovett, J. V., Hoult, A. H. C. and Christen, O. 1994. Biologically active secondary metabolites of barley. IV. Hordenine production by different barley lines. *Journal of Chemical Ecology* 20(8), 1945–1954.

Macías, F. A., Marín, D., Oliveros-Bastidas, A., Castellano, D., Simonet, A. M. and Molinillo, J. M. 2005a. Structure–activity relationships (SAR) studies of benzoxazinones, their degradation products and analogues. Phytotoxicity on standard target species (STS). *Journal of Agricultural and Food Chemistry* 53(3), 538–548.

Macías, F. A., Oliveros-Bastidas, A., Marín, D., Castellano, D., Simonet, A. M. and Molinillo, J. M. 2005b. Degradation studies on benzoxazinoids. Soil degradation dynamics of (2 R)-2-O-β-D-glucopyranosyl-4-hydroxy-(2 H)-1, 4-benzoxazin-3 (4 H)-one

(DIBOA-Glc) and its degradation products, phytotoxic allelochemicals from Gramineae. *Journal of Agricultural and Food Chemistry* 53(3), 554–561.

Macías, F. A., Mejías, F. J. and Molinillo, J. M. 2019. Recent advances in allelopathy for weed control: from knowledge to applications. *Pest Management Science* 75(9), 2413–2436.

Mahmood, A., Cheema, Z. A., Mushtaq, M. N. and Farooq, M. 2013a. Maize-sorghum intercropping systems for purple nutsedge management. *Archives of Agronomy and Soil Science* 59(9), 1279–1288.

Mahmood, K., Khan, M. B., Ijaz, M., Ullah, H., Mahmood, A., Song, Y. Y., Luo, S. M. and Zeng, R. S. 2013b. Exogenous application of signaling compounds enhances rice allelopathic potential in rhizosphere soil. *International Journal of Agriculture and Biology* 15(6), 1319–1324.

Marcinkevičienė, A., Kriaučiūnienė, Z., Bogužas, V. and Velička, R. 2013. Allelopathic effects of cover crops on spring barley germination and establishment. *Journal of Food, Agriculture and Environment* 11(3/4), 684–688.

Matsuo, H., Taniguchi, K., Hiramoto, T., Yamada, T., Ichinose, Y., Toyoda, K., Takeda, K. and Shiraishi, T. 2001. Gramine increase associated with rapid and transient systemic resistance in barley seedlings induced by mechanical and biological stresses. *Plant and Cell Physiology* 42(10), 1103–1111.

Maver, M., Miras-Moreno, B., Lucini, L., Trevisan, M., Pii, Y., Cesco, S. and Mimmo, T. 2020. New insights in the allelopathic traits of different barley genotypes: middle eastern and Tibetan wild-relative accessions vs. cultivated modern barley. *PLoS ONE* 15(4), e0231976.

May, F. and Ash, J. 1990. An assessment of the allelopathic potential of Eucalyptus. *Australian Journal of Botany* 38(3), 245–254.

Mirsky, S. B., Ryan, M. R., Teasdale, J. R., Curran, W. S., Reberg-Horton, C. S., Spargo, J. T., Scott Wells, M. S., Keene, C. L. and Moyer, J. W. 2013. Overcoming weed management challenges in cover crop-based organic rotational no-till soybean production in the Eastern United States. *Weed Technology* 27(1), 193–203.

Mori, M., Tomita, C., Sugimoto, K., Hasegawa, M., Hayashi, N., Dubouzet, J. G., Ochiai, H., Sekimoto, H., Hirochika, H. and Kikuchi, S. 2007. Isolation and molecular characterization of a Spotted leaf 18 mutant by modified activation-tagging in rice. *Plant Molecular Biology* 63(6), 847–860.

Muller, C. H. and Del Moral, R. 1966. Soil toxicity induced by terpenes from Salvia leucophylla. *Bulletin of the Torrey Botanical Club* 93(2), 130–137.

Muller, W. H. and Muller, C. H. 1964. Volatile growth inhibitors produced by Salvia species. *Bulletin of the Torrey Botanical Club* 91(4), 327–330.

Nair, M. G., Whitenack, C. J. and Putnam, A. R. 1990. 2,2′-OXO-1, 1 ′-azobenzene A microbially transformed allelochemical from 2,3-Benzoxazolinone: I. *Journal of Chemical Ecology* 16(2), 353–364.

Neal, A. L., Ahmad, S., Gordon-Weeks, R. and Ton, J. 2012. Benzoxazinoids in root exudates of maize attract Pseudomonas putida to the rhizosphere. *PLoS ONE* 7(4), e35498.

Nichols, V., Verhulst, N., Cox, R. and Govaerts, B. 2015. Weed dynamics and conservation agriculture principles: a review. *Field Crops Research* 183, 56–68.

Ninkovic, V. 2003. Volatile communication between barley plants affects biomass allocation. *Journal of Experimental Botany* 54(389), 1931–1939.

Ninkovic, V., Markovic, D. and Rensing, M. 2021. Plant volatiles as cues and signals in plant communication. *Plant, Cell and Environment* 44(4), 1030–1043.

Novoplansky, A. 2009. Picking battles wisely: plant behaviour under competition. *Plant, Cell and Environment* 32(6), 726–741.

Ohno, T., Doolan, K., Zibilske, L. M., Liebman, M., Gallandt, E. R. and Berube, C. 2000. Phytotoxic effects of red clover amended soils on wild mustard seedling growth. *Agriculture, Ecosystems and Environment* 78(2), 187–192.

Ohsawa, R. and Nakatani, C. 2005. Inter- and intra-cultivar variations in the allelopathic effect of leaf aqueous extract of buckwheat (Fagopyrum esculentum Moench) on the growth of lettuce seedling. *Fagopyrum* 22, 21–24.

Omezzine, F., Bouaziz, M., Simmonds, M. S. J. and Haouala, R. 2014. Variation in chemical composition and allelopathic potential of mixoploid *Trigonella foenum-Graecum L.* with developmental stages. *Food Chemistry* 148, 188–195.

Otte, B. A., Rice, C. P., Davis, B. W., Schomberg, H. H., Mirsky, S. B. and Tully, K. L. 2020. Phenolic acids released to soil during cereal rye cover crop decomposition. *Chemoecology* 30(1), 25–34.

Petersen, J., Belz, R., Walker, F. and Hurle, K. 2001. Weed suppression by release of isothiocyanates from turnip-rape mulch. *Agronomy Journal* 93(1), 37–43.

Pheng, S., Olofsdotter, M., Jahn, G. and Adkins, S. W. 2009. Potential allelopathic rice lines for weed management in Cambodian rice production. *Weed Biology and Management* 9(4), 259–266.

Pierik, R., Visser, E. J. W., De Kroon, H. and Voesenek, L. A. C. J. 2003. Ethylene is required in tobacco to successfully compete with proximate neighbours. *Plant, Cell and Environment* 26(8), 1229–1234.

Pouryousef, M., Yousefi, A. R., Oveisi, M. and Asadi, F. 2015. Intercropping of fenugreek as living mulch at different densities for weed suppression in coriander. *Crop Protection* 69, 60–64.

Raman, H., Shamaya, N. and Pratley, J. 2018. Genetic variation for weed competition and allelopathy in rapeseed (*Brassica napus L.*). *Biological Approaches for Controlling Weeds* 27.

Rehman, S., Shahzad, B., Bajwa, A. A., Hussain, S., Rehman, A., Cheema, S. A., Abbas, T., Ali, A., Shah, L., Adkins, S. and Li, P. 2019. Utilizing the allelopathic potential of Brassica species for sustainable crop production: a review. *Journal of Plant Growth Regulation* 38(1), 343–356.

Reigosa, M. J., Pedrol, N. and González, L. 2006. *Allelopathy: A Physiological Process with Ecological Implications*. Springer Science & Business Media Dordrecht, Netherlands.

Rice, E. L. 1984. *Allelopathy*, 2nd. ed. Academic Press, Orlando, FL.

Santonja, M., Bousquet-Mélou, A., Greff, S., Ormeño, E. and Fernandez, C. 2019. Allelopathic effects of volatile organic compounds released from *Pinus halepensis* needles and roots. *Ecology and Evolution* 9(14), 8201–8213.

Sawada, K., Hasegawa, M., Tokuda, L., Kameyama, J., Kodama, O., Kohchi, T., Yoshida, K. and Shinmyo, A. 2004. Enhanced resistance to blast fungus and bacterial blight in transgenic rice constitutively expressing OsSBP, a rice homologue of mammalian selenium-binding proteins. *Bioscience, Biotechnology, and Biochemistry* 68(4), 873–880.

Schandry, N. and Becker, C. 2020. Allelopathic plants: models for studying plant-interkingdom interactions. *Trends in Plant Science* 25(2), 176–185.

Schmelz, E. A., Huffaker, A., Sims, J. W., Christensen, S. A., Lu, X., Okada, K. and Peters, R. J. 2014. Biosynthesis, elicitation and roles of monocot terpenoid phytoalexins. *The Plant Journal: For Cell and Molecular Biology* 79(4), 659–678.

Schmidt, A., Mack, G., Möhring, A., Mann, S. and El Benni, N. 2019. Stricter cross-compliance standards in Switzerland: economic and environmental impacts at farm- and sector-level. *Agricultural Systems* 176, 102664.

Schreiner, O. and Reed, H. S. 1907. The production of deleterious excretions by roots. *Bulletin of the Torrey Botanical Club* 34(6), 279-303.

Schulz, M., Marocco, A., Tabaglio, V., Macias, F. A. and Molinillo, J. M. G. 2013. Benzoxazinoids in rye allelopathy–from discovery to application in sustainable weed control and organic farming. *Journal of Chemical Ecology* 39(2), 154-174.

Schütz, V., Bigler, L., Girel, S., Laschke, L., Sicker, D. and Schulz, M. 2019. Conversions of benzoxazinoids and downstream metabolites by soil microorganisms. *Frontiers in Ecology and Evolution* 7, 238.

Sepulveda, B. A. and Corcuera, L. J. 1990. Effect of gramine on the susceptibility of barley leaves to *Pseudomonas syringae*. *Phytochemistry* 29(2), 465-467.

Shamaya, N., Raman, H., Rohan, M., Pratley, J. and Wu, H. 2018. Natural variation for interference traits against annual ryegrass in canola. In: *AusCanola 2018, 20th Australian Research Asesmbly on Brassicas*, 2018 Perth, 174-178.

Shelef, O., Hahn, P. G., Getman-Pickering, Z. and Medina, A. M. 2019. Coming to common ground: the challenges of applying ecological theory developed aboveground to rhizosphere interactions. *Frontiers in Ecology and Evolution* 7, 58.

Shimura, K., Okada, A., Okada, K., Jikumaru, Y., Ko, K. W., Toyomasu, T., Sassa, T., Hasegawa, M., Kodama, O., Shibuya, N., Koga, J., Nojiri, H. and Yamane, H. 2007. Identification of a biosynthetic gene cluster in rice for momilactones. *Journal of Biological Chemistry* 282(47), 34013-34018.

Smith, H., Casal, J. J. and Jackson, G. M. 1990. Reflection signals and the perception by phytochrome of the proximity of neighbouring vegetation. *Plant, Cell and Environment* 13(1), 73-78.

Song, Q., Qin, F., He, H., Wang, H. and Yu, S. 2019. Allelopathic potential of rain leachates from Eucalyptus urophylla on four tree species. *Agroforestry Systems* 93(4), 1307-1318.

Sturm, D. J., Peteinatos, G. and Gerhards, R. 2018. Contribution of allelopathic effects to the overall weed suppression by different cover crops. *Weed Research* 58(5), 331-337.

Subrahmaniam, H. J., Libourel, C., Journet, E. P., Morel, J. B., Muños, S., Niebel, A., Raffaele, S. and Roux, F. 2018. The genetics underlying natural variation of plant-plant interactions, a beloved but forgotten member of the family of biotic interactions. *Plant Journal: For Cell and Molecular Biology* 93(4), 747-770.

Swanton, C. J., Nkoa, R. and Blackshaw, R. E. 2015. Experimental methods for crop-weed competition studies. *Weed Science* 63(SP1), 2-11.

Szwed, M., Mitrus, J., Wiczkowski, W., Dębski, H. and Horbowicz, M. 2020. If phenolic compounds in the soil with buckwheat residues affect the emergence and growth of weed seedlings? *Acta Physiologiae Plantarum* 42(9), 1-11.

Szwed, M., Wiczkowski, W., SzawaraNowak, D., Obendorf, R. L. and Horbowicz, M. 2019. Allelopathic influence of common buckwheat root residues on selected weed species. *Acta Physiologiae Plantarum* 41(6), 92.

Teasdale, J. R. and Mohler, C. L. 1993. Light transmittance, soil-temperature, and soil-moisture under residue of hairy vetch and rye. *Agronomy Journal* 85(3), 673-680.

Tibugari, H., Chiduza, C., Mashingaidze, A. B. and Mabasa, S. 2019. Quantification of sorgoleone in sorghum accessions from eight southern African countries. *South African Journal of Plant and Soil* 36(1), 41-50.

Tin, W. W. T., Hayashi, H., Otomatsu, T., Hirose, K., Hasegawa, K. and Shigemori, H. 2009. Caprolactam, an inhibitory allelochemical exuded from germinating buckwheat (*Fagopyrum esculentum*) seeds. *Heterocycles* 78(5), 1217–1222.

Tominaga, T. and Uezu, T. 1995. Weed suppresion by buckwheat. *Current Advances in Buckwheat Research* 2, 693–697.

Tukey, H. B. 1966. Leaching of metabolites from Above-ground plant parts and its implications. *Bulletin of the Torrey Botanical Club* 93(6), 385–401.

Uddin, M. R., Park, K. W., Kim, Y. K., Park, S. U. and Pyon, J. Y. 2010. Enhancing sorgoleone levels in grain sorghum root exudates. *Journal of Chemical Ecology* 36(8), 914–922.

Uddin, M. R., Thwe, A. A., Kim, Y. B., Park, W. T., Chae, S. C. and Park, S. U. 2013. Effects of jasmonates on sorgoleone accumulation and expression of genes for sorgoleone biosynthesis in sorghum roots. *Journal of Chemical Ecology* 39(6), 712–722.

Uesugi, A., Johnson, R. and Kessler, A. 2019. Context-dependent induction of allelopathy in plants under competition. *Oikos* 128(10), 1492–1502.

Ullah, R., Aslam, Z., Maitah, M., Zaman, Qu, Bashir, S., Hassan, W. and Chen, Z. 2020. Sustainable weed control and enhancing nutrient use effciency in crops through *Brassica* (*Brassica compestris L.*) allelopathy. *Sustainability* 12(14), 5763.

Van Dam, N. M. and Bouwmeester, H. J. 2016. Metabolomics in the rhizosphere: tapping into belowground chemical communication. *Trends in Plant Science* 21(3), 256–265.

Venturi, V. and Keel, C. 2016. Signaling in the rhizosphere. *Trends in Plant Science* 21(3), 187–198.

Wang, N. Q., Kong, C. H., Wang, P. and Meiners, S. J. 2021. Root exudate signals in plant-plant interactions. *Plant, Cell and Environment* 44(4), 1044–1058.

Wari, D., Alamgir, K. M., Mujiono, K., Hojo, Y., Tani, A., Shinya, T., Nakatani, H. and Galis, I. 2019. Brown planthopper honeydew-associated symbiotic microbes elicit momilactones in rice. *Plant Signaling and Behavior* 14(11), 1655335.

Weidenhamer, J. D., Mohney, B. K., Shihada, N. and Rupasinghe, M. 2014. Spatial and temporal dynamics of root exudation: how important is heterogeneity in allelopathic interactions? *Journal of Chemical Ecology* 40(8), 940–952.

Weston, L. A., Alsaadawi, I. S. and Baerson, S. R. 2013. Sorghum allelopathy–from ecosystem to molecule. *Journal of Chemical Ecology* 39(2), 142–153.

Weston, L. A. and Mathesius, U. 2013. Flavonoids: their structure, biosynthesis and role in the rhizosphere, including allelopathy. *Journal of Chemical Ecology* 39(2), 283–297.

Wink, M. 2003. Evolution of secondary metabolites from an ecological and molecular phylogenetic perspective. *Phytochemistry* 64(1), 3–19.

Wippich, C. and Wink, M. 1985. Biological properties of alkaloids. Influence of quinolizidine alkaloids and gramine on the germination and development of powdery mildew, *Erysiphe graminis* f.sp. *hordei*. *Experientia* 41(11), 1477–1479.

Wu, H., Pratley, J., Lemerle, D. and Haig, T. 1999. Crop cultivars with allelopathic capability. *Weed Research* 39(3), 171–180.

Xie, Y., Wu, L., Zhu, B., Wu, H., Gu, Q., Rajer, F. U. and Gao, X. 2017. Digital gene expression profiling of the pathogen-resistance mechanism of *Oryza sativa* 9311 in response to *Bacillus amyloliquefaciens* FZB42 induction. *Biological Control* 110, 89–97.

Xu, M., Galhano, R., Wiemann, P., Bueno, E., Tiernan, M., Wu, W., Chung, I. M., Gershenzon, J., Tudzynski, B., Sesma, A. and Peters, R. J. 2012. Genetic evidence for natural product-mediated plant-plant allelopathy in rice (*Oryza sativa*). *New Phytologist* 193(3), 570–575.

Xuan, T. D., Minh, T. N., Anh, L. H. and Khanh, T. D. 2016. Allelopathic momilactones A and B are implied in rice drought and salinity tolerance, not weed resistance. *Agronomy for Sustainable Development* 36(3), 52.

Yang, X. F. and Kong, C. H. 2017. Interference of allelopathic rice with paddy weeds at the root level. *Plant Biology* 19(4), 584–591.

Yoshida, Y., Miyamoto, K., Yamane, H., Nishizawa, Y., Minami, E., Nojiri, H. and Okada, K. 2017. OsTGAP1 is responsible for JA-inducible diterpenoid phytoalexin biosynthesis in rice roots with biological impacts on allelopathic interaction. *Physiologia Plantarum* 161(4), 532–544.

Zhang, S. Z., Li, Y. H., Kong, C. H. and Xu, X. H. 2016. Interference of allelopathic wheat with different weeds. *Pest Management Science* 72(1), 172–178.

Zhang, T., Fan, B. and Wang, P. 2018 Barnyardgrass root recognition behaviour for rice allelopathy. *Agronomy* 8(4), 39.

Zhang, Z., Liu, Y., Yuan, L., Weber, E. and Van Kleunen, M. 2021. Effect of allelopathy on plant performance: a meta-analysis. *Ecology Letters* 24(2), 348–362.

Zhao, H., Li, H. B., Kong, C. H., Xu, X. H. and Liang, W. J. 2005. Chemical response of allelopathic rice seedlings under varying environmental conditions. *Allelopathy Journal* 15(1), 105–110.

Zhao, H. H., Kong, C. H. and Xu, X. H. 2019. Herbicidal efficacy and ecological safety of an allelochemical-based benzothiazine derivative. *Pest Management Science* 75(10), 2690–2697.

Zhao, M., Cheng, J., Guo, B., Duan, J. and Che, C. T. 2018. Momilactone and related diterpenoids as potential agricultural chemicals. *Journal of Agricultural and Food Chemistry* 66(30), 7859–7872.

Zubair, H. M., Pratley, J. E., Sandral, G. A. and Humphries, A. 2017. Allelopathic interference of alfalfa (Medicago sativa L.) genotypes to annual ryegrass (*Lolium rigidum*). *Journal of Plant Research* 130(4), 647–658.

Chapter 5

Advances in understanding invasive characteristics in weed species

Ahmet Uludağ, Çanakkale Onsekiz Mart University, Turkey; Mehmet Arslan, Erciyes University, Turkey; İlhan Üremiş, Hatay Mustafa Kemal University, Turkey; and Necmi Aksoy, Düzce University, Turkey

1 Introduction

Invasive alien species (IAS) are organisms including plants introduced either accidentally or deliberately into a natural environment where they are not normally found, resulting in serious negative consequences for their new environment. Invasive alien plants (IAP) are a growing issue in weed science. Because a particular ecosystem has not had the time to adapt to and develop ways to regulate a new species (e.g. through the evolution of diseases and pests targeting it), a new species can outcompete native species and local crops, spreading rapidly and disrupting both agricultural production and local ecosystems with consequences such as biodiversity loss and loss of ecosystem services (Ward et al., 2014; Zimdahl, 2018).

Since the definition of a weed is a plant that grows in a place where it is not wanted, with resulting negative economic and ecological effects, it is not easy to draw a line between weeds and IAP because the definitions of both are context-dependent as are concepts such as invasion biology and weed science (Daehler, 1998; Müller-Schärer et al., 2018; Witt et al., 2018; Shah et al., 2020). For example, *Ambrosia artemisiifolia* (common ragweed), which is native to Central and North America, is considered an agricultural weed in its native range (Chauvel et al., 2021) as well as introduced ranges such as Hungary and other central European countries (Pinke et al., 2011; Ramona, 2017), but

http://dx.doi.org/10.19103/AS.2021.0098.05

is classified as an IAP in other parts of Europe and on other continents (van Boheemen et al., 2017; Kropf et al., 2018; Zambak and Uludag, 2019; Chauvel et al., 2021).

Weeds and IAPs also need to be seen in a broader ecological context. Both weediness and invasiveness, for example are a result of disturbance of natural habitats by man (Ellstrand et al., 2010). An IAP can be a problem in both agricultural landscapes and natural habitats as shown in the case of *Ipomoea triloba*, originally from the Caribbean but now a widespread tropical weed that can overwhelm both crops and natural vegetation (Fig. 1) (Yazlık et al., 2018). In addition, re-vegetation of semi-disturbed areas such as pastures can result in new weed species in riparian and agricultural areas (Miyawaki and Washitani, 2004). The concepts of weeds and IAP as potential drivers of biodiversity and ecosystem service loss has brought new understanding to aspects of weed science such as weed ecology, weed management, herbicide resistance, and so on (Ward et al., 2014; Neve et al., 2018; Markus et al., 2018; Darmency, 2019; Sun et al., 2020).

One of the main features of IAP is a rapid adaptation in new habitats but little is known about what the causes of this rapid adaptation are. The evolutionary success of weeds has been attributed to genetic and physiological features (Grant, 1967; Baker, 1974; Pyšek et al., 2019; Ni et al., 2021). The evolution of a new weed species is explained as an introduction of an alien species passing geographic barriers followed by recombination through hybridization (McNeill, 1976; Ellstrand et al., 2010). Other suggested changes causing rapid evolution

Figure 1 *Ipomoea triloba*: a problem in both agricultural and natural areas.

of IAP and weeds include genetic bottlenecks, polyploidy and stress-induced modification of the genome (Oka and Morishima, 1982; Prentis et al., 2008; Dlugosch and Parker, 2008).

Lowry et al. (2013) have compiled invasion-related hypotheses for all organism groups and identified 17 hypotheses common to all groups. However, others have suggested that there is no one theory that can explain all invasive plant and weed invasions (Dai et al., 2020). Ten hypotheses explaining weed evolution and adaptation have been suggested (Clements and Jones, 2021). They are:

1. general-purpose genotypes,
2. life-history strategies,
3. ability to evolve rapidly,
4. epigenetic capacity,
5. hybridization,
6. herbicide resistance,
7. herbicide tolerance,
8. cropping system vulnerability,
9. co-evolution of weeds with human management, and
10. the ability of weeds to adapt to climate change.

In this chapter, we focus on epigenetic modifications. However, we start with genetic modifications to show the importance of epigenetics in invasiveness and weediness. We do not distinguish between weeds and IAPs due to the contextual nature of these concepts.

2 Genetic modifications as a factor in invasiveness

Several hypotheses have been proposed to explain how invasive plants successfully establish themselves and become dominant in new environments (Mack, 1996; Catford et al., 2009; Inderjit et al., 2005). Numerous studies have been conducted to show how genetically regulated phenotypic variations in invasive plants play a critical role in increased competitiveness, successful establishment, superior fitness, greater range expansion and invasion success (Facon et al., 2006; Lavergne and Molofsky, 2007; Roman and Darling, 2007; Barrett, 2015). The proposed mechanisms behind invasion success include: genetic diversity, enemy release, resource fluctuation, increased competitive ability and interspecific hybridization. It is hard to say when or where a particular mechanism is likely to be important (Perrins et al., 1992; Mack, 1996). For example, both genetic and epigenetic variation play a role in explaining the invasion success of *Phragmites australis* (Fig. 2) in the environments where the plants were introduced (Liu et al., 2018).

Figure 2 *Phragmites australis*: both genetic and epigenetic variation played a similar role for adaptation in the environments where the plants were introduced.

Among the proposed mechanisms of invasion success, genetic diversity has been well studied and documented (Facon et al., 2006; Lavergne and Molofsky, 2007; Roman and Darling, 2007; Barrett, 2015). Genetic diversity among individuals of a species reflects the presence of different alleles in the population that determine its capacity to adapt to new environmental conditions. Low genetic diversity can decrease the ability of a population to respond to selection pressures. The genetic diversity of a species, therefore, plays a significant role in determining its potential to become invasive.

There are several scenarios in explaining invasive potential due to genetic variation. These include a single introduction with a few genotypes from one source population or multiple introductions with many genotypes from different source populations. *Imperata cylindrica*, a serious threat in timber plantations, is a good example of multiple introductions (Fig. 3). Multiple introduced genetic lineages of parental material from East Asia contributed to the establishment of *I. cylindrica* in the southern states of the US (Lucardi et al., 2020). *I. cylindrica* has both sexual and clonal reproduction potential with high genetic diversity that facilitates its establishment and invasiveness (Capo-Chichi et al., 2008; Lucardi et al., 2014a). The current range expansion of *I. cylindrica* has also been attributed to interspecific hybridization (Lucardi et al., 2014b).

Ambrosia artemisiifolia (Table 1), which was introduced to South America, Australia and Asia from various source populations originating from its native North America, is also a good example of multiple introductions as a factor in

Figure 3 *Imperata cylindrica*: is a good example for multiple introductions.

its invasive success (Oswalt and Marshall, 2008; Gaudeul et al., 2011; Martin et al., 2014; van Boheemen et al., 2017). Another good example of multiple introductions is *Phragmites australis*. Introduced populations of *P. australis* in North America had a higher level of genetic diversity and higher phenotypic variation than populations in its native range due to multiple introductions from the Mediterranean region, sub-Saharan Africa and the Middle East (Lavergne and Molofsky, 2007). However, multiple introductions are not always responsible for high genetic diversity in invasive populations. For example, low genetic diversity in an invasive population of *Impatiens glandulifera* was detected, despite its multiple introductions, with phenotypic plasticity seen as a more significant factor in invasiveness (Hagenblad et al., 2015).

Gene expression of invasive plant species is under the influence of environmental factors. Advanced molecular techniques are now helping us to understand differentially expressed genes that may contribute to the success

Table 1 Native and introduced ranges of weeds and invasive alien plants mentioned in the chapter

Species	Family	Native Areas	Introduced Areas	Additional References
Alternanthera philoxeroides (Mart) Griseb.	Amaranthaceae	South America	Asia, North America, Australia	Pan et al. (2007)
Ambrosia artemisiifolia L.	Asteraceae	North and Central America	South America, Australia, Europe and Asia	Zambak and Uludag (2019)
Eichhornia crassipes (Mart.) Solms.	Pontederiaceae	South America	North America, Europe, Asia, Africa, Australia	Tetik and Uremis (2015)
Impatiens glandulifera Royle	Balsaminaceae	Himalayan, North-West Pakistan and Northern India	North America, Asia, Europe, Australia	Tanner and Gange (2020)
Imperata cylindrica (L.) P.Beauv.	Poaceae	Europe, Asia, Africa, Australia	North and South America, New Zealand	Rusdy (2020)
Ipomoea triloba L.	Convolvulaceae	Caribbean (Tropical America)	North and South America, Asia, Africa, Europe (Spain), Australia (pantropical)	Yazlık et al. (2018)
Phragmites australis (Cav.) Trin. ex Steud	Poaceae	Cryptogenic		Pyšek et al. (2019)
Reynoutria japonica Houtt.	Polygonaceae	Asia	North and South America, Europe, Australia	Del Tredici (2017)
Solanum elaeagnifolium Cav.	Solanaceae	North and South America	Europe, Asia, Africa, Australia	Uludag et al. (2016)
Spartina anglica C.E. Hubb.	Poaceae	North America	North America, Australia, Europe, Asia (China)	Proença et al. (2019)
Triadica sebifera (L.) Small	Euphorbiaceae	China	North America, Asia, Africa, Europe, Australia	Chapman et al. (2019)

Source: CABI, 2021.

of invasive species in recipient environments. A genome-wide gene expression study on *A. artemisiifolia* showed that only 180 out of 45 000 genes were differentially expressed between native plants and invasive plants in their new environment (Hodgins et al., 2013). These findings confirmed expectations that only a few genes play a significant role in the adaptation of invasive species in the environments where they are introduced (Prentis et al., 2010).

Genetically controlled phenotypic variations in introduced invasive species also depend on the number of introduced individuals and the number of locations of introduction. Lower phenotypic variations are expected in communities with low population sizes. When a limited number of invader species are introduced into a new environment, inbreeding leads to decreased genetic variation that considerably reduces the fitness of the species (Estoup et al., 2016; Uller and Leimu, 2011; Schrieber and Lachmuth, 2017). However, reductions in genetic variation can be alleviated by intra- and interspecific hybridization, new mutations and genome duplications which create new variants or species (Estoup et al., 2016).

Increased global transportation of plant species increases the occurrence of interspecific hybridizations that result in the formation of new hybrid taxa in the regions where the new plant species are introduced. Inter-species hybridization can mask deleterious genes or transfer favourable genes that promote invasions (Abbott, 1992; Petit et al., 2003; Estoup et al., 2016; Wendel et al., 2016). Hybridization can increase invasiveness, enhance establishment and survival rates in new habitats by creating novel phenotypes relative to parental taxa. The importance of hybridization has been confirmed, for example by the study of 35 hybrid taxa (Schierenbeck and Ellstrand, 2009). Environmental conditions in new habitats can be more appropriate for hybrids because they have been shown to cope better with environmental extremes (Williams et al., 2001). Increased reproductive ability of new hybrids is an indication of natural selection playing a significant role in shaping hybrid performance and invasiveness over time (Hovick and Whitney, 2014).

The successful establishment of invasive plants has been attributed to biotic factors such as the absence of specialist herbivores in new environments (Blossey and Nötzold, 1995). Invaders escaping from native herbivores that are absent in newly introduced habitats can become more successful because they can allocate more resources to evolve into more vigorous ecotypes rather than allocating defensive resources to survive. An example is *Triadica sebifera* (Syn. *Sapium sebiferum*) which has a higher competitive ability by allocating more resources to producing biomass rather than using those resources for defence compared to its native environment (Huang et al., 2012; Siemann and Rogers, 2001, 2003a,b). It has been shown that herbivore attacks can alter secondary metabolite synthesis and hormone signalling pathways (Pieterse and Dicke, 2007). Herbivory-induced changes in the genes controlling plant secondary metabolism, hormone signalling pathways and plant defence have been found in *Arabidopsis thaliana* (Ehlting et al., 2008; Davila Olivas et al., 2017).

3 Epigenetic modifications as a factor in invasiveness

Significant attention has recently been given to epigenetic modifications in weed ecological genetics since they can govern the ecological-adaptive potential of weed species. The term epigenetics has been defined by Waddington (1953) as relating to the alteration of gene expression in a cell during development. Epigenetic modifications are heritable but reversible changes in gene expression without a change in the DNA sequence that persist through one or more generations (Richards, 2006; Holliday, 2006; Jones, 2012; Richards et al., 2017; Shi et al., 2019). In natural plant populations, the remarkable phenotypic variance within and among populations can be seen due to epigenetic modifications (Steward et al., 2002; Aina et al., 2004; Medrano et al., 2014; Foust et al., 2016; Kooke et al., 2015; Zhang et al., 2010; Liu et al., 2015; Dubin et al., 2015; Wibowo et al., 2016; Asensi-Fabado et al., 2017). The four main types of epigenetic modification are (Fig. 4) (Sahu et al., 2013; Lamke and Baurle, 2017):

1 DNA methylation,
2 histone modifications,
3 chromatin configuration, and
4 actions of non-coding RNA species that affect messenger RNA availability.

Among the known types of epigenetic modification, DNA methylation is the most common type that controls gene expression, DNA conformation, DNA stability, transposon silencing and gene imprinting independent of heritable

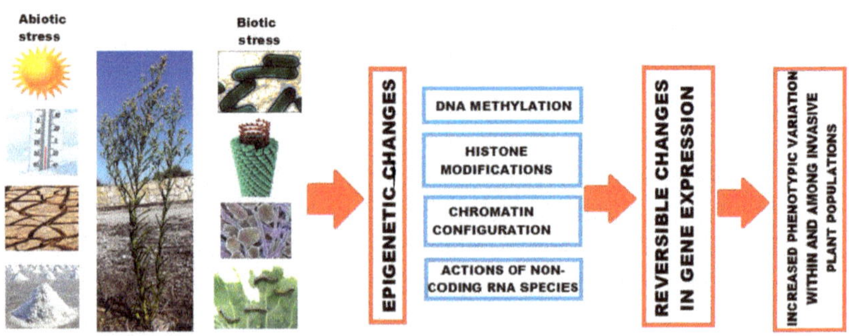

Figure 4 Epigenetic mechanisms of invasive alien plants' response to biotic and abiotic stresses.

genes (Jaenisch and Bird, 2003; Goll and Bestor, 2005; Takeda and Paszkowski, 2006; Hackett et al., 2013). DNA methylation occurs through the addition of a methyl group to the fifth carbon of the pyrimidine ring of cytosine nucleotides (Akimoto et al., 2007; Jones, 2012; Ni et al., 2018). In plants, DNA methylation occurs in three methylation contexts, CG, CHG and CHH, where H is A, C or T. CHG and CHH contexts are unique to plants (Law and Jacobsen, 2010).

For some invasive species, the development of genetic diversity through gene mutations, gene drift and selection is too slow a process to cope with environmental extremes compared with gene regulation by epigenetic modification (Zhang et al., 2013; Medrano et al., 2014). Epigenetic variation, therefore, confers a significant potential advantage to invasive plants in more extreme or unstable environments (Lele et al., 2018). Invasive weed species can evolve more rapidly through epimutations that create phenotypic diversity to allow adaptation.

Climate change influences the direction and degree of evolution of invasive plants since factors such as distribution, phenotypic plasticity, natural enemies and epigenetic modifications are significantly influenced by environmental factors such as increased temperature, CO_2 and altered precipitation regimes. Significant phenotypic variations can result from epigenetic responses to environmental stimuli (Labra et al., 2002a; Aina et al., 2004; Dowen et al., 2012; Sani et al., 2013; Radford et al., 2014; Ni et al., 2018). An increasing number of studies have shown how abiotic stress factors, such as exposure to drought (Steward et al., 2002; Labra et al., 2002b), high temperature (Dubin et al., 2015) or cold (Steward et al., 2002) govern the extent of adaptability (Castonguay and Angers, 2012; Douhovnikoff and Dodd, 2014; Dubin et al., 2015). It has been suggested that environmental factors may interact with specific loci that alter gene expression through epigenetic mechanisms and increase the ability of invasive weeds to adapt to extreme abiotic stresses (Jones, 2012; Dowen et al., 2012; Radford et al., 2014). It is known that abrupt climatic change applies selection pressure on plants that can affect the direction and degree of adaptation (Jump et al., 2006; Eveno et al., 2008).

Epigenetic modifications can therefore play a crucial role in the adaptation and evolution of natural plant populations. For example, the adaptive divergence of wild populations of *Viola cazorlensis* has been associated with methylation-based epigenetic modifications (Herrera and Bazaga, 2010). Paun et al. (2010) also reported that phenotypic variation among geographically and ecologically diverse allotetraploid sibling orchid taxa (*Dactylorhiza majalis, D. traunsteineri* and *D. majalis* ssp. *ebudensis*) was mostly due to epigenetic factors regulating gene expression in response to environmental stimulus, mainly water availability and temperature, which changed DNA methylation profiles. Two populations of *Laguncularia racemosae* grown in saline and non-saline areas showed that populations grown in saline sites were hypomethylated in comparison to non-saline sites and that populations grown in saline and non-saline areas exhibited different phenotypic characteristics due to epigenetic modifications (Lira-Medeiros et al., 2010). Epigenetic variation can, in turn, regulate differential gene expression and transposable element activation or repression. Altered expression patterns of genes in *Triticum* sp., for example were found to be caused by abiotic stress-activated retrotransposons that silenced genes adjacent to the insertion sites of retrotransposons (Kashkush et al., 2003).

Plant species reproducing asexually have very low levels of genetic diversity due to a lack of genetic variations through recombination and gene flow, limiting their adaptive potential in fluctuating environments. This means that successful invasions by clonal plant species can occur, despite low levels of genetic diversity. Studies on extremely invasive clonal plant species, such as *Reynoutria japonica* (syn: *Fallopia japonica*), *Alternanthera philoxeroides*, *Spartina anglica* and *Eichhornia crassipes* (Fig. 5) showed insignificant or no genetic variations (Baumel et al., 2001; Xu et al., 2003; Ainouche et al., 2004; Zhang et al., 2010; Richards et al., 2012, 2017; Geng et al., 2007; Parepa et al., 2014; Tetik and Uremis, 2015; Zhang et al., 2016; Banta and Richards, 2018; Holm et al., 2018; Kooke et al., 2019). Phenotypic plasticity and epigenetic mutations were the major sources for phenotypic variation to adapt to abiotic extremes. Epigenetic studies conducted with asexually reproduced *A. philoxeroides* under saline conditions revealed significantly higher epigenetic variations within the population (Shi et al., 2018) and between populations,

Figure 5 *Eichhornia crassipes*: a clonal reproducing plant species.

despite sharing almost identical DNA sequences (Shi et al., 2019). The adaptive and expansion potential of *A. philoxeroides* under novel and fluctuating environmental conditions was attributed to epigenetic variations due to progressive accumulation of epimutations. Epigenetic variations detected in asexually propagated species in extreme environments may compensate for narrow genetic variation (Schlichting, 1986; Sultan, 2004).

Invasive alien weed species have a great advantage compared to other species in terms of response to climate extremes since they demonstrate higher genetic or phenotypic diversity. Invasive species have higher phenotypic plasticity that enables them to optimize access to resources such as moisture, nutrition and sunlight (Davidson et al., 2011). The adaption capacity of invasive weeds to climate extremes can be attributed to rapid evolution to novel conditions (Oduor et al., 2016). This suggests the impact of invasiveness and weediness of species is expected to increase with climate change (Kriticos et al., 2004; Thuiller et al., 2006). This suggests the need for more epigenetic studies of invasive alien plants such as *Solanum elaeagnifolium* (Fig. 6) (Chiarini, 2014; Uludag et al., 2016; Chavana et al., 2021).

4 Conclusion

Both weediness and invasiveness are the results of human activities in disrupting natural environments. Global climate change has exacerbated invasions by alien species and caused new weeds to emerge in many areas, not only because of genetics but also due to epigenetics. The role of epigenetics is not only limited to understanding the invasiveness or weediness of a species but also affects weed management. By better understanding both genetic and epigenetic mechanisms which allow some plants to become invasive, we can better identify invasive species as well as predict and track future invasions. We will also be in a better position, for example to identify how to enable crops and native species to compete more effectively with invasive rivals. We believe that understanding the role of all aspects of epigenetics in plants can therefore help in creating more sustainable agriculture.

5 Future trends in research

This chapter covers key themes relevant to the evolution of weeds and invasive alien plants, with a focus on epigenetic modifications as a factor in invasiveness. Recently more attention has been given to existing heritable epigenetic variance in plant populations since epigenetic modifications can govern the ecological-adaptive potential of weed species. Advances in sequencing technologies, accessibility of sequencing data, bioinformatics methods and biological interpretations are needed to bring exciting new discoveries and challenges on

Figure 6 *Solanum elaeagnifolium*: an aggressive weed and invasive alien plant for future studies.

the role of epigenetics with the incorporation of whole epigenetic information jointly with environment and huge DNA sequence.

6 References

Abbott, R. J. (1992). Plant invasions, interspecific hybridization and the evolution of new plant taxa. Trends in Ecology and Evolution 7(12): 401–405.

Aina, R., Sgorbati, S., Santagostino, A., Labra, M., Ghiani, A. and Citterio, S. (2004). Specific hypomethylation of DNA is induced by heavy metals in white clover and industrial hemp. Physiologia Plantarum 121(3): 472–480.

Ainouche, M. L., Baumel, A. and Salmon, A. (2004). Biological relevance of polyploidy, ecology to genomics: *Spartina anglica* Schreb. a natural model system for analysing early evolutionary changes that affect allopolyploid genomes. Biological Journal of the Linnean Society 82(4): 475–484.

Akimoto, K., Katakami, H., Kim, H. J., Ogawa, E., Sano, C. M., Wada, Y. and Sano, H. (2007). Epigenetic inheritance in rice plants. Annals of Botany 100(2): 205-217.

Asensi-Fabado, M. A., Amtmann, A. and Perrella, G. (2017). Plant responses to abiotic stress: the chromatin context of transcriptional regulation. Biochimica et Biophysica Acta. Gene Regulatory Mechanisms 1860(1): 106-122.

Baker, H. G. (1974). The evolution of weeds. Annual Review of Ecology and Systematics 5(1): 1-24.

Banta, J. A. and Richards, C. L. (2018). Quantitative epigenetics and evolution. Heredity 121(3): 210-224.

Barrett, S. C. (2015). Foundations of invasion genetics: the Baker and Stebbins legacy. Molecular Ecology 24(9): 1927-1941.

Baumel, A., Ainouche, M. L. and Levasseur, J. E. (2001). Molecular investigations in populations of *Spartina anglica* C. E. Hubbard (Poaceae) invading coastal Brittany (France). Molecular Ecology 10(7): 1689-1701.

Blossey, B. and Nötzold, R. (1995). Evolution of increased competitive ability in invasive nonindigenous plants: a hypothesis. Journal of Ecology 83(5): 887.

CABI (2021). Invasive Species Compendium. Wallingford, UK: CAB International. Available at: www.cabi.org/isc

Capo-Chichi, L. J. A., Faircloth, W. H., Williamson, A. G., Patterson, M. G., Miller, J. H. and Van Santen, E. (2008). Invasion dynamics and genotypic diversity of cogongrass (*Imperata cylindrica*) at the point of introduction in the Southeastern United States. Invasive Plant Science and Management 1(2): 133-141.

Castonguay, E. and Angers, B. (2012). The key role of epigenetics in the persistence of asexual lineages. Genetics Research International 2012: 534289. DOI: 10.1155/2012/534289.

Catford, J. A., Jansson, R. and Nilsson, C. (2009). Reducing redundancy in invasion ecology by integrating hypotheses into a single theoretical framework. Diversity and Distributions 15: 22-40.

Chapman, D., Pescott, O. L., Roy, H. E. and Tanner, R. (2019). Improving species distribution models for invasiven non- native species with biologically informed pseudo-absence selection. Journal of Biogeography 46: 1029-1040.

Chauvel, B., Fried, G., Follak, S., Chapman, D., Kulakova, Y., Le Bourgeois, T., Marisavljevic, D., Monty, A., Rossi, J. P., Starfinger, U., Tanner, R., Tassus, X., Van Valkenburg, J. and Regnier, E. (2021). Monographs on invasive plants in Europe N° 5: *Ambrosia trifida* L. Botany Letters 168(2): 167-190.

Chavana, J., Singh, S., Vazquez, A., Christoffersen, B., Racelis, A. and Kariyat, R. R. (2021). Local adaptation to continuous mowing makes the noxious weed *Solanum elaeagnifolium* a superweed candidate by improving fitness and defence traits. Scientific Reports 11(1): 1.

Chiarini, F. E. (2014). Variation in rDNA loci of polyploid *Solanum elaeagnifolium* (Solanace). New Zealand Journal of Botany 52(3): 277-284.

Clements, D. R. and Jones, V. L. (2021). Ten ways that weed evolution defies human management efforts amidst a changing climate. Agronomy 11(2): 284.

Daehler, C. C. (1998). The taxonomic distribution of invasive angiosperm plants: ecological insights and comparison to agricultural weeds. Biological Conservation 84(2): 167-180.

Dai, Z. C., Wan, L. Y., Qi, S. S., Rutherford, S., Ren, G. Q., Wan, J. S. H. and Du, D. L. (2020). Synergy among hypotheses in the invasion process of alien plants: a road map within a timeline. Perspectives in Plant Ecology, Evolution and Systematics 47, 125575.

Darmency, H. (2019). Does genetic variability in weeds respond to non-chemical selection pressure in arable fields? Weed Research 59(4): 260–264.

Davidson, A. M., Jennions, M. and Nicotra, A. B. (2011). Do invasive species show higher phenotypic plasticity than native species and, if so, is it adaptive? A meta-analysis. Ecology Letters 14(4): 419–431.

Davila Olivas, N. H., Kruijer, W., Gort, G., Wijnen, C. L., van Loon, J. J. and Dicke, M. (2017). Genome-wide association analysis reveals distinct genetic architectures for single and combined stress responses in *Arabidopsis thaliana*. New Phytologist 213(2): 838–851.

Del Tredici, P. (2017). The introduction of Japanese knotweed, *Reynoutria japonica*, into North America. The Journal of the Torrey Botanical Society 144(4): 406–416.

Dlugosch, K. M. and Parker, I. M. (2008). Founding events in species invasions: genetic variation, adaptive evolution, and the role of multiple introductions. Molecular Ecology 17(1): 431–449.

Douhovnikoff , V. and Dodd, R. S. (2014). Epigenetics: a potential mechanism for clonal plant success. Plant Ecology 216(2): 227–233.

Dowen, R. H., Pelizzola, M., Schmitz, R. J., Lister, R., Nery, J. M. D., Dixon, J. E. and Ecker, J. R. (2012). Widespread methylation in response to biotic stress. Proceedings of the National Academy of Sciences of the United States of America 109: e2183–e2191.

Dubin, M. J., Zhang, P., Meng, D., Remigereau, M. S., Osborne, E. J., Paolo Casale, F., Drewe, P., Kahles, A., Jean, G., Vilhjálmsson, B., Jagoda, J., Irez, S., Voronin, V., Song, Q., Long, Q., Rätsch, G., Stegle, O., Clark, R. M. and Nordborg, M. (2015). DNA methylation in *Arabidopsis* has a genetic basis and shows evidence of local adaptation. eLife 4: e05255 DOI: 10.7554/eLife.05255.

Ehlting, J., Chowrira, S. G., Mattheus, N., Aeschliman, D. S., Arimura, G. I. and Bohlmann, J. (2008). Comparative transcriptome analysis of *Arabidopsis thaliana* infested by diamond back moth (*Plutella xylostella*) larvae reveals signatures of stress response, secondary metabolism, and signalling. BMC Genomics 9: 154. DOI: 10.1186/1471-2164-9-154.

Ellstrand, N. C., Heredia, S. M., Leak-Garcia, J. A., Heraty, J. M., Burger, J. C., Yao, L., Nohzadeh-Malakshah, S. and Ridley, C. E. (2010). Crops gone wild: evolution of weeds and invasives from domesticated ancestors. Evolutionary Applications 3(5–6): 494–504.

Estoup, A., Ravigné, V., Hufbauer, R. A., Vitalis, R., Gautier, M. and Facon, B. (2016). Is there a genetic paradox of biological invasion? Annual Review of Ecology, Evolution, and Systematics 47(1): 51–72.

Eveno, E., Collada, C., Guevara, M. A., Léger, V., Soto, A., Díaz, L., Léger, P., González Martínez, S. C., Cervera, M. T., Plomion, C. and Garnier-Géré, P. H. (2008). Contrasting patterns of selection at *Pinus pinaster* Ait. drought stress candidate genes as revealed by genetic differentiation analyses. Molecular Biology Evolution 25: 417–437.

Facon, B., Genton, B. J., Shykoff, J., Jarne, P., Estoup, A. and David., P. (2006). A general eco-evolutionary framework for understanding bioinvasions. Trends in Ecology and Evolution 21(3): 130–135.

Foust, C. M., Preite, V., Schrey, A. W., Verhoeven, K. J. F. and Richards, C. L. (2016). Genetic and epigenetic differences associated with environmental gradients in replicate populations of two salt marsh perennials. Molecular Ecology 25(8): 1639–1652.

Gaudeul, M., Giraud, T., Kiss, L. and Shykoff, J. A. (2011). Nuclear and chloroplast microsatellites show multiple introductions in the worldwide invasion history of common ragweed, *Ambrosia artemisiifolia*. PLoS ONE 6(3): e17658.

Geng, Y., Pan, X., Xu, C., Zhang, W., Li, B., Chen, J., Lu, B. and Song, Z. (2007). Phenotypic plasticity rather than locally adapted ecotypes allows the invasive alligator weed to colonize a wide range of habitats. Biological Invasions 9(3): 245-256.

Goll, M. G. and Bestor, T. H. (2005). Eukaryotic cytosine methyl transferases. Annual Review of Biochemistry 74: 481-514.

Grant, W. F. (1967). Cytogenetic factors associated with the evolution of weeds. Taxon 16(4): 283-293.

Hackett, J. A., Sengupta, R., Zylicz, J. J., Murakami, K., Lee, C., Down, T. A. and Surani, M. A. (2013). Germline DNA demethylation dynamics and imprint erasure through 5-hydroxymethylcytosine. Science 339(6118): 448-452.

Hagenblad, J., Hülskötter, J., Acharya, K. P., Brunet, J., Chabrerie, O., Cousins, S. A., Dar, P. A., Diekmann, M., De Frenne, P., Hermy, M., Jamoneau, A., Kolb, A., Lemke, I., Plue, J., Reshi, Z. A. and Graae, B. J. (2015). Low genetic diversity despite multiple introductions of the invasive plant species *Impatiens glandulifera* in Europe. BMC Genetics 16: 103.

Herrera, C. M. and Bazaga, P. (2010). Epigenetic differentiation and relationship to adaptive genetic divergence in discrete populations of the violet *Viola cazorlensis*. New Phytologist 187(3): 867-876.

Hodgins, K. A., Lai, Z., Nurkowski, K., Huang, J. and Rieseberg, L. H. (2013). The molecular basis of invasiveness: differences in gene expression of native and introduced common ragweed (*Ambrosia artemisiifolia*) in stressful and benign environments. Molecular Ecology 22(9): 2496-2510.

Holliday, R. (2006). Epigenetics: a historical overview. Epigenetics 1(2): 76-80.

Holm, A. K., Elameen, A., Oliver, B. W., Brandsæter, L. O., Floistad, I. S. and Brurberg, M. B. (2018). Low genetic variation of invasive *Fallopia* spp. in their northernmost European distribution range. Ecology and Evolution 8(1): 755-764.

Hovick, S. M. and Whitney, K. D. (2014). Hybridisation is associated with increased fecundity and size in 2014, 17. Ecology Letters 17(11): 1464-1477.

Huang, W., Carrillo, J., Ding, J. Q. and Siemann, E. (2012). Invader partitions ecological and evolutionary responses to above- and belowground herbivory. Ecology 93(11): 2343-2352.

Inderjit, Cadotte, M. W. and Colautti, R. I. (2005). The ecology of biological invasions: past, present and future. In: Inderjit (Ed.), Invasive Plants: Ecological and Agricultural Aspects. Basel: Birkhäuser Verlag, 19-43.

Jaenisch, R. and Bird, A. (2003). Epigenetic regulation of gene expression: how the genome integrates intrinsic and environmental signals. Nature Genetics 33 (Suppl.): 245-254.

Jones, P. A. (2012). Functions of DNA methylation: islands, start sites, gene bodies and beyond. Nature Reviews. Genetics 13(7): 484-492.

Jump, A. S., Hunt, J. M., Martinez-Izquierdo, J. A. and Penuelas, J. (2006). Natural selection and climate change: temperature-linked spatial and temporal trends in gene frequency in *Fagus sylvatica*. Molecular Ecology 15(11): 3469-3480.

Kashkush, K., Feldman, M. and Levy, A. A. (2003). Transcriptional activation of retrotransposons alters the expression of adjacent genes in wheat. Nature Genetics 33(1): 102-106.

Kooke, R., Johannes, F., Wardenaar, R., Becker, F., Etcheverry, M., Colot, V., Vreugdenhil, D., Keurentjes, J.J.B. (2015). Epigenetic basis of morphological variation and phenotypic plasticity in *Arabidopsis thaliana*. Plant Cell 27 (2): 337-348.

Kooke, R., Morgado, L., Becker, F., van Eekelen, H., Hazarika, R., Zheng, Q., de Vos, R.C.H., Johannes, F., Keurentjes, J.J.B. (2019). Epigenetic mapping of the Arabidopsis metabolome reveals mediators of the epigenotypephenotype map. Genome Research 29 (1), 96-106.

Kriticos, D. J., Yonow, T. and Mcfadyen, R. E. (2004). The potential distribution of *Chromolaena odorata* (Siam weed) in relation to climate. Weed Research 45(4): 246-254.

Kropf, M., Huppenberger, A. S. and Karrer, G. (2018). Genetic structuring and diversity patterns along rivers-local invasion history of *Ambrosia artemisiifolia* (Asteraceae) along the Danube River in Vienna (Austria) shows non-linear pattern. Weed Research 58(2): 131-140.

Labra, M., Ghiani, A., Citterio, S., Sgorbati, S., Sala, F., Vannini, C., Ruffini-Castiglione, M. and Bracale, M. (2002a). Analysis of cytosine methylation pattern in response to water deficit in pea root tips. Plant Biology 4(6): 694-699.

Labra, M., Vannini, C., Bracale, M. and Sala, F. (2002b). Methylation changes in specific sequences in response to water deficit. Plant Biosystems - An International Journal Dealing with All Aspects of Plant Biology 136(3): 269-275.

Lamke, J. and Baurle, I. (2017). Epigenetic and chromatin based mechanisms in environmental stress adaptation and stress memory in plants. Genome Biology 18(1): 124.

Lavergne, S. and Molofsky, J. (2007). Increased genetic variation and evolutionary potential drive the success of an invasive grass. Proceedings of the National Academy of Sciences of the United States of America 104(10): 3883-3888.

Law, J. A. and Jacobsen, S. E. (2010). Establishing, maintaining and modifying DNA methylation patterns in plants and animals. Nature Reviews. Genetics 11(3): 204-220.

Lele, L., Ning, D., Cuiping, P., Xiao, G. and Weihua, G. (2018). Genetic and epigenetic variations associated with adaptation to heterogeneous habitat conditions in a deciduous shrub. Ecology and Evolution 8(5): 2594-2606.

Lira-Medeiros, C. F., Parisod, C., Fernandes, R. A., Mata, C. S., Cardoso, M. A. and Ferreira, P. C. G. (2010). Epigenetic variation in mangrove plants occurring in contrasting natural environment. PLoS ONE 5(4): e10326.

Liu, J., Feng, L., Li, J. and He, Z. (2015). Genetic and epigenetic control of plant heat responses. Frontiers in Plant Science 6: 267.

Liu, L., Pei, C., Liu, S., Guo, X., Du, N. and Guo, W. (2018).Genetic and epigenetic changes during the invasion of a cosmopolitan species (*Phragmites australis*). Ecology and Evolution 8(13): 6615-6624.

Lowry, E., Rollinson, E. J., Laybourn, A. J., Scott, T. E., Aiello-Lammens, M. E., Gray, S. M., Mickley, J. and Gurevitch, J. (2013). Biological invasions: a field synopsis, systematic review, and database of the literature. Ecology and Evolution 3(1): 182-196.

Lucardi, R. D., Wallace, L. E. and Ervin, G. N. (2014a). Invasion success in cogongrass (*Imperata cylindrica*): A population genetic approach exploring genetic diversity and historical introductions. Invasive Plant Science and Management 7(1): 59-75.

Lucardi, R. D., Wallace, L. E. and Ervin, G. N. (2014b). Evaluating hybridization as a potential facilitator of successful cogongrass (*Imperata cylindrica*) invasion in Florida, USA. Biological Invasions 16(10): 2147-2161.

Lucardi, R. D., Wallace, L. E. and Ervin, G. N. (2020). Patterns of genetic diversity in highly invasive species: cogongrass (*Imperata cylindrica*) expansion in the invaded range of the Southern United States (US). Plants 9(4): 423. DOI: 10.3390/plants9040423.

Mack, R. N. (1996). Predicting the identity and fate of plant invaders: emergent and emerging approaches. Biological Conservation 78(1-2): 107-121.

Markus, C., Pecinka, A., Karan, R., Barney, J. N. and Merotto Jr., A. (2018). Epigenetic regulation-contribution to herbicide resistance in weeds? Pest Management Science 74(2): 275-281.

Martin, L. B., Coon, C. A. C., Liebl, A. L. and Schrey, A. W. (2014). Surveillance for microbes and range expansion in house sparrows. Proceedings of the Royal Society B: Biological Sciences 281: 20132690.

McNeill, J. (1976). The taxonomy and evolution of weeds. Weed Research 16(6): 399-413.

Medrano, M., Herrera, C. M. and Bazaga, P. (2014). Epigenetic variation predicts regional and local intraspecific functional diversity in a perennial herb. Molecular Ecology 23(20): 4926-4938.

Miyawaki, S. and Washitani, I. (2004). Invasive alien plant species in riparian areas of Japan: the contribution of agricultural weeds, revegetation species and aquacultural species. Global Environmental Research (English Edition) 8(1): 89-101.

Müller-Schärer, H., Sun, Y., Chauvel, B., Karrer, G., Kazinczi, G., Kudsk, P., Oude, A. G. J. M. L., Schaffner, U., Skjoth, C. A., Smith, M., Vurro, M., de Weger, L. A. and Lommen, S. T. E. (2018). Cross-fertilizing weed science and plant invasion science to improve efficient management: a European challenge. Basic and Applied Ecology 33: 1-13.

Neve, P., Barney, J. N., Buckley, Y., Cousens, R. D., Graham, S., Jordan, N. R., Lawton-Rauh, A., Liebman, M., Mesgaran, M. B., Schut, M., Shaw, J., Storkey, J., Baraibar, B., Baucom, R. S., Chalak, M., Childs, D. Z., Christensen, S., Eizenberg, H., Fernández-Quintanilla, C., French, K., Harsch, M., Heijting, S., Harrison, L., Loddo, D., Macel, M., Maczey, N., Merotto, A., Mortensen, D., Necajeva, J., Peltzer, D. A., Recasens, J., Renton, M., Riemens, M., Sønderskov, M. and Williams, M. (2018). Reviewing research priorities in weed ecology, evolution and management: a horizon scan. Weed Research 58(4): 250-258.

Ni, B., You, J., Li, J., Du, Y., Zhao, W. and Chen, X. (2021). Genetic and epigenetic changes during the upward expansion of *Deyeuxia angustifolia* Kom. in the Alpine Tundra of the Changbai Mountains, China. Plants 10: 291.

Ni, P., Li, S. G., Lin, Y. P., Xiong, W., Huang, X. N. and Zhan, A. B. (2018). Methylation divergence of invasive Ciona ascidians: significant population structure and local environmental influence. Ecology and Evolution 8(20): 10272-10287.

Oduor, A. M. O., Leimu, R. and van Kleunen, M. (2016). Invasive plant species are locally adapted just as frequently and at least as strongly as native plant species. Journal of Ecology 104(4): 957-968.

Oka, H. I. and Morishima, H. (1982). Ecological genetics and the evolution of weeds. In: Holzner W, Numata N, eds. Biology and ecology of weeds. The Hague: Dr W Junk Publishers, 73-89.

Oswalt, M. L. and Marshall, G. D. (2008). Ragweed as an example of worldwide allergen expansion. Allergy, Asthma, and Clinical Immunology 4(3): 130-135.

Pan, X. Y., Geng, Y. P., Sosa, A. J., Zhang, W. J., Li, B. and Chen, J. K. (2007). Invasive Alternanthera philoxeroides: biology, ecology and management. Acta Phytotaxonomica Sinica 45: 884-900.

Parepa, M., Fischer, M., Krebs, C. and Bossdorf, O. (2014). Hybridization increases invasive knotweed success. Evolutionary Applications 7(3): 413–420.
Paun, O., Bateman, R. M., Fay, M. F., Hedren, M., Civeyrel, L. and Chase, M. W. (2010). Stable epigenetic effects impact adaptation in allopolyploid orchids (Dactylorhiza: Orchidaceae). Molecular Biology Evolution 27(11): 2465–2473.
Perrins, J., Williamson, M. and Fitter, A. (1992). Do annual weeds have predictable characters? Acta Oecologica - International Journal of Ecology 13: 517–533.
Petit, R. J., Bodénès, C., Ducousso, A., Roussel, G. and Kremer, A. (2003). Hybridization as a mechanism of invasion in oaks. New Phytologist 161(1): 151–164.
Pieterse, C. M. J. and Dicke, M. (2007). Plant interactions with microbes and insects: from molecular mechanisms to ecology. Trends in Plant Science 12(12): 564–569.
Pinke, G., Karácsony, P., Czúcz, B. and Botta-Dukát, Z. (2011). Environmental and land-use variables determining the abundance of *Ambrosia artemisiifolia* in arable fields in Hungary. Preslia 83: 219–235.
Prentis, P. J., Wilson, J. R., Dormontt, E. E., Richardson, D. M. and Lowe, A. J. (2008). Adaptive evolution in invasive species. Trends in Plant Science 13(6): 288–294.
Prentis, P. J., Woolfit, M., Thomas-Hall, S. R., Ortiz-Barrientos, D., Pavasovic, A., Lowe, A. J. and Schenk, P. M. (2010). Massively parallel sequencing and analysis of expressed sequence tags in a successful invasive plant. Annals of Botany 106(6): 1009–1017.
Proença, B., Nez, T., Poli, A., Ciutat, A., Devaux, L., Sottolichio, A., Montaudouin, X. and Michalet, R. (2019). Intraspecific facilitation explains the spread of the invasive engineer *Spartina anglica* in Atlantic salt marshes. Journal of Vegetation Science 30: 212–223.
Pyšek, P., Skalova, H., Čuda, J., Guo, W. Y., Doležal, J., Kauzál, O., Lambertini, C., Pyšková, K., Brix, H. and Meyerson, L. A. (2019). Physiology of a plant invasion: biomass production, growth and tissue chemistry of invasive and native Phragmites australis populations. Preslia 91(1): 51–75.
Radford, E. J., Ito, M., Shi, H., Corish, J. A., Yamazawa, K., Isganaitis, E., Seisenberger, S., Hore, T. A., Reik, W., Erkek, S., Peters, A. H., Patti, M. E. and Ferguson-Smith, A. C. (2014). In utero effects. In utero undernourishment perturbs the adult sperm methylome and intergenerational metabolism. Science 345: 1255903.
Ramona, S. (2017). Chemical control of the invasive species *Ambrosia artemisiifolia* L. in sun flower agroecosystem. International Multidisciplinary Scientific Geoconference: SGEM 17: 161–167.
Richards, E. J. (2006). Inherited epigenetic variation – revisiting softinheritance. Nature Reviews Genetics 7: 395–401.
Richards, C. L., Alonso, C., Becker, C., Bossdorf, O., Bucher, E. and Colomé-Tatché, M. and Verhoeven, K. J. (2017). Ecological plant epigenetics: evidence from model and non-model species, and the way forward. Ecology Letters 20: 1576–1590.
Richards, C. L., Schrey, A. W. and Pigliucci ,M. (2012). Invasion of diverse habitats by few Japanese knotweed genotypes is correlated with epigenetic differentiation. Ecology Letters 15(9): 1016–1025.
Roman, J. and Darling, J. A. (2007). Paradox lost: genetic diversity and the success of aquatic invasions. Trends in Ecology and Evolution 22(9): 454–464.
Rusdy, M. (2020). Imperata cylindrica: reproduction, dispersal, and controls. CAB Reviews 15(38): 76–80. CAB International 2020 (Online ISSN 1749-8848).

Sahu, P. P., Pandey, G., Sharma, N., Puranik, S., Muthamilarasan, M. and Prasad, M. (2013). Epigenetic mechanisms of plant stress responses and adaptation. Plant Cell Reports 32: 1151-1159. DOI: 10.1007/s00299-013-1462-x.

Sani, E., Herzyk, P., Perrella, G., Colot, V. and Amtmann, A. (2013). Hyperosmotic priming of *Arabidopsis* seedlings establish a long-term somatic memory accompanied by specific changes of the epigenome. Genome Biology 14: R59.

Schierenbeck, K. A. and Ellstrand, N. C. (2009). Hybridization and the evolution of invasiveness in plants and other organisms. Biological Invasions 11(5): 1093-1105.

Schlichting, C. D. (1986). The evolution of phenotypic plasticity in plants. Annual Review of Ecology and Systematics 17(1): 667-693.

Schrieber, K. and Lachmuth, S. (2017). The genetic paradox of invasions revisited: The potential role of inbreeding × environment interactions in invasion success. Biological Reviews 92(2): 939-952.

Shah, K. K., Tiwari, I., Tripathi, S., Subedi, S. and Shrestha, J. (2020). Invasive alien plant species: A threat to biodiversity and agriculture in Nepal. Agriways 08(1): 62-73.

Shi, W., Chen, X., Gao, L., Xu, C. Y., Ou, X., Bossdorf, O., Yang, J. and Geng, Y. (2019). Transient stability of epigenetic population differentiation in a clonal invader. Frontiers in Plant Science 9: 1851.

Shi, W., Hu, X., Ou, X., Yang, J. and Geng, Y. (2018). Increased population epigenetic diversity of the clonal invasive species *Alternanthera philoxeroides* in response to salinity stress. Genes and Genetic Systems 93: 259-269.

Siemann, E. and Rogers, W. E. (2001). Genetic differences in growth of an invasive tree species. Ecology Letters 4(6): 514-518.

Siemann, E. and Rogers, W. E. (2003a). Increased competitive ability of an invasive tree limited by an invasive beetle. Ecological Applications 13(6): 1503-1507.

Siemann, E. and Rogers, W. E. (2003b). Reduced resistance of invasive varieties of the alien tree *Sapium sebiferum* to a generalist herbivore. Oecologia 135(3): 451-457.

Steward, N., Ito, M., Yamaguchi, Y., Koizumi, N. and Sano, H. (2002). Periodic DNA methylation in maize nucleosomes and demethylation by environmental stress. Journal of Biological Chemistry 277(40): 37741-37746.

Sultan, S. E. (2004). Promising directions in plant phenotypic plasticity. Perspectives in Plant Ecology, Evolution and Systematics 6(4): 227-233.

Sun, Y., Kaleibar, B. P., Oveisi, M. and Müller-Schärer, H. (2020). Addressing climate change: what can plant invasion science and weed science learn from each other? Frontiers in Agronomy 2: 40.

Takeda, S. and Paszkowski, J. (2006). DNA methylation and epigenetic inheritance during plant gametogenesis. Chromosoma 115(1): 27-35.

Tanner, R. A. and Gange, A. C. (2020). Himalayan balsam, *Impatiens glandulifera*: its ecology, invasion and management. Weed Research 60(1): 4-7.

Tetik, O. and Uremis, I. (2015). Eichhornia crassipes. In: Onen, H. (Ed.), Türkiye İstilacı Bitkiler Katalogu. Ankara, Turkey: Gıda Tarım ve Hayvancılık Bakanlığı, 303-315.

Thuiller, W., Richardson, D. M., Rouget, M., Proches, S. and Wilson, J. R. U. (2006). Interactions between environment, species traits, and human uses describe patterns of plant invasions. Ecology 87(7): 1755-1769.

Uller, T. and Leimu, R. (2011). Founder events predict changes in genetic diversity during hu-man-mediated range expansions. Global Change Biology 17: 3478-3485. DOI: 10.1111/j.1365-2486.2011.02509.x.

Uludag, A., Gbehounou, G., Kashefi, J., Bouhache, M., Bon, M. C., Bell, C. and Lagopodi, A. L. (2016). Review of the current situation for solanum elaeagnifolium in the mediterranean basin. EPPO Bulletin 46(1): 139–147.

van Boheemen, L. A., Lombaert, E., Nurkowski, K. A., Gauffre, B., Rieseberg, L. H. and Hodgins, K. A. (2017). Multiple introductions, admixture and bridgehead invasion characterize the introduction history of *Ambrosia artemisiifolia* in Europe and Australia. Molecular Ecology 26(20): 5421–5434.

Waddington, C. H. (1953). Genetic assimilation of an acquired character. Evolution 7(2): 118–126.

Ward, S. M., Cousens, R. D., Bagavathiannan, M. V., Barney, J. N., Beckie, H. J., Busi, R., Davis, A. S., Dukes, J. S., Forcella, F., Freckleton, R. P., Gallandt, E. R., Hall, L. M., Jasieniuk, M., Lawton-Rauh, A., Lehnhoff, E. A., Liebman, M., Maxwell, B. D., Mesgaran, M. B., Murray, J. V., Neve, P., Nuñez, M. A., Pauchard, A., Queenborough, S. A. and Webber, B. L. (2014). Agricultural weed research: a critique and two proposals. Weed Science 62(4): 672–678.

Wendel, J. F., Jackson, S. A., Meyers, B. C. and Wing, R. A. (2016). Evolution of plant genome architecture. Genome Biology 17: 37.

Wibowo, A., Becker, C., Marconi, G., Durr, J., Price, J., Hagmann, J., Papareddy, R., Putra, H., Kageyama, J., Becker, J., Weigel, D. and Gutierrez-Marcos, J. (2016). Hyperosmotic stress memory in *Arabidopsis* is mediated by distinct epigenetically labile sites in the genome and is restricted in the male germline by DNA glycosylase activity. eLife 5: e13546.

Williams, J., Boecklen, W. and Howard, D. (2001). Reproductive processes in two oak (*Quercus*) contact zones with different levels of hybridization. Heredity 87: 680–690. DOI: 10.1046/j.1365-2540.2001.00968.x.

Witt, A., Beale, T. and van Wilgen, B. W. (2018). An assessment of the distribution and potential ecological impacts of invasive alien plant species in eastern Africa. Transactions of the Royal Society of South Africa 73(3): 217–236.

Xu, C. Y., Zhang, W. J., Fu, C. Z. and Lu, B. R. (2003). Genetic diversity of alligator weed in China by RAPD analysis. Biodiversity and Conservation 12(4): 637–645.

Yazlık, A., Uremis, İ, Uludag, A., Uzun, K. and Senol, S. G. (2018). *Ipomoea triloba*: an alien plant threatening many habitats in Turkey. EPPO Bulletin 48(3): 589–594.

Zambak, Ş. and Uludag, A. (2019). Düzce ilindeki arsız zaylan (*Ambrosia artemisiifolia* L.) populasyonlarının durumu. Turkish Journal of Weed Science 22(1): 67–80.

Zhang, Y. Y., Fischer, M., Colot, V. and Bossdorf, O. (2013). Epigenetic variation creates potential for evolution of plant phenotypic plasticity. New Phytologist 197(1): 314–322.

Zhang, Y. Y., Parepa, M., Fischer, M. and Bossdorf, O. (2016). Epigenetics of colonizing species. A study of Japanese knotweed in Central Europe. In: Barrett, S. C. H., Colautti, R. I., Dlugosch, K. M. and Rieseberg, L. H. (Eds), Invasion Genetics. Chichester, UK: John Wiley & Sons, Ltd, 328–340.

Zhang, Y. Y., Zhang, D. Y. and Barrett, S. C. H. (2010). Genetic uniformity characterizes the invasive spread of water hyacinth (*Eichhornia crassipes*), a clonal aquatic plant. Molecular Ecology 19(9): 1774–1786.

Zimdahl, R. L. (2018). Fundamentals of Weed Science. Academic Press. San Diego, California.

Part 2

Intelligent weed control technologies

Chapter 6

Modelling the effects of cropping systems on weed dynamics: the trade-off between process analysis and decision support

Nathalie Colbach, AgroSup Dijon, INRAE, Université de Bourgogne, France

1 Introduction

Non-chemical weed control must be based on combining many partially efficient and mostly preventive techniques, with an emphasis on biological regulation (Liebman and Gallandt, 1997). This complex management is not made any easier by the dozens of contrasting weed species which differ in response to these management techniques (Fried et al., 2010). Moreover, weed management must not only aim at controlling harmful weeds. It must also promote ecosystem services such as the contribution of weeds to wild plant diversity and as a resource for fauna (Blaix et al., 2018).

These complexities mean that we need tools that synthesise our knowledge on weeds and, particularly, quantify the effects of cropping systems on these pests. Models have been increasingly used to evaluate and design cropping systems since the 1980s (Ould-Sidi and Lescourret, 2011, Dury et al., 2012, Martin et al., 2013). These models are a simplified and relatively abstract representation of a process or system which are used to describe, explain or predict that process or system. Their complexity can vary greatly, ranging from

http://dx.doi.org/10.19103/AS.2021.0098.07

models consisting of a single equation or function, to highly complex models aggregating a large number of equations, contingency tables etc. Ecology was probably the first scientific discipline to use formalised models such as Leslie's matrix models (Leslie, 1945). These first models were deterministic, that is, a given input set only gave one possible set of output values. The more recent stochastic (or probabilistic) models have at least one parameter value that is chosen randomly, and each input set is associated with a distribution of output variables after repeated simulation runs. Stochastic models aim at predicting the probability of events rather than at understanding the underlying processes (Legay, 1996).

Predictive weed models are of potential value to a wide range of users. Scientists can use them to understand processes and events. However, they also need methodologies to evaluate and design multifunctional cropping systems. Technical institutes and extension services need to produce advice for their farmers to optimise production and financial returns. Farmers want guidance on what action to take, both for short-term tactical decisions (To spray or not to spray? When to spray?) and for strategic decisions (Which rotation? Which tillage strategy is best for this rotation?). Government decision-makers and regulators need to evaluate risks and to decide if and how to regulate these risks.

The models that best answer the scientists' requirements are weed dynamics models. These synthesise and quantify effects of cropping system components on weed life cycles and can then be used to test particular scenarios ('virtual experiments'). Though these weed models have been around for quite some time, they are rarely used for making actual predictions or to make decisions (Freckleton and Stephens, 2009). The earliest models did integrate a few cropping techniques, usually to determine the control level necessary to keep a weed population stable; but they did not attempt to evaluate more complex cropping systems (Colbach and Debaeke, 1998). However, designing weed control strategies not only requires predictions of average effects of cropping techniques. It also requires predictions of how these effects vary, in order to estimate the probability of success or failure of a given management strategy. Farmers are notably risk-adverse and often prefer a strategy that produces a stable income over a number of years to a strategy with a higher average income but which varies from year to year (Wossink et al., 1997, Doohan et al., 2010, Ridier et al., 2013).

Different approaches are available to address these points. For instance, variability can be predicted with deterministic models of the interactions between cropping techniques and environmental conditions (Colbach et al., 2005). However, complex deterministic models are disadvantaged by their large number of parameters, with the risk of an unstable model that is very sensitive to small variations in parameter values (Gressel, 2005,

Freckleton and Stephens, 2009). The alternative is to tackle uncertainty with stochastic functions to obtain a frequency distribution of expected results (Holst et al., 2007, Freckleton and Stephens, 2009). The resulting mean is not necessarily identical to the output obtained with mean parameter values (Freckleton and Watkinson, 1998). Stochastic functions are usually empirical (i.e. descriptive instead of explanatory) and cannot therefore be used in conditions other than those in which their parameter sets were estimated.

The objective of the present chapter is to critically evaluate the main characteristics of these *a priori* contradictory approaches:

- deterministic vs. stochastic; and
- mechanistic vs. empirical.

The chapter does not attempt to be a generic modelling paper or to provide a comprehensive list of models. It focuses on modelling cropping system effects on weed infestation and/or dynamics, with the ultimate aim of evaluating and designing better cropping systems. This requires a multi-disciplinary approach, balancing different experimental and philosophical concepts. Another challenge is the duality of model uses, requiring a compromise between representing processes for scientists and providing predictions for decision-makers. The chapter shows that modelling approaches depend on the modeller's objective. It analyses the merits and drawbacks of different approaches, in order to propose a compromise for both representing cropping system × environment interactions correctly and the ability to run numerous simulations accurately, easily and rapidly.

This chapter is based on an earlier paper (Colbach, 2010) which is updated here and focuses specifically on weeds. We will not cover models focusing on a particular weed stage, for which recent reviews can be found elsewhere (seed germination and dormancy, Batlla et al., 2020, emergence, Royo-Esnal et al., 2020, weed-crop interference, Singh et al., 2020). The chapter also does not focus on decision support systems, which include many aspects other than weed dynamics, for example, weed identification or the ergonomics of graphic interfaces for uses (González-Andújar, 2020, Kanatas et al., 2020).

2 Comparing models: case studies

Models have multiple uses, that is to describe, explain or predict a process or a system. Defining the potential future use of a model is a first crucial step in a modelling process. Here, we focus on models for designing cropping systems for integrated crop production. This requires a good ranking ability of cropping systems in a large range of situations (e.g. 'system A is less infested by weeds

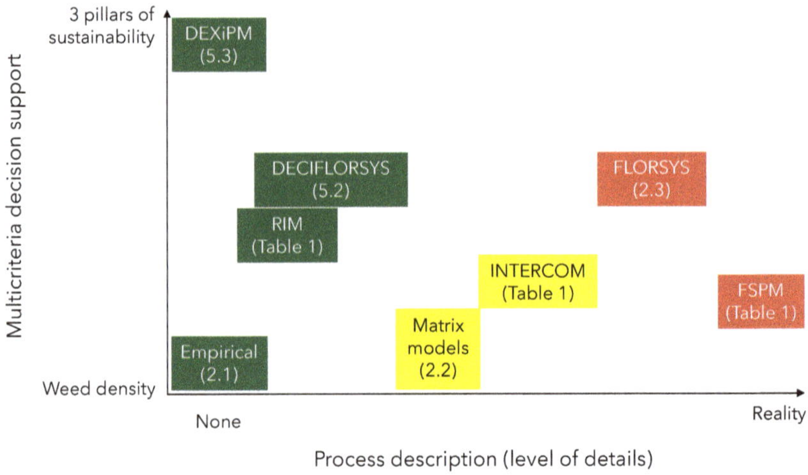

Figure 1 Trade-off between the extent of multicriteria evaluation vs. the detail of process representation in a few major contrasting weed dynamics models (see Table 1 and subsequent sections for further details). Colours illustrate easiness of use (green = easy, yellow = intermediate, red = difficult) (Nathalie Colbach, 2020 CC BY NC ND).

than system B') and the ability to take account of variability (e.g. 'system A is less infested by weeds than system B for 90% of the weather scenarios). We also want to understand effects to delimit the domain of validity of the model and to avoid statistical relationships devoid of any reality.

This section assesses three contrasting models developed to quantify the effect of cropping systems on weed dynamics (hence 'weeds = f(cropping system)' models) at the field level. The aim is to illustrate three very different approaches to identify their respective merits and disadvantages before moving on to more general and theoretical considerations in the following sections. The case studies were chosen to maximise structural differences in terms of precision of biophysical processes vs. range of evaluation criteria (Fig. 1) with a preference for validated models. Some of the models include bioeconomic submodels (e.g. RIM, Pannell et al., 2004, Lacoste and Powles, 2017), which we will not discuss here. Similarly, we will not consider multicriteria evaluation tools such as DEXiPM (DEXi Pest Management, Pelzer et al., 2012) which mostly do not cover weeds.

2.1 A single-equation static model

In contrast to other areas of modelling (e.g. crop diseases, Ennaïfar et al., 2007), there are no models that use an empirical approach to directly predict weed incidence from a set of input variables describing crop and management

Table 1 Short presentation of the most recent and most complete weed dynamics models (focusing on 'validated' models). Green cells show the best approaches for predicting and understanding the effects of cropping systems on crop production, weed dynamics and yield loss due to weeds

Model	Cropping practices	Weather	Scale	Process-based model	Canopy model	Yield model	Yield loss model	Weed-weed competition	Resources for which plants compete	Crops	Weeds	Reference[a]
INTERCOM	Crop sowing techniques	Yes	Post-emergence	Yes	Homogeneous two-species	Ecophysiological	Difference from simulations with vs without weeds	No	Light, water	Wheat	*Alopecurus myosuroides*	(Kropff and van Laar, 1993, Andrew and Storkey, 2017)
RIM	Main operations and options	No	Multi-annual	No	No	No	f(weed vs crop density)	No	Not specified	Arable crops	Single species (e.g. *Lolium rigidum*)	(Pannell et al., 2004, Lacoste and Powles, 2017, Torra and Monjardino, 2020)
No name	Crop sowing & Weed control options	Yes	Multi-annual	No	No	No	f(weed vs crop density)	No	Not specified	Cereals	Single species (e.g. *Avena fatua)*	(Molinari et al., 2020)
APSIM	Main operations and options	Yes	Multi-annual	Yes	Homogeneous	Ecophysiological	Difference from simulations with vs without weeds	No	Light, water and nitrogen	Arable crops	Four groups	(Keating et al., 2003)

(*Continued*)

Table 1 (*Continued*)

Model	Cropping practices	Weather	Scale	Process-based model	Canopy model	Yield model	Yield loss model	Weed-weed competition	Resources for which plants compete	Crops	Weeds	Reference[a]
Weed Manager	Main operations and options	Yes	Multi-annual	Yes	Homogeneous two-species	Ecophysiological	f(green area index)	No	Light	Wheat	Multispecies weed flora	(Parsons et al., 2009, Benjamin et al., 2010)
FLORSYS	Detailed list of operations	Yes	Multi-annual	Yes	3D individual-based	Ecophysiological	Difference from simulations with vs without weeds	Yes	Light, nitrogen and water (only during pre-emergent growth)	Arable crops (including perennial grasses and legumes)	Multispecies weed flora	(Gardarin et al., 2012, Munier-Jolain et al., 2013, Colbach et al., 2014, Colbach et al., 2021)
FSPM (functional-structural plant models)	Crop sowing operations	Yes	Post-emergence	Yes	3D individual-based with detailed species-specific plant architecture	Ecophysiological	Difference from simulations with vs without weeds	Yes	Light	Species-specific	Species-specific	(Evers and Bastiaans, 2016)

[a] References synthesizing models rather than giving all the details.

practices. Primot et al. (Primot et al., 2006) used a similar approach to predict the probability that weed biomass in oilseed rape exceeded a particular threshold, from inputs describing crop management practices (e.g. tillage, fertilisation). However, they also included variables based on field observations of type of weed flora, weed and crop plant densities (Table 2), which are typical for tactical rather than strategic decisions. They fitted a single equation to observations from fields monitored in four French regions (Eure et Loir, Yvelines, Yonne and Puy de Dôme) during three years (Primot et al., 2006), testing different weed-biomass thresholds and equation formats (linear vs logistic) and input selection methods.

Despite the simplicity of the model structure and the low number of parameters, the overall model accuracy (AUC, based on a receiver operating characteristic curve analysis estimating the frequencies of correct and incorrect model-based decisions) evaluated by cross-validation was high (0.79, Table 2). An AUC of 0.50 indicates a model only as good as a random decision and 1 indicates a perfect prediction (Murtaugh, 1966, Swets, 1988). However, this high predictive ability was only true as long as the model comprised input variables describing early crop and weed densities. Without these variables (as is more typical for strategic decision support), model accuracy dropped to 0.55 (for this same biomass threshold and a logistic model, Primot et al., 2006). Despite its higher predictive accuracy, a model relying on early crop and weed inputs is less useful for decision support, as it can only be used for choosing a limited number of management techniques, for example, only application of spring herbicides.

Deciding whether to spray based on this model, combined with field observations, is a tactical decision and requires a relevant biomass threshold to assess weed harmfulness for crop production. Such harmfulness thresholds are highly questionable (Oliver, 1988, O'Donovan, 1996, Swanton et al., 1999). Even the best of these thresholds usually disregard variability in water and nutrient resources, rarely quantify yield losses due to weed assemblages (Swinton et al., 1994), and only consider annual effects (McDonald and Riha, 1999, Munier-Jolain et al., 2002). The latter is particularly troubling as weed seeds survive for several years in the soil (Lewis, 1973). Whilst a single weed plant surviving in a given year does not affect crop production, the hundreds or thousands of seeds that might emerge in later years may harm future crops. Primot's threshold has the advantage of being based on weed biomass (rather than density), which has been shown to be a much better indicator of weed-related yield loss (Milberg and Hallgren, 2004, Colbach and Cordeau, 2018).

If observation-based inputs are eliminated from the model, they can be used for strategic decisions, for example, to decide which tillage strategy would be best. However, the model only estimates average effects whereas farmers are also interested in the probabilities of risk and success in different scenarios.

Table 2 Empirical model predicting the probability $P(y>y_T)$ that weed biomass exceeds a certain threshold in oilseed rape crops at winter onset from inputs x_i describing crop management as well as observations on the type of weed flora, crop and weed plant densities (Primot et al., 2006). Example of $y_T = 0.1$ t ha^{-1} and a logistic model; model accuracy AUC estimated by cross-validation= 0.79

A. Equation

$$P(y > y_T) = \frac{\exp\left(\alpha_0 + \sum_{i=1}^{p} \alpha_i x_i\right)}{1 + \exp\left(\alpha_0 + \sum_{i=1}^{p} \alpha_i x_i\right)}$$

B. Values of regression coefficients α_i

Input variable	Unit	Type	Estimate	Standard-error
Intercept (α_0)			−3.94	3.17
Soil mineral nitrogen available at crop emergence	kg N ha^{-1}	Measured	−0.0326	0.00216
Sowing date	Julian days	Practice	0.0252	0.0143
Shallow tillage (vs mouldboard ploughing)	Yes or no	Practice	0.88	0.371
Oilseed rape density 4-6 weeks after sowing (OD)	Plants m^{-2}	Measured	−0.00163	0.0107
Weed plant 4-6 weeks after sowing (WD)	Plants m^{-2}	Measured	−0.0085	0.007
OD ×WD	Plants^2 m^{-4}	Measured	0.00033	0.00015
Type of weed flora		Measured		
Type1			0	0
Type2			−0.87	0.705
Type3			−3.226	0.860
Type4			−4.077	0.821
Type5			−2.063	0.846
Type6			−2.485	0.589
Type7			0.573	1.163

C. Typology of the weed flora

Type	Group	Time of weed emergence	Height of weed relative to oilseed rape
1	Monocotyledonous	Summer	Equal or shorter
2	Monocotyledonous	Indifferent[a]	Equal or shorter
3	Dicotyledonous	Summer	Equal or shorter
4	Dicotyledonous	Summer	Taller
5	Dicotyledonous	Autumn	Equal or shorter
6	Dicotyledonous	Indifferent[a]	Equal or shorter
7	Dicotyledonous	Indifferent[a]	Taller

[a] Indifferent: weed emerging regardless of the season of the year.

Empirical models including the effects of pedoclimatic variables make this possible, as illustrated by Ennaïfar's disease model (Schoeny et al., 2001, Ennaïfar et al., 2007). The performance of each cropping system is predicted for several weather scenarios chosen randomly from available databases to obtain distributions of disease incidence and yield loss. For instance, when winter wheat was sown on 15 November in the moist and temperate climate of Brittany, significant yield loss (i.e. a reduction in yield of at least 5%) occurred in 3 years out of 30 compared to every year when sowing on 15 October. In the drier climate of the Paris Basin, the frequency of a 5% yield loss decreased from 19 out of 30 to 0 years when sowing was delayed.

2.2 Matrix-based models

Matrix-based models, such as Leslie's matrix (Leslie, 1945), follow a dynamic approach instead of the static predictions characterising empirical models. They account for a life cycle consisting of successive stages of weed development over time. Among the earliest models, González-Andújar and Fernández-Quintanilla (1991) describe the population dynamics of *Avena sterilis* L. ssp *ludoviciana* (Durieu) Nyman in the cereal-based systems of Central Spain (Fig. 2). Simulation length is chosen by the user. For each year, the model predicts the densities of three successive seed banks according to season, and two cohorts of seedlings according to their emergence period, since some of these seedlings survive to become adults producing new seeds. The successive

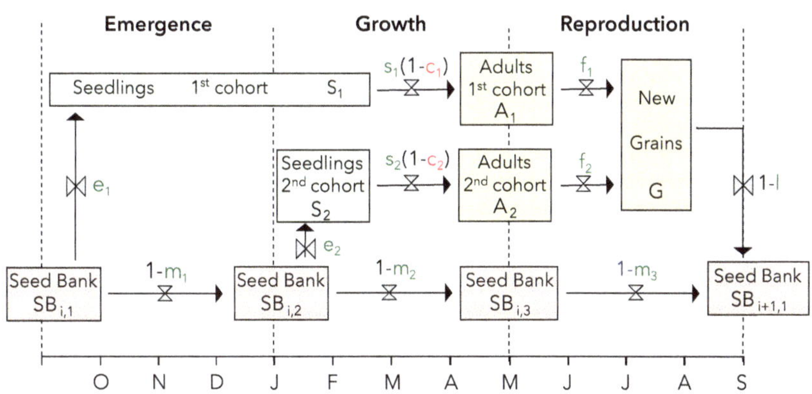

Figure 2 Diagram of the annual life cycle of the population of *Avena sterilis* used in the model of González-Andújar and Fernández-Quintanilla (González-Andújar and Fernández-Quintanilla, 1991). Life stages: three successive seed banks ($SB_{i,1}$, $SB_{i,2}$, $SB_{i,3}$), two cohorts of seedings (S_1, S_2) and adults (A_1, A_2), and seed production (G). Demographic rates: seedling emergence *e*, seedling survival in the absence of control measures *s*, plant mortality due to control measures *c*, fecundity *f*, loss of newly produced seeds *l*, mortality of seeds in the soil *m*.

stages are related by demographic rates. Some of these rates are constants (e.g. seedling emergence, seed mortality in soil and surface seed loss), while others are density-dependent (e.g. seedling survival, fecundity). More recent models make these demographic rates vary with management, for example, emergence rates depending on tillage type (see review Colbach and Debaeke, 1998).

Such models are not designed to model weed biology and ecology in detail, but can model biology well enough to improve decision-making (Colbach and Debaeke, 1998). They make it possible to compare average weed dynamics of contrasting cropping systems in the region where their parameters were estimated. When linked to yield loss functions and economic submodels, they can calculate crop production and profitability, as González-Andújar and Fernández-Quintanilla did in their follow-up paper (1993). In these respects, they are useful in decision support.

Despite its simplicity, González-Andújar and Fernández-Quintanilla's model produces realistic results. It was evaluated by comparing results with independent observations from three scenarios monitored over four years, showing that simulations and field observations were highly correlated ($r = 0.94$). However, the model disregards interactions with pedoclimatic conditions and does not represent biophysical processes underpinning management practices. This makes it very difficult to use the model outside the region and type of cropping system where the model functions and parameters were established. Without considering weather effects, it is impossible to run frequency analysis to estimate probabilities of success.

2.3 A model built from process-based submodels

The most complete weed dynamics models, such as FlorSys, follow the principles of crop models (Gardarin et al., 2012, Munier-Jolain et al., 2013, Colbach et al., 2014, Colbach et al., 2021). FlorSys can be considered as a virtual experimental field on which many and diverse cropping systems can be tested and evaluated, in terms of crop production, benefits (e.g. biodiversity) and adverse effects of weeds (e.g. impact on production) (Mézière et al., 2015). The user inputs the complete list of crop management operations lasting for several years, similar to the management operations applied to a real field in an experimental station or a farmer's field, together with latitude, daily weather data and soil characteristics (Fig. 3). The list includes all management operations (i.e. tillage, sowing, mechanical weeding, fertilisation, pesticide spraying, mowing/cutting and harvesting operations). These must be described in detail in terms of dates and options (e.g. date, density, depth, interrow, orientation, equipment, species and cultivars, seed treatments and impurity rate of the seed lot for a sowing operation) over the years or decades that the simulation is meant to run.

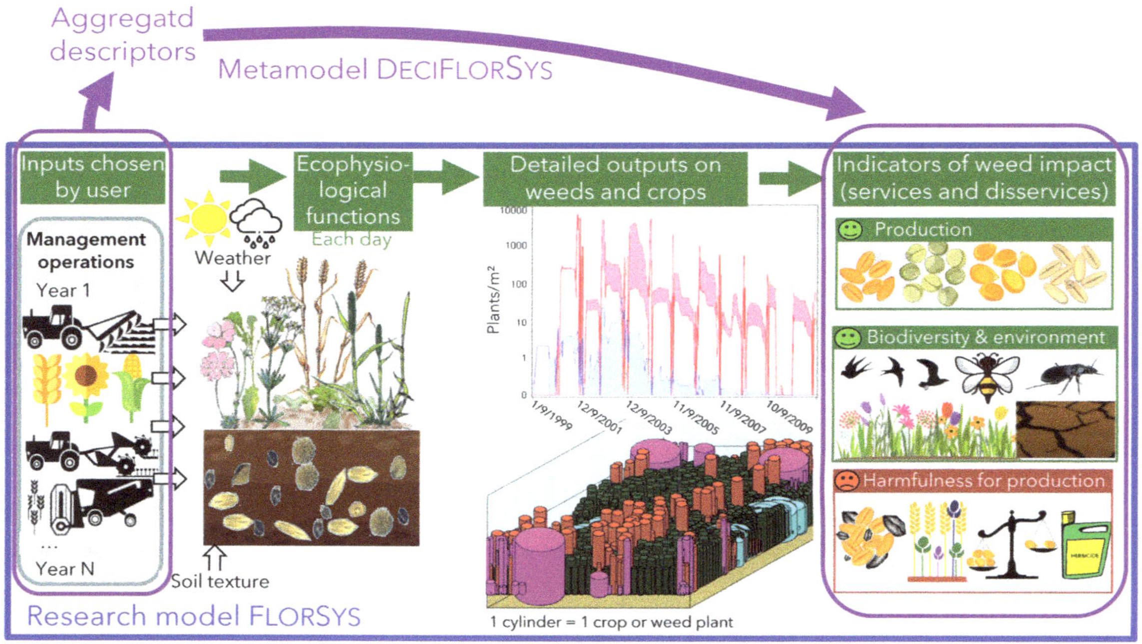

Figure 3 General representation of the (1) research model FLORSYS (blue rectangle) which simulates crop growth and weed dynamics from a detailed description of cropping system, weather and soil inputs based on a mechanistic representation of biophysical processes at a daily time step and in 3D (details in Figure 4), and then aggregates these data into indicators of weed services and disservices (Mézière et al., 2015, Colbach et al., 2014, Munier-Jolain et al., 2013, Gardarin et al., 2012, Pointurier et al., 2021); and the (2) metamodel DECIFLORSYS (purple boxes and arrows) which directly estimates weed (dis)services from aggregated cropping system inputs based on machine learning (Colas et al., 2020, Colas, 2018) (Nathalie Colbach, 2020 CC BY NC ND).

Seeds and plants of multiple crop and weed species follow a succession of life stages (e.g. dormant seeds, seedlings) driven by ecophysiological processes (e.g. germination, photosynthesis, respiration) and cropping techniques (Fig. 4). Each function and its associated parameters were individually estimated in specific experiments in growth chambers, greenhouses, field trials or *in silico* experiments with other models. The total number of species-dependent parameters in the model is enormous (211), to which are added

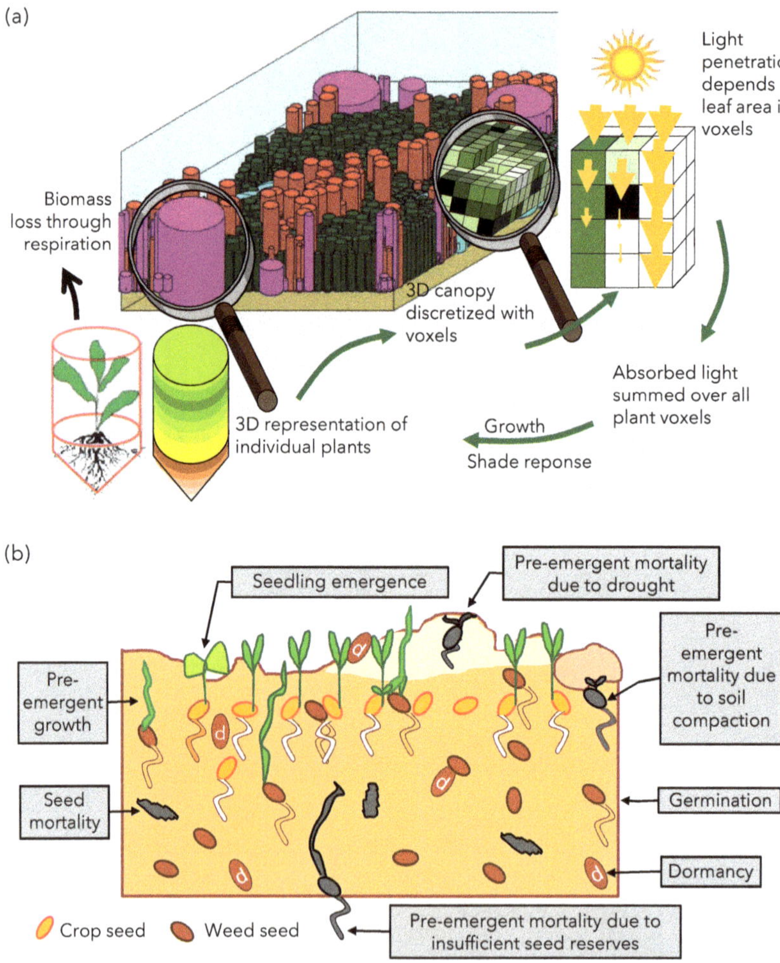

Figure 4 Simplified representation of the main ecophysiological functions and processes included in the FLORSYS model (Colbach et al., 2014, Munier-Jolain et al., 2013, Gardarin et al., 2012, Pointurier et al., 2021). (a) Above-ground 3D individual-based representation of the crop-weed canopy, focusing on plant-plant competition for light. (b) Below-ground processes related to seed dynamics (Nathalie Colbach, 2020 CC BY NC ND).

species-independent parameters for describing physical processes (e.g. seed movements during tillage, soil structure dynamics). The model is currently parametrised for 30 common and contrasting annual weed species and 33 cash and cover crop species, including several varieties of pea, wheat and faba bean.

The model was evaluated with independent data from field observations (Colbach et al., 2016, Colbach et al., 2021, Pointurier et al., 2021). This is rare for weed dynamics models because it is so difficult to find well-documented multiannual fields from different pedoclimates. To date, FLORSYS has been evaluated using expert knowledge an annual trial measuring light interception in heterogeneous plant canopies, a bi-annual trial monitoring weekly weed emergence, several cropping-system trials assessing crop and weed variables (e.g. plant densities, biomass, yield) several times a year over a dozen years and the annual data from several hundred fields recorded by the Biovigilance-Flore network (Fried et al., 2008). The evaluation showed that crop biomass and yields, weed seed banks, daily weed species densities and, particularly, densities averaged over the years were generally well predicted and ranked (e.g. modelling efficiency of 0.55 for crop yield, 85% of observed weed species density included inside the simulated confidence interval) as long as a corrective function was added to keep weeds from flowering during winter at more southern latitudes (Colbach et al., 2016, Pointurier et al., 2021). The prediction quality was usually better at the species than at the community scale, and better at the rotation than at the daily scale.

The thorough representation of biophysical processes produces very detailed outputs, by day and in 3D, which are essential to understand why a given technique or cropping system results in a given performance. The possible applications of the FLORSYS model (see compilations by Colbach et al., 2019, Colbach, 2020, Colbach et al., 2021) are greater than for the models presented in the two previous sections. The model is used for a wide range of purposes, such as:

- field networks simulating cropping systems recorded in farms to track innovations in weed management strategies using few or no herbicides;
- sensitivity analyses simulating virtual varieties and cropping systems resulting from random choice of species traits and management techniques in order to identify crop ideotypes and ideal cropping systems which reconcile crop production with low herbicide use and high biodiversity;
- virtual experiments to optimise choices for individual techniques (e.g. Which tillage tool is best in a given rotation? How frequently and when to carry out mechanical weeding?) and to test prototypes proposed by experts and stakeholders; and
- participatory workshops with farmers.

Whatever the approach and objective, cropping systems are always run with several series of weather scenarios to estimate probabilities of success and failure.

2.4 Major differences

The three models discussed have several similarities:

- They work at the field level.
- They predict one or several weed variables as a function of cropping system and weather variables.
- Their prediction quality has been evaluated.

However, despite modelling the same object and system, the chosen types of models and the associated experimental studies are completely different. The equations constituting the static single-equation model (section 2.1) were each estimated in a single step, from a large data set relating input and output variables (e.g. tillage tools and weed biomass). The shape of the equation is very similar to a basic ANOVA used for statistically separating treatments in experiments. Linking inputs and outputs can result in surprising results, such as the increase in weed biomass with increasing crop density in Table 2 at high weed density (resulting from the OD × WD interaction). This is not so much the result of the expected cause-to-effect relationship (i.e. increased crop density leaves less space for weeds) but shows rather that favourable conditions increase both crop and weed growth, resulting in a positive correlation.

In contrast, the matrix model (Section 2.2) and, particularly, the mechanistic model (Section 2.3) are an aggregation of equations, with only ten for the former and several thousand for the latter. These are usually independently estimated from small and diverse data sets covering intermediate environmental and weed variables (e.g. soil water potential and weed seed germination). The shapes of the equations of the mechanistic model are often as simple as that of empirical models (e.g. linear or logistic regressions), and the parameters are estimated with the same method (i.e. minimising the sum of error squares). At a first glance, FLORSYS could be considered as a more detailed version of the matrix models discussed in Section 2.2. However, there is a fundamental difference insofar as matrix models focus on weeds and include crops, at best, via functions linking weed-plant survival to crop-sowing density or, vice-versa, crop yield loss to weed state variables (e.g. plant density or biomass). In contrast, FLORSYS does not discern between crop and weed individuals, except that crop seeds are sown and newly produced seeds exported to calculate yield.

The input variables, weed stages, cropping system effects and interactions are described in much more detail in FLORSYS. The resulting model is very complex, parameter-hungry, process-based and multi-annual, synthesising and organising knowledge, in addition to predicting effects. The subsequent sections evaluate whether this additional complexity is necessary and sufficient for 'weed = f(cropping system)' models for designing integrated cropping systems.

3 Limiting the modelled system: temporal, spatial and species scales

Models are a representation of a process or a system of which the extent and the resolution (time-step, spatial unit...) must be limited. These choices depend not only on the model aim, here to evaluate and design cropping systems, but also on the modelled organism, here weeds. The spatiotemporal limitations are crucial because they determine the extent and detail of the processes to be studied experimentally and the way these processes will be modelled.

3.1 Temporal scales

The temporal extent of the modelled system should depend on the longevity of the modelled organism. As most weed seeds survive for several years (Lewis, 1973), weed dynamics models are usually multi-annual (see previous reviews by Colbach and Debaeke, 1998, Holst et al., 2007, Freckleton and Stephens, 2009, Bagavathiannan et al., 2020). Conversely, for pests without year-to-year survival, annual models (e.g. Minogue, 1989) are sufficient as there is no (significant) effect of field history on pest dynamics. Models aiming mainly at quantification of effects and decision-making (rather than at a representation of processes) sometimes summarize past field history, expressing weed risk in a given year as a function of favourable crops grown in the past (e.g. Munier-Jolain et al., 2005).

The empirical model discussed in Section 2.1 is static and produces a snapshot of weed incidence. The matrix models such as those discussed in Section 2.2 work with an annual time-step (i.e. temporal resolution), calculating a limited number of key life-stages such as mature plants and newly produced seeds once a year. Such an approach is insufficient when attempting to optimise complex operations such as tillage strategies. For instance, plant mortality after harrowing varies with environmental conditions (e.g. soil moisture) and weed stage (Kurstjens and Kropff, 2001), which both vary over time. Some crop models therefore use hourly or even minute-based time-steps, particularly in functional-structural plant models (FSPM) which explicitly describe the development over time of the 3D architecture or structure of plants including a detailed representation of individual organs (Vos et al., 2010). However, we focus here on models for designing cropping systems and not for predicting detailed weed architecture or spread in a field. As farmers make daily decisions and manage daily operations, a daily time-step appears to be the most adequate solution.

3.2 Spatial scales

The spatial extent of a model must be chosen according to the dispersal ability of the analysed organism and the model's objective. Though weed seeds disperse in space and weed communities in a given field depend on the neighbourhood of the analysed field (Petit et al., 2013), most weed models are limited to a single field, focusing on the effect of field history and current practices (see previous reviews by Colbach and Debaeke, 1998, Holst et al., 2007, Freckleton and Stephens, 2009, Bagavathiannan et al., 2020). If the objective, though, is to evaluate whether contrasting ecosystem services can be better reconciled at the landscape scale (land sparing) rather than inside each field (land sharing), then the relevant weed model must be extended to the landscape level (as was done for FLORSYS, Colbach et al., 2018). This is also the case when aiming to monitor gene flow in weed populations, particularly of phenotypes strongly interacting with cropping system components. This is, for instance, the case for models focused on the spread of herbicide-resistance genes in weed populations (Maxwell et al., 1990, Colbach, 2009). These models consider landscapes as a mosaic of fields and semi-natural areas. In each spatial unit, the dynamics depend on the crop grown in the field and its cropping techniques (or the type of semi-natural area and its management) as well as weed phenotype (herbicide resistant vs. susceptible). During flowering, the various units exchange pollen. Exchanges depend on field areas, shapes and distances. The genotypes of the newly produced seeds are calculated as a function of the proportions of the genotypes of the seed-producing plants and of the pollen cloud.

Crop-weed canopies are inherently heterogeneous because weeds are usually distributed in patches within fields (e.g. Bigwood and Inouye, 1988), contrasting species co-exist (Fried et al., 2010), successive emergence flushes occur for each weed species (e.g. *Alopecurus myosuroides* Huds., Colbach et al., 2006a) and plants exhibit a high morphological plasticity (Colbach et al., 2020b). The resources available for each plant thus not only depend on its own morphology and ability to take up and use resources, but also on those of its neighbours. As a consequence, some modellers (including for FLORSYS) now include individual 3D canopy representations (Renton, 2013, Colbach et al., 2021).

However, the question remains whether integrating intra-field variability into models is always necessary. The monospecies parent model of FLORSYS disregards this variability and its evaluation has shown that models can correctly rank situations and cropping systems (Colbach et al., 2007) despite not covering intra-field variability. However, modelling intra-field variability is crucial if the model's objective is to evaluate spatially variable techniques such as herbicide applications in precision agriculture (Fernández-Quintanilla et al., 2020) or sowing patterns in intercropping (Gaudio et al., 2019).

Different approaches exist to integrate intra-field variability. The most detailed approach is the use of functional-structural plant models (FSPM), which describe the location and architecture of each plant in detailed 3D, focusing on competition (examples in Vos et al., 2010, Gaudio et al., 2019). Because of their detail, these models are usually annual and bi-species at best, disregarding any pre-emergence processes. To reconcile a multi-annual approach with intra-field variability, some weed dynamics models divide the field into homogenous subunits and simulate the weed life cycle within each of these. The various units can exchange pollen and/or seeds (e.g. Ballaré et al., 1987) and influence neighbour units via competition (e.g. Gonzalez-Andujar et al., 2000). FLORSYS' approach is a compromise insofar as it represents each individual plant but does so in a very simple manner, as included in other weed dynamics models (e.g. Röhrig et al., 1999).

3.3 Species scales

Most weed models only consider one species and neglect intra-species variability. This can be acceptable in some intensive and simplified cropping systems where there is often one main weed species (e.g. *A. myosuroides* in intensive winter-crop rotation (Van Himme and Bulcke, 1975)). However, particularly in systems with one dominant selection pressure, intra-species variability can quickly become an issue. The first weed models considering intra-species variability looked at the development of herbicide resistance in weed populations selected by the repetitive use of a single active ingredient. One possibility is to consider separate self-pollinating resistant and susceptible populations where resistant seeds can only be produced by a resistant parent or after a mutation on a susceptible parent (Gressel and Segel, 1978). A more mechanistic approach consists in simulating the possible plant genotypes and cross-pollination where resistant seeds can also be produced by susceptible plants pollinated by resistant pollen in case of a dominant resistance allele (Maxwell et al., 1990). This is the approach also used by FLORSYS, which considers both target-site resistance depending on a single gene and two alleles (Colbach et al., 2017) and non-target-site resistance depending on several quantitative genes using probabilistic functions (Délye et al., 2020). There are, to date, many weed models focusing on herbicide resistance, genotypes and heredity (see review by Renton et al., 2014) but none include as many cropping techniques other than herbicides and in as much detail as FLORSYS.

Variability interacting with less drastic selection pressure is less easy to detect and model, though no less important (e.g. selection of different dormancy patterns in rotations). For instance, seed dormancy is related to a large number of genes (Jana and Thai, 1987), and the same approach as for non-target-site herbicide resistance could be used. However, intra-species variability is not always due to different genotypes but can also result from maternal effects, that is, conditions during seed production. The monospecies parent model of FLORSYS uses empirical functions to relate seed dormancy to nitrogen availability and water deficit during seed production (Colbach et al., 2006b), two variables that strongly depend on crop management (in interaction with climate).

With the reduction in herbicide use and increased crop diversification, weed communities could again become more diverse (Jastrzebska et al., 2019, Neyret et al., 2020). Weed models must thus be multi-species though, to date, few are (even among the simpler matrix-based models) and none include as many species as FLORSYS. To make this possible, plant morphology needs to be generic and applicable to all types of species (e.g. dicots and monocots, erect and rosette-shaped plants). Similarly, only processes pertinent for all species type were included (e.g. tillering specific to grass weeds was neglected) (Colbach et al., 2021). Despite these simplifications, mechanistic models such as FLORSYS still require an enormous number of parameters, which hinders the addition of new species to the model. This is the reason why Gardarin et al. (Gardarin et al., 2012, Gardarin et al., 2016, Colbach et al., 2020b) developed a new methodology based on functional relationships to estimate difficult-to-measure model parameters from easily measured species traits, trait databases and/or expert opinion. The approach was recently extended to crop species (Gardarin et al., 2016, Colbach et al., 2020b).

4 Modelling approaches: empirical versus mechanistic models

The previous section emphasised the key role of the chosen modelling approach for determining the resolution of a model and the experiments needed to build it. Even after considering temporal, spatial and species-related causes of variability, the larger question remains of how to predict the variability of the effect of a given cropping technique between years, fields and regions. This variability results from the interactions between cropping system components (e.g. the effect of mouldboard ploughing vs. no ploughing depends on previous crops) as well as between these components and environmental conditions (e.g. the efficacy of mechanical weeding depends on soil moisture). The choice of a modelling approach depends initially on individual functions and submodels, which can be empirical or mechanistic, stochastic or deterministic. The final model can be a combination of different approaches.

The choice of the modelling approach is also influenced by fashions and by traditions in different scientific disciplines. Here, we consider a multi-annual model of cropping system effects on weed dynamics. The latter are the domain of ecology and related disciplines, which tend to use more empirical and stochastic models. Cropping systems are mainly the domain of agronomists who, at least in the past, were more inclined towards process-based and deterministic models. The following section critically analyses these various models, their strengths and weaknesses, starting by comparing empirical and mechanistic models.

4.1 Testing scenarios with mechanistic models

One way to take account of interactions is to develop mechanistic models which are based on explanations of biophysical processes, instead of simply quantifying relationships between input and output variables with empirical functions. The evaluation of FLORSYS (Colbach et al., 2016, 2021, Pointurier et al., 2021) showed that this mechanistic model of the interactions between cropping techniques and environmental conditions correctly ranked scenarios (different options of a cropping technique, different cropping systems). The model also satisfactorily predicted the variations of effects of a given cropping technique. Mechanistic models can be used in a large range of situations without re-estimating parameter values as they aim to represent universal processes instead of describing an event in a particular situation. This is important for the objective of designing cropping systems in different pedoclimatic conditions.

Mechanistic models such as FLORSYS can be considered as a virtual experimental field where existing and virtual cropping systems can be tested over many years and repeated with different weather series (Section 2.3). This could also be done with empirical or matrix-based models (Sections 2.1 and 2.2) as long as these include pedoclimatic effects. However, only mechanistic models can perform a diagnostic function to understand why a given cropping technique or system results in a given performance.

4.2 The mechanistic model to organise research and synthesise knowledge

Another major advantage of mechanistic models over empirical ones is their contribution to organising research. As an aggregation of submodels describing individual processes or techniques, mechanistic models synthesise existing knowledge produced by different teams and disciplines. For instance, FLORSYS includes submodels based on data from the literature (e.g. seed mortality in soil, Gardarin et al., 2010b), virtual experiments using simulation models (e.g. to link pre-emergent seedling mortality to soil structure, Gardarin et al., 2010a), experiments specifically set up for modelling purposes in controlled conditions (e.g. temperature and light effects on germination, Colbach et al., 2002) and using garden plots (e.g. plant morphology response to shading, Colbach et al., 2020b), submodels taken from other models after field testing (e.g. seed movements during mouldboard ploughing, Colbach et al., 2000) and other connected submodels (e.g. the STICS soil submodel, Colbach et al., 2020a).

Once a first model version is built, knowledge gaps can be identified when no adequate data can be found to fit a crucial function. Once simulations are compared to independent field observations, missing or deficient submodels can be identified (e.g. seed predation in continuously untilled fields, Colbach et al., 2016), which then sets off another round of modelling to include new functions (e.g. seed predation, Perthame et al., 2018).

Mechanistic models are also better adapted to changing conditions. It is much easier to include new processes and/or techniques. For instance, FLORSYS was initially developed for conventional agriculture based on tillage and high mineral fertiliser input (Gardarin et al., 2012, Munier-Jolain et al., 2013, Colbach et al., 2014). It was recently extended by adding submodels for seed predation to adapt to no-till systems where predation is much more frequent than in tilled systems (Perthame et al., 2018), and plant-plant competition for nitrogen to adapt to low-input systems (Moreau et al., 2021).

4.3 Which processes should be detailed?

The choice between empirical and mechanistic models is not as clear-cut as stated above. Attention has been drawn to the danger of over-parametrisation of mechanistic models and to the problem of data input (Grundy, 2003). Indeed, each additional parameter increases the prediction error because of estimation error. This explains why, when looking at a single species, prediction quality was better for the monospecies FLORSYS parent model (Colbach et al., 2007) than the multispecies FLORSYS (Colbach et al., 2016, Colbach et al., 2021, Pointurier et al., 2021). Moreover, mechanistic models often need large and complicated input data sets which are not always complete and reliable, and make the model difficult to use for outsiders and/or affect prediction quality (Colbach et al., 2016). For instance, weed dynamics models use the viable seed bank initially present in the simulated field as input, which is difficult and expensive to estimate (Ambrosio et al., 2004).

Mechanistic models do not always necessarily produce better predictions than more robust empirical models. For instance, dormancy variations in weed seeds can be predicted with a physiological model based on the action of a phytochrome in the seed (Vleeshouwers and Bouwmeester, 2001). The model evaluation showed the prediction of release from dormancy in spring to be largely overestimated (Vleeshouwers and Bouwmeester, 2001). Conversely, many weed dynamics models simply predict dormancy as a seasonal variation with time (see previous reviews in Colbach and Debaeke, 1998, Holst et al., 2007, Bagavathiannan et al., 2020), and the few models to be evaluated showed that both emergence and multi-annual dynamics were satisfactorily predicted (e.g. FLORSYS, Section 2.3).

Moreover, the distinction of an empirical vs. a mechanistic approach depends on the modelling scale and the scientific discipline. For instance, the authors of FLORSYS consider weed seed movement during mouldboard ploughing to be modelled mechanistically because the model describes the rotation of the furrow and its subsequent break-up, resulting in soil translating onto the plough pan as a function of soil compaction, ploughing depth and width (Colbach et al., 2000). However, any physicist would consider this model to be empirical as it does not consider processes such as friction between the soil clods and the plough share or the fissuring of soil clod aggregates during furrow break-up (Kouwenhoven and Terpstra, 1972). In addition, many functions in FLORSYS cannot be considered as mechanistic, even at the field and plant scale at which the model was developed. Priority was given to the mechanistic decomposition of processes interacting the most with cropping system components. Other processes were simply described by empirical functions (e.g. seasonal seed dormancy).

Consequently, a suitable compromise would be a model combining both mechanistic and empirical approaches:

- giving priority to processes interacting with the key driving variables, i.e. cropping system components; and
- limiting the mechanistic decomposition to biophysical processes that are consistent with the temporal, spatial and species scales of the model.

For instance, FLORSYS can integrate interactions between tillage and soil structure because it is limited to a small field cluster. Actual landscape models simulating hundreds of fields neglect such interactions and focus on dispersal processes at the landscape scale; they integrate the effect of mouldboard ploughing with empirical transition matrices, giving probabilities of infectious crop residues or weed seeds moving between soil layers, calculated with a mechanistic model or with random functions (Colbach, 2009).

5 Modelling approaches: stochastic versus deterministic models

This section looks at issues in selecting stochastic and deterministic models. It also looks at the broader choice of empirical versus mechanistic or stochastic versus deterministic models.

5.1 How to tackle residual variability?

The previous section considered "macroscopic" variability in weed dynamics, which can be easily represented both mechanistically (by describing biophysical processes) and empirically (by quantifying interactions between input variables). Matrix-based models often predict the density of emerged weeds using weed seed bank density multiplied by an emergence rate and make this emergence rate depend on tillage type (e.g. plough vs. no-plough) (Cousens et al., 1986). The same effect is represented mechanistically in FLORSYS by several successive functions interacting with environmental conditions, that is, seed movements depending on tillage characteristics and soil structure, seed dormancy varying with season, seed germination and pre-emergent seedling growth depending on seed depth as well as soil temperature, moisture and structure (Fig. 4b). Despite this detailed mechanistic approach, FLORSYS is unable to predict all the variability in weed emergence, as shown by model evaluation using experiments either controlling (e.g. seed density, depth, age) or closely monitoring variables (e.g. weather, soil hydrothermal conditions and structure) (Colbach et al., 2006a).

The use of stochastic functions for randomly choosing parameter values from a set distribution is a simple approach to account for variability in processes, both in mechanistic and empirical models. This approach is often used in ecology, for instance, to evaluate the long-term dynamics of different species strategies with matrix-based models, combined with elasticity analyses (Caswell, 2001). In agronomic models, which are often more mechanistic, the use of stochastic functions is more recent and can occur at a very small scale, for example, the random placing of seeds between soil clods in a specific soil layer (e.g. between 1 cm and 3 cm deep) when sowing a crop (Dürr et al., 2001).

The opposite approach would be to remain deterministic, that is, working with constant parameter values, and to further study the process to understand and model the causes of variability. Such an approach is more complex. It requires additional experiments that identify these causes instead of describing the variation of a parameter. It probably also requires moving to a finer scale of organisation (e.g. from the plant level to the physiological or molecular level). It also adds a large number of parameters to the model, thus increasing the risk of over-parametrisation.

5.2 Variability in input variables

As mentioned, the main objective of the chapter is to test scenarios of weed management for strategic decision-making in integrated crop production. In such a situation, agricultural operations cannot be considered as stochastic events as is frequently done in ecological models (Claessen et al., 2005,

Garnier et al., 2006). These operations result from the farmer's decisions and can be modelled with decision rules (Aubry et al., 1998). A decision rule usually consists of a conditional function ('if ... then...; else...') relating the decision to objectives and constraints, with an evaluation criterion of the result. These rules apply both at short-term scales (e.g. choose the timing of mechanical weeding to maximise weed seedling mortality, as a function of weed seedling stage and soil moisture) and long-term scales (e.g. choose tillage tools to maximise weed seed burial, as a function of previous and pre-previous crops).

These decision rules can be represented deterministically. They can rapidly become very complicated particularly when upscaling from the field to the landscape. In that case, variability in management options at the field level is often disregarded, usually by using fixed management programmes. Simulations then focus on variability in the landscape management, mainly the location of the different crops in the field pattern to study dispersal processes. Crops can be placed deterministically to compare contrasted patterns (Colbach et al., 2018), or randomly with probabilistic functions (Colbach, 2009).

Stochasticity is usually applied to variables that cannot be controlled by the farmer, such as weather. In addition, this variable cannot be predicted sufficiently well or over the long term to improve farming decisions. Consequently, weather sets are often chosen randomly in simulations, with cropping system scenarios being repeated several times to obtain the probabilities of success (see example in Section 2.1).

5.3 The choice depends on the objective and system

The choice of a stochastic versus a deterministic approach often depends on the objective of the modeller. For instance, ecologists evaluating long-term strategies of species dynamics will prefer a stochastic approach to predict weed biomass and seed production because of its simplicity, amongst other things. In contrast, agronomists aiming at optimising the choice and timing of agricultural operations to control weeds will prefer an ecophysiological approach predicting weed biomass and seed production from plant state variables (e.g. leaf area, plant height, architecture) driven by environmental conditions (e.g. temperature, radiation, rainfall) (Fig. 4a).

However, the deterministic description of an event often requires describing the process at a very small level of organisation, which is not necessarily compatible with the spatiotemporal scale of the modelled system. For instance, the Brownian movement of pollen grains in a liquid appears stochastic at the macroscopic scale but can be described with deterministic collisions of particles at the molecular level (Brown, 1828). This level of representation is not compatible with a cropping system model working on a daily basis over several years, either at the field or the landscape level.

5.4 The contribution of model evaluation

Sometimes, the choice of empirical vs. mechanistic, or stochastic vs. deterministic is not voluntary but due to a lack of knowledge of the biophysical processes. For instance, because no prior knowledge existed on between-field seed dispersal by agricultural tools, this process can be described with stochastic functions based on Poisson distributions (Colbach et al., 2004). Such stop-gap measures make sensitivity analyses and model evaluation even more crucial in deciding which parameters and functions need to be refined. The main component of evaluation is the comparison of model output with independent observations ('validation'), a step that is still comparatively rare, even for the numerous weed dynamics models (Holst et al., 2007). Model evaluation is particularly important in the case of sensitive mechanistic models aggregated from numerous and diverse sources, but is an essential step for any model to check assumptions and parameter estimates. Indeed, no modeller should forget the main modelling rule of 'garbage in, garbage out'.

Model evaluation also plays a key role in developing simulations by identifying:

- the domain of validity of the model, i.e. the conditions in which the model can be used given its underlying assumptions;
- the prediction error; and
- the major input variables necessary for a good prediction.

Various approaches and statistical methods have been presented for evaluating crop models (Wallach, 2006) that also apply to 'weeds = f(cropping system)' models.

6 How to bridge the gap between process analysis and decision support

6.1 Support decision-making or synthesise knowledge on processes?

The main results of the comparison of different 'weeds = f(cropping system)' models in the previous sections are summarised in Table 3, contrasting the features specific to each approach. It is apparent that mechanistic models are usually deterministic whereas stochastic functions are more often applied to empirical models. Models are not necessarily entirely empirical or mechanistic and strictly stochastic or deterministic; they are situated on a continuum, depending on their objectives and experimental constraints.

This summary also shows that empirical models mostly focus on a single objective – to quantify and predict effects. Conversely, mechanistic modellers

Table 3 Caricatural summary of contrasting modelling approaches for 'weeds = f(cropping system)' models (based on Colbach, 2010)

	Modelling approach		
	Empirical[1] / stochastic[2]	⇔	Mechanistic[3] /deterministic[4]
A. Developing models (choice of model structure and parameter estimation)			
Necessary level of knowledge	Limited	⇔	Detailed
Number of experimental situations	Numerous and diverse	⇔	Few, well-chosen
Type of experiment	Surveys, data bases, *in silico* experiments with mechanistic models	⇔	Growth chambers, greenhouses, field experiments
Aim of experiment	Quantify and rank effects	⇔	Understand and identify processes
Characterising state variables on experiments	In general	⇔	In detail
Model structure	Simple, small number of parameters	⇔	Complicated, large number of parameters
Temporal scale	Static	⇔	Dynamic
Crop modelling	Focus on weeds only	⇔	Weeds and crops are represented with similar detail
B. Using models			
Extrapolability	Limited (to conditions and combinations present in the data base)	⇔	Large (if the relevant biophysical processes are integrated in the model)
Sensitivity	Robust	⇔	Sensitive
Testing scenarios	Comparing scenarios	⇔	Optimising the choice and timing of cultivation technique
	Long-term (years, decades)	⇔	Short and medium-term (days, years)
Diagnosis	Difficult or impossible	⇔	Easy
Ease of use	Fewer inputs, faster simulations	⇔	Many complex inputs, slow simulations
Targeted users	Farmers, crop advisors, extension services, technical institutes,	⇔	Scientists, technical institutes

[1] Empirical functions directly relate observations to input variables, without attempting to identify or explain the underlying biophysical processes.
[2] In stochastic models, at least one parameter is chosen randomly; to each input set (scenario) is associated a distribution of output values produced by several simulation runs (repetitions).
[3] Mechanistic models are based on sub-models that are proposals for explaining biophysical processes.
[4] Deterministic functions associate to each input set (scenario) a single output set.

usually have a double objective - knowledge aggregation on processes and predicting effects of cropping systems for decision support. The model complexity necessary for the former objective might impede achieving the latter objective. Indeed, models such as FLORSYS were initially developed for scientists. Though they are essential tools for cropping system design (see Sections 2.3 and 4.1.1), their use is often too complicated and slow for end users such as farmers or government decision makers. In particular, this type of model is much too slow to use in participatory workshops with farmers and other stakeholders, where the objective is to run a continuous loop of cropping system design, evaluation, analysis and redesign as well as to foster discussions among participants about ideal options.

All the models presented here evaluate the performance of existing cropping systems rather than proposing alternative cropping systems to meet the user's goals and constraints. To find a series of cropping systems meeting these goals and constraints would take time, as the range of possible cropping system candidates is enormous. The search for cropping system solutions would be considerably faster if 'weeds = f(cropping system)' models could be inverted into 'cropping system = f(weeds)' models, that is, if cropping system variables were determined as a function of weed-management objectives. Then, optimal cropping systems could be estimated as a function of a set of objectives and constraints (Peyrard et al., 2007). This would be possible for a simple model as the one discussed in section 2.1 but the more realistic mechanistic models are too complicated to invert in this way.

One solution to this problem is to automate the design-evaluation loop used in participatory workshops by connecting weed models to optimisation algorithms (Press et al., 2007, Venter, 2010). This also has the advantage that potential cropping systems are not limited by the experts' imagination. FLORSYS was, for instance, used with genetic optimisation algorithms to identify cereal-based rotations that reconcile crop production with weed-based food for bees (Colbach et al., 2021). However, this approach is limited if simulations are slow, as for FLORSYS (several hours to several days for one system over 30 years and 10 weather repetitions).

6.2 Metamodel to support decision-making and synthesise knowledge on processes

We therefore propose a different approach based on transforming complex mechanistic models into simpler empirical models for prediction ('meta-models' or, as in computer science, 'emulators', i.e. a computer program or electronic device imitating another program or device). This approach has already been used in other disciplines to simplify complex mechanistic models, for instance, for yield predictions (e.g. Brooks et al., 2001, Baumann et al., 2002) or pesticide leaching (e.g. Holman et al., 2004). These emulators are not only easier to use and combine with other models; they also reduce the risk of bad predictions caused by badly estimated complex input variables.

Meta-modelling is different from the traditional methods of model simplification (where a family of alternative models ranging from simple to complex is tested to identify the simplest model with the best prediction ability), or nested models (where complex models are increasingly simplified by deleting or aggregating state variables and/or process functions) (Haefner, 2005). The approach proposed here consists of using mechanistic models to run *in silico* experiments (i.e. simulations or virtual experiments) in all the conditions relevant for future end users and then fitting regressions and other models to correlate the simulation output directly to its input.

This is the approach that Colas et al. (2020) chose in developing a new tool – DeciFlorSys – to support cropping-system design. To ensure that the new tool would not only be useful but actually used, it was co-developed with future users, namely crop advisors and farmers. The first step consisted in building a large set of simulation results by running FlorSys over several thousand virtual farm fields, including both actual systems (from farm surveys etc.) and systems based on random choices of techniques. The latter were essential to ensure that the domain of validity of the metamodel was not limited by current practices or experts' imagination, particularly in terms of crop diversity, rotational patterns or crop mixtures.

Data mining was then used to link the weed impacts on crop production and biodiversity simulated by FlorSys to cropping-system variables, as if developing an empirical model from field data. Multivariate regression trees (Breiman et al., 1984, De'ath, 2001) provide a synthetic graphical representation ('decision tree') to identify candidate changes in cropping systems that improve their performance, through navigation among branches (see example in Fig. 5). In addition, multivariate random forests (Breiman, 2001, Segal and Xiao, 2011), which function as black-box models ('predictor'), directly predict weed impact indicators from cropping-system variables, thus emulating FlorSys, but with a much faster response time and in a way that is easier to handle than the parent model.

While the DeciFlorSys predictor is as good as FlorSys in ranking cropping systems, it cannot adequately evaluate effects that strongly interact with pedoclimatic conditions, such as the effect of tillage timing with respect to soil moisture (Colas, 2018, Colas et al., 2019). To fine-tune cropping systems in terms of operation dates, for example, and to run a diagnosis on a cropping system, it is still necessary to go back to FlorSys.

Basing the decision-support system on simulations rather than field observations makes it possible to benefit from the huge amount of knowledge included in FlorSys and to browse a huge number of possible cropping systems

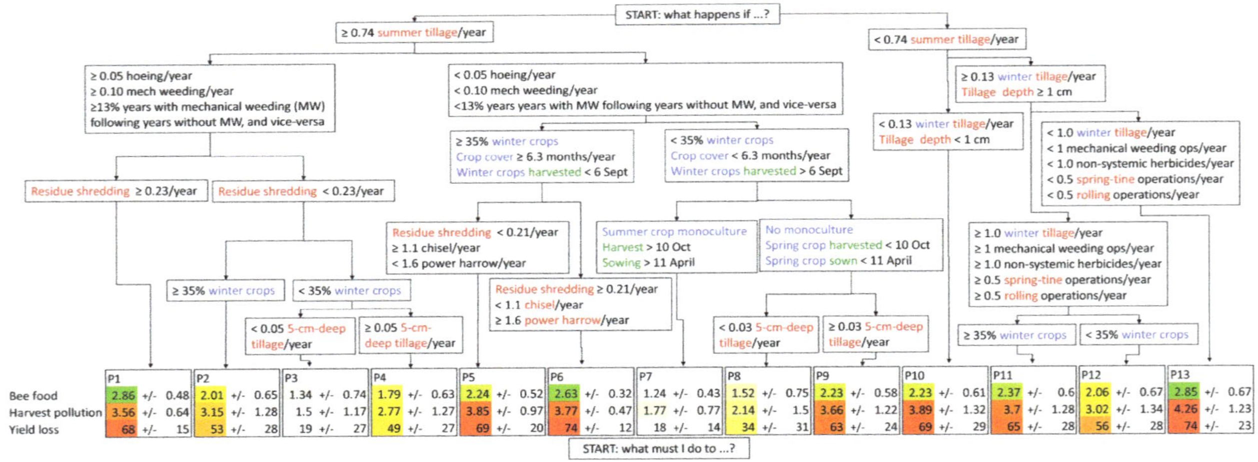

Figure 5 Example of decision tree from the decision support system DeciFlorSys, identifying cropping system rules for reaching a series of 13 weed impact profiles based on a multivariate regression tree linking weed impact indicators simulated by FlorSys to combinations of cropping-system variables (blue: rotation, green: sowing/harvest, brown: tillage, black: weed control), based on 4350 cropping systems resulting from a random combination of cultural techniques from 20 French regions. Indicator values were rescaled to [0,1], averaged over 30 years and 10 weather repetitions, and coloured from white (minimum) to green (maximum) for biodiversity and from white to red (maximum) for harmfulness to crop production. Uncoloured cells show standard error including weather effects and variability among systems in a branch. The cross-validation error of the tree was 0.106. The tree can be read top-down to get an idea of the performance of a proposed combination of management practices, or bottom-up to identify the practices corresponding to a target performance (Colas et al., 2020, based on data and methods from Colas, 2018) (Floriane Colas, 2019).

that could never be investigated *in situ*. The quality of a meta-model depends particularly on the experimental design of the *in silico* experiments which need careful planning (Conti and O'Hagan, 2010). Meta-models are usually used for facilitating model analysis, that is, sensitivity analysis, uncertainty analysis or calibration. Here, the focus is on facilitating use by farmers, advisors and others. Simplifying complex models initially developed for scientists into more simple and easier-to-use software packages is one way of disseminating scientific knowledge to end users and contributes to the adoption of integrated weed management. When developing the meta-models, it is necessary to work directly with the end users to choose an easy-to-use model structure, and particularly, input variables that can be easily collected. Just as in the initial detailed models, the output variables must also be chosen with care to be pertinent relative to the future use of the model. For instance, FLORSYS calculates the indicators of weed impact on crop production and biodiversity that were developed together with stakeholders, including farmers and crop advisors (Mézière et al., 2015).

6.3 Moving from multicriteria evaluation of weed floras to multicriteria evaluation of cropping systems

Even the very complicated FLORSYS model, with its many weed-impact indicators, does not predict all the output variables that farmers are interested in. Farmers are not only interested in weed infestation and crop yield but also in profitability. Moreover, fields are usually infested with other pests, and herbicides are not the only cropping technique that adversely affects the environment or human health. This would mean that prospective cropping systems need to be evaluated by running several models simultaneously, each focusing on different criteria, such as weeds, other pests (e.g. Minogue, 1989, Ennaïfar et al., 2007), soil erosion (Evrard et al., 2010) etc. However, this would very soon become unmanageable in terms of simulation management and time. The advantage of meta-models as a synthesis of numerous complex models thus becomes even more evident.

Other teams have used a very different approach, developing multicriteria evaluation tools based on expert opinion (e.g. DEXiPM, Pelzer et al., 2012) and then gradually improving the individual, deficient branches of these tools, for instance with the help of mechanistic models. Decision trees based on FLORSYS simulations have been introduced into the multicriteria evaluation tool DEXiPM (Pelzer et al., 2012) to replace the current biodiversity 'branch' (Cavan et al., 2019). DEXiPM was developed for *ex ante* assessment of the sustainability of integrated crop management systems. In addition to a series of basic indicators describing the cropping system and the context of the assessment, this tool includes 86 aggregated indicators assessing the three dimensions of sustainability, in terms of social, environmental and economic issues.

7 Conclusion and future trends

Managing weeds to limit both crop production loss and environmental impacts is a major challenge of agriculture. Because of the large number of factors and the complexity of interactions, models are essential to synthesise our knowledge of weeds and to quantify the effects of cropping systems on weed dynamics. These models must be able to rank candidate cropping systems as a function of weed incidence, and to account for variability in effects to estimate the risk of success or failure of a particular system. Two contrasting approaches are possible. Mechanistic models describe variability with process-based, usually deterministic sub-models quantifying interactions between cropping system components and environmental conditions. Empirical models directly relate observations to input variables, using few parameters, and usually quantify variability with probabilistic (stochastic) functions.

This chapter has critically evaluated these *a priori* contradictory approaches, that is, the deterministic vs. stochastic and the mechanistic vs. empirical representations of cropping system effects in weed models. Model structure and modelling approaches must be chosen depending on the objective, the modelled system/organism and the available knowledge. However, even when considering only a single object (weeds) and objective (cropping system design and evaluation), contrasting model structures and modelling approaches exist, and some models combine different approaches.

Modelling choices sometimes also depend on the modelling history of the relevant scientific discipline as well as methods for data collection and analysis. For weed dynamics models, two divergent trends coexist, starting from simple matrix models. One trend goes towards increasingly complicated mechanistic models, illustrating the accumulation of knowledge on the functioning of the agroecosystem. The other trend goes towards empirical models, benefiting from recent improvements in data collection and management (e.g. remote sensing, database management) as well as statistical methods (e.g. data mining, machine learning). We argue that models using a mechanistic representation of the cropping system × environment interactions are best for quantifying effects and accounting for their variability, combined with a subsequent transformation with *in silico* experiments into empirical models of the major cropping system components.

Following two basic modelling principles, i.e. 'All models are wrong, but some are useful (Box, 1976) and 'Garbage in, garbage out', it is essential to evaluate ('validate') models relative to their objective, to determine what is needed to improve them and where and how to use them to ensure their usefulness.

8 Where to look for further information

The latest compilation of models for decision support in weed management:

- Chantre, G. R. and González-Andújar, J. L., eds. (2020). *Decision Support Systems for Weed Management*. Springer International Publishing, Cham.

Previous reviews on weed dynamics models:

- Colbach, N. and Debaeke, P. (1998). Integrating crop management and crop rotation effects into models of weed population dynamics: a review. *Weed Science* 46:717-728.
- Holst, N., Rasmussen, I. A. and Bastiaans, L. (2007). Field weed population dynamics: a review of model approaches and applications. *Weed Research* 47:1-14.
- Freckleton, R. P. and Stephens, P. A. (2009). Predictive models of weed population dynamics. *Weed Research* 49:225-232.
- Bagavathiannan, M., Beckie, H., Chantre, G., González-Andújar, J., León, R., Neve, P., Poggio, S., Schutte, B., Somerville, G., Werle, R. and Van Acker, R. (2020). Simulation models on the ecology and management of arable weeds: structure, quantitative insights, and applications. *Agronomy* 10:1611, 10.3390/agronomy10101611.

9 References

Ambrosio, L., Iglesias, L., Mari, C. and Del Monte, J. P. (2004). Evaluation of sampling methods and assessment of the sample size to estimate the weed seedbank in soil, taking into account spatial variability. *Weed Research* 44(3), 224-236.

Andrew, I. K. S. and Storkey, J. (2017). Using simulation models to investigate the cumulative effects of sowing rate, sowing date and cultivar choice on weed competition. *Crop Protection* 95, 109-115.

Aubry, C., Papy, F. and Capillon, A. (1998). Modelling decision-making processes for annual crop management. *Agricultural Systems* 56(1), 45-65.

Bagavathiannan, M. V., Beckie, H. J., Chantre, G. R., González-Andújar, J. L., León, R. G., Neve, P., Poggio, S. L., Schutte, B. J., Somerville, G. J., Werle, R. and Van Acker, R. V. (2020). Simulation models on the ecology and management of arable weeds: structure, quantitative insights, and applications. *Agronomy* 10(10), 1611.

Ballaré, C. L., Scopel, A. L., Ghersa, C. M. and Sánchez, R. A. (1987). The population ecology of Datura ferox in soybean crops: a simulation approach to incorporating seed dispersal. *Agriculture, Ecosystems and Environment* 19(2), 177-188.

Batlla, D., Malavert, C., Farnocchia, R. B. F. and Benech-Arnold, R. (2020). Modelling weed Seedbank dormancy and germination. In: Chantre, G. R. and González-Andújar, J. L. (Eds) *Decision Support Systems for Weed Management*. Cham: Springer International Publishing, 61-83.

Baumann, D. T., Bastiaans, L. and Kropff, M. J. (2002). Intercropping system optimization for yield, quality, and weed suppression combining mechanistic and descriptive models. *Agronomy Journal* 94(4), 734–742.
Benjamin, L. R., Milne, A. E., Parsons, D. J. and Lutman, P. J. W. (2010). A model to simulate yield losses in winter wheat caused by weeds, for use in a weed management decision support system. *Crop Protection* 29(11), 1264–1273.
Bigwood, D. W. and Inouye, D. W. (1988). Spatial pattern analysis of seed banks : an improved method and optimized sampling. *Ecology* 69(2), 497–507.
Blaix, C., Moonen, A. C., Dostatny, D. F., Izquierdo, J., Le Corff, J., Morrison, J., Von Redwitz, C., Schumacher, M. and Westerman, P. R. (2018). Quantification of regulating ecosystem services provided by weeds in annual cropping systems using a systematic map approach. *Weed Research* 58(3), 151–164.
Box, G. E. P. (1976). Science and statistics. *Journal of the American Statistical Association* 71(356), 791–799.
Breiman, L. (2001). Random forests. *Machine Learning* 45(1), 5–32.
Breiman, L., Friedman, J. H., Olshen, R. A. and Stone, C. J. (1984). *Classification and Regression Trees*. New York: Chapman and Hall (Wadsworth, Inc.), 254 p.
Brooks, R. J., Semenov, M. A. and Jamieson, P. D. (2001). Simplifying Sirius: sensitivity analysis and development of a meta-model for wheat yield prediction. *European Journal of Agronomy* 14(1), 43–60.
Brown, R. (1828). A brief account of microscopical observations made in the months of June, July and August, 1827, on the particles contained in the pollen of plants; and on the general existence of active molecules in organic and inorganic bodies. *The Philosophical Magazine* 4(21), 161–173.
Caswell, H. (2001). *Matrix Population Models*. Sunderland: Sinauer, Associates, Incorp. Publishers.
Cavan, N., Omon, B., Tailleur, A., Dubois, S., Queyrel, W., Van Inghelandt, B., Colbach, N. and Angevin, F. (2019). Comparaison de méthodes de conception de systèmes de culture innovants pour la gestion durable des adventices. In: Colbach, N., Angevin, F., Bockstaller, C., Chauvel, B., Denieul, C., Moreau, D., Omon, B., Pellet, D., Rodriguez, A., Trannoy, L., et al. (Eds) *Gestion des adventices dans un contexte de changement - Séminaire CoSAC*. Paris, France, 46–49.
Claessen, D., Gilligan, C. A., Lutman, P. J. W. and Van Den Bosch, Fvd (2005). Which traits promote persistence of feral GM crops? Part 1: implications of environmental stochasticity. *Oikos* 110(1), 20–29.
Colas, F. (2018). *Co-développement d'un modèle d'aide à la décision pour la gestion intégrée de la flore adventice. Métamodélisation et analyse de sensibilité d'un modèle mécaniste complexe (FLORSYS) des effets des systèmes de culture sur les services et disservices écosystémiques de la flore adventice*. Dijon, France: University Bourgogne Franche-Comté, 334 p.
Colas, F., Cordeau, S., Granger, S., Jeuffroy, M.-H., Pointurier, O., Queyrel, W., Rodriguez, A., Villerd, J. and Colbach, N. (2020). Co-development of a decision support system for integrated weed management: contribution from future users. *European Journal of Agronomy* 114, 126010.
Colas, F., Villerd, J. and Colbach, N. (2019). Simplification d'un modèle complexe pour le développement d'un modèle d'aide à la décision pour la gestion agroécologique de la flore adventice. *Végéphyl - 24e conférence du COLUMA - Journées Internationales sur la Lutte Contre Les Mauvaises Herbes*. Orléans, France.

Colbach, N. (2009). How to model and simulate the effects of cropping systems on population dynamics and gene flow at the landscape level. Example of oilseed rape volunteers and their role for co-existence of GM and non-GM crops. *Environmental Science and Pollution Research* 16(3), 348–360.

Colbach, N. (2010). Modelling cropping system effects on crop pest dynamics: how to compromise between process analysis and decision aid. *Plant Science* 179(1-2), 1–13.

Colbach, N. (2020). How to use a "virtual field" to evaluate and design integrated weed management strategies at different spatial and temporal scales. In: Chantre, G. R. and González-Andújar, J. L. (Eds) *Decision Support Systems for Weed Management*. Springer International Publishing, Cham, Switzerland, 227–248.

Colbach, N., Beaudoin, N., Guyot, S., Villerd, J. and Moreau, D. (2020a). How to Model Crop-Weed Competition for Soil Resources: Connecting the STICS Soil Submodel to the FLORSYS Weed Dynamics Model. *XIIth Stics Users Seminar*. Montpellier, France, 24–25.

Colbach, N., Moreau, D., Dugué, F., Gardarin, A., Strbik, F. and Munier-Jolain, N. (2020b). The response of weed and crop species to shading. How to predict their morphology and plasticity from species traits and ecological indexes? *European Journal of Agronomy* 121, 126158.

Colbach, N., Bertrand, M., Busset, H., Colas, F., Dugué, F., Farcy, P., Fried, G., Granger, S., Meunier, D., Munier-Jolain, N. M., Noilhan, C., Strbik, F. and Gardarin, A. (2016). Uncertainty analysis and evaluation of a complex, multi-specific weed dynamics model with diverse and incomplete data sets. *Environmental Modelling and Software* 86, 184–203.

Colbach, N., Busset, H., Yamada, O., Dürr, C. and Caneill, J. (2006a). ALOMYSYS: modelling black-grass (*Alopecurus myosuroides* Huds.) germination and emergence, in interaction with seed characteristics, tillage and soil climate. II. *European Journal of Agronomy* 24(2), 113–128.

Colbach, N., Dürr, C., Roger-Estrade, J., Chauvel, B. and Caneill, J. (2006b). ALOMYSYS: modelling black-grass (*Alopecurus myosuroides* Huds.) germination and emergence, in interaction with seed characteristics, tillage and soil climate. I. *European Journal of Agronomy* 24(2), 95–112.

Colbach, N., Chauvel, B., Dürr, C. and Richard, G. (2002). Effect of environmental conditions on *Alopecurus myosuroides* germination. I. Effect of temperature and light. *Weed Research* 42(3), 210–221.

Colbach, N., Chauvel, B., Gauvrit, C. and Munier-Jolain, N. M. (2007). Construction and evaluation of ALOMYSYS, modelling the effects of cropping systems on the blackgrass life-cycle. From seedling to seed production. *Ecological Modelling* 201(3–4), 283–300.

Colbach, N., Colas, F., Cordeau, S., Maillot, T., Queyrel, W., Villerd, J. and Moreau, D. (2021). The FLORSYS crop-weed canopy model, a tool to investigate and promote agroecological weed management. *Field Crops Research* 261, 108006.

Colbach, N., Collard, A., Guyot, S. H. M., Mézière, D. and Munier-Jolain, N. M. (2014). Assessing innovative sowing patterns for integrated weed management with a 3D crop:weed competition model. *European Journal of Agronomy* 53, 74–89.

Colbach, N. and Cordeau, S. (2018). Reduced herbicide use does not increase crop yield loss if it is compensated by alternative preventive and curative measures. *European Journal of Agronomy* 94, 67–78.

Colbach, N., Cordeau, S., Garrido, A., Granger, S., Laughlin, D., Ricci, B., Thomson, F. and Messéan, A. (2018). Landsharing vs landsparing: how to reconcile crop production and biodiversity? A simulation study focusing on weed impacts. *Agriculture, Ecosystems and Environment* 251, 203-217.

Colbach, N., Cordeau, S., Queyrel, W., Maillot, T., Villerd, J. and Moreau, D. (2019). Du champ virtuel au champ réel - ou comment utiliser un modèle de simulation pour diagnostiquer des stratégies de gestion durables des adventices? *Agronomie, Environnement et Sociétés* 9, 111-128.

Colbach, N. and Debaeke, P. (1998). Integrating crop management and crop rotation effects into models of weed population dynamics: a review. *Weed Science* 46(6), 717-728.

Colbach, N., Dürr, C., Roger-Estrade, J. and Caneill, J. (2005). How to model the effects of farming practices on weed emergence. *Weed Research* 45(1), 2-17.

Colbach, N., Fernier, A., Le Corre, V., Messéan, A. and Darmency, H. (2017). Simulating changes in cropping practices in conventional and glyphosate-resistant maize. I. Effects on weeds. *Environmental Science and Pollution Research International* 24(12), 11582-11600.

Colbach, N., Meynard, J. M., Clermont-Dauphin, C. and Messéan, A. (2004). Genesys-COLZA: un modèle des effets à moyen et long terme des systèmes de culture sur les flux de gènes entre champs de colza et repousses dans un espace agricole. In: Monestiez, P., Lardon, S. and Seguin, B. (Eds) *Organisation spatiale des activités agricoles et processus environnementaux.* Paris: INRA Editions, 193-215 .

Colbach, N., Roger-Estrade, J., Chauvel, B. and Caneill, J. (2000). Modelling vertical and lateral seed bank movements during moulboard ploughing. *European Journal of Agronomy* 13(2-3), 111-124.

Conti, S. and O'Hagan, A. (2010). Bayesian emulation of complex multi-output and dynamic computer models. *Journal of Statistical Planning and Inference* 140(3), 640-651.

Cousens, R., Doyle, C. J., Wilson, B. J. and Cussans, G. W. (1986). Modelling the economics of controlling *Avena fatua* in winter wheat. *Pesticide Science* 17(1), 1-12.

De'ath, G. (2001). Multivariate regression trees: A new technique for modeling species-environment relationships. *Ecology* 83, 1105-1117.

Délye, C., Colbach, N. and Le Corre, V. (2020). Résistances aux herbicides : mécanismes, situation en France et bonnes pratiques. *Innovations Agronomiques* 81, 33-49.

Doohan, D., Wilson, R., Canales, E. and Parker, J. (2010). Investigating the human dimension of weed management: new tools of the trade. *Weed Science* 58(4), 503-510.

Dürr, C., Aubertot, J. -N., Richard, G., Dubrulle, P., Duval, Y. and Boiffin, J. (2001). Simple: A model for SIMulation of PLant Emergence prediction the effects of soil tillage and sowing operations. *Soil Science Society of America Journal* 65(2), 414-423.

Dury, J., Schaller, N., Garcia, F., Reynaud, A. and Bergez, J. E. (2012). Models to support cropping plan and crop rotation decisions: a review. *Agronomy for Sustainable Development* 32(2), 567-580.

Ennaïfar, S., Makowski, D., Meynard, J.-M. and Lucas, P. (2007). Evaluation of models to predict take-all incidence in winter wheat as a function of cropping practices, soil, and climate. *European Journal of Plant Pathology* 118(2), 127-143.

Evers, J. B. and Bastiaans, L. (2016). Quantifying the effect of crop spatial arrangement on weed suppression using functional-structural plant modelling. *Journal of Plant Research* 129(3), 339-351.

Evrard, O., Nord, G., Cerdan, O., Souchere, V., Le Bissonnais, Y. and Bonte, P. (2010). Modelling the impact of land use change and rainfall seasonality on sediment export from an agricultural catchment of the northwestern European loess belt. *Agriculture, Ecosystems and Environment* 138(1-2), 83-94.

Fernández-Quintanilla, C., Dorado, J., Andújar, D. and Peña, J. M. (2020). Site-specific based models. In: Chantre, G. R. and González-Andújar, J. L. (Eds) *Decision Support Systems for Weed Management*. Cham: Springer International Publishing, 143-157.

Freckleton, R. P. and Stephens, P. A. (2009). Predictive models of weed population dynamics. *Weed Research* 49(3), 225-232.

Freckleton, R. P. and Watkinson, A. R. (1998). Predicting the determinants of weed abundance: a model for the population dynamics of *Chenopodium album* in sugar beet. *Journal of Applied Ecology* 35(6), 904-920.

Fried, G., Norton, L. R. and Reboud, X. (2008). Environmental and management factors determining weed species composition and diversity in France. *Agriculture, Ecosystems and Environment* 128(1-2), 68-76.

Fried, G., Petit, S. and Reboud, X. (2010). A specialist-generalist classification of the arable flora and its response to changes in agricultural practices. *BMC Ecology* 10, 20.

Gardarin, A., Coste, F., Wagner, M.-H. and Dürr, C. (2016). How do seed and seedling traits influence germination and emergence parameters in crop species? A comparative analysis. *Seed Science Research* 26(4), 317-331.

Gardarin, A., Dürr, C. and Colbach, N. (2010a). Effects of seed depth and soil structure on the emergence of weeds with contrasted seed traits. *Weed Research* 50(2), 91-101.

Gardarin, A., Dürr, C., Mannino, M. R., Busset, H. and Colbach, N. (2010b). Seed mortality in the soil is related to the seed coat thickness. *Seed Science Research* 20(4), 243-256.

Gardarin, A., Dürr, C. and Colbach, N. (2012). Modeling the dynamics and emergence of a multispecies weed seed bank with species traits. *Ecological Modelling* 240, 123-138.

Garnier, A., Deville, A. and Lecomte, J. (2006). Stochastic modelling of feral plant populations with seed immigration and road verge management. *Ecological Modelling* 197(3-4), 373-382.

Gaudio, N., Escobar-Gutierrez, A. J., Casadebaig, P., Evers, J. B., Gérard, F., Louarn, G., Colbach, N., Munz, S., Launay, M., Marrou, H., Barillot, R., Hinsinger, P., Bergez, J.-E., Combes, D., Durand, J.-L., Frak, E., Pagès, L., Pradal, C., Saint-Jean, S., Van Der Werf, W. and Justes, E. (2019). Current knowledge and future research opportunities for modeling annual crop mixtures: a review. *Agronomy for Sustainable Development* 39(2), 20.

González-Andújar, J. L. (2020). Introduction to decision support systems. In: Chantre, G. R. and González-Andújar, J. L. (Eds) *Decision Support Systems for Weed Management*. Cham: Springer International Publishing, 25-38.

González-Andújar, J. L. and Fernández-Quintanilla, C. (1991). Modelling the population dynamics of *Avena sterilis* under dry-land cereal cropping systems. *Journal of Applied Ecology* 28(1), 16-27.

González-Andújar, J. L. and Fernández-Quintanilla, C. (1993). Strategies for the control of *Avena sterilis* in winter wheat production systems in central Spain. *Crop Protection* 12(8), 617-623.

Gonzalez-Andujar, J. L., Perry, J. N. and Moss, S. R. (2000). Modelling effects of spatial patterns on the seed bank dynamics of *Alopecurus myosoroides*. *Weed Science* 47, 697-705.

Gressel, J. (2005). Problems in qualifying and quantifying assumptions in plant protection models: resultant simulations can be mistaken by a factor of million. *Crop Protection* 24(11), 1007–1015.

Gressel, J. and Segel, L. A. (1978). The paucity of plants evolving genetic resistance to herbicides : possible reasons and implications. *Journal of Theoretical Biology* 75(3), 349-371.

Grundy, A. C. (2003). Predicting weed emergence: a review of approaches and future challenges. *Weed Research* 43(1), 1-11.

Haefner, J. W. (2005). *Modeling Biological Systems: Principles and Applications*. Springer US, New York 475 p .

Holman, I. P., Dubus, I. G., Hollis, J. M. and Brown, C. D. (2004). Using a linked soil model emulator and unsaturated zone leaching model to account for preferential flow when assessing the spatially distributed risk of pesticide leaching to groundwater in England and Wales. *The Science of the Total Environment* 318(1-3), 73-88.

Holst, N., Rasmussen, I. A. and Bastiaans, L. (2007). Field weed population dynamics: a review of model approaches and applications. *Weed Research* 47(1), 1-14.

Jana, S. and Thai, K. M. (1987). Patterns of changes of dormant genotypes in *Avena fatua* populations under different agricultural conditions. *Canadian Journal of Botany* 65(8), 1741-1745.

Jastrzebska, M., Kostrzewska, M. K., Marks, M., Jastrzebski, W. P., Treder, K. and Makowski, P. (2019). Crop rotation compared with continuous rye cropping for weed biodiversity and rye yield: a case study of a long-term experiment in Poland. *Agronomy-Basel* 9, 17.

Kanatas, P., Traylos, I. S., Gazoulis, I., Tataridas, A., Tsekoura, A. and Antonopoulos, N. (2020). Benefits and limitations of decision support systems (DSS) with a special emphasis on weeds. *Agronomy-Basel* 10, 10.

Keating, B. A., Carberry, P. S., Hammer, G. L., Probert, M. E., Robertson, M. J., Holzworth, D., Huth, N. I., Hargreaves, J. N. G., Meinke, H., Hochman, Z., Mclean, G., Verburg, K., Snow, V., Dimes, J. P., Silburn, M., Wang, E., Brown, S., Bristow, K. L., Asseng, S., Chapman, S., Mccown, R. L., Freebairn, D. M. and Smith, C. J. (2003). An overview of APSIM, a model designed for farming systems simulation. *European Journal of Agronomy* 18(3-4), 267-288.

Kouwenhoven, J. K. and Terpstra, R. (1972). Characterization of soil handling with mould-board ploughs. *Netherlands Journal of Agricultural Science* 20(3), 180-192.

Kropff, M. J. and van Laar, H. H. (Eds.) (1993). *Modelling Crop-Weed Interactions*. CAB International, Wallingford, Oxfordshire, 274 p.

Kurstjens, D. A. G. and Kropff, M. J. (2001). The impact of uprooting and soil-covering on the effectiveness of weed harrowing. *Weed Research* 41(3), 211-228.

Lacoste, M. and Powles, S. (2017). RIM: anatomy of a weed management decision support system for adaptation and wider application. *Weed Science* 63(3), 676-689.

Legay, J. M. (1996). *L'expérience et le modèle. Un discours sur la méthode*. Paris: INRA.

Leslie, P. H. (1945). On the use of matrices in population mathematics. *Biometrika* 33, 183-212.

Lewis, J. (1973). Longevity of crop and weed seeds: survival after 20 years in soil. *Weed Research* 13(2), 179-191.

Liebman, M. and Gallandt, E. R. (1997). Many little hammers: ecological management of crop-weed interactions. In: Jackson, L. E. (Ed) *Ecology in Agriculture*. Academic Press San Diego, CA, 291-343 .

Martin, G., Martin-Clouaire, R. and Duru, M. (2013). Farming system design to feed the changing world: a review. *Agronomy for Sustainable Development* 33(1), 131–149.
Maxwell, B. D., Roush, M. L. and Radosevich, S. R. (1990). Predicting the evolution and dynamics of herbicide resistance in weed populations. *Weed Technology* 4(1), 2–13.
McDonald, A. and Riha, S. (1999). Model of crop:weed competition applied to maize: *Abutilon theophrasti* interactions. I. Model description and evaluation. *Weed Research* 39(5), 355–369.
Mézière, D., Petit, S., Granger, S., Biju-Duval, L. and Colbach, N. (2015). Developing a set of simulation-based indicators to assess harmfulness and contribution to biodiversity of weed communities in cropping systems. *Ecological Indicators* 48, 157–170.
Milberg, P. and Hallgren, E. (2004). Yield loss due to weeds in cereals and its large-scale variability in Sweden. *Field Crops Research* 86(2–3), 199–209.
Minogue, K. P. (1989). Diffusion and spatial probability models for disease spread. In: Jeger, M. J. (Ed) *Spatial Components of Plant Disease Epidemics*. Prentice Hall, Englewood Cliffs , New Jersey, 127–143.
Molinari, F. A., Blanco, A. M., Vigna, M. R. and Chantre, G. R. (2020). Towards an integrated weed management decision support system: A simulation model for weed-crop competition and control. *Computers and Electronics in Agriculture* 175, 105597.
Moreau, D., Pointurier, O., Perthame, L., Beaudoin, N., Villerd, J. and Colbach, N. (2021). Integrating plant-plant competition for nitrogen into a 3D individual-based model simulating the effects of cropping systems on weed dynamics. *Field Crops Research* 268, 108166.
Munier-Jolain, N. M., Chauvel, B. and Gasquez, J. (2002). Long-term modelling of weed control strategies: analysis of threshold-based options for weed species with contrasted competitive abilities. *Weed Research* 42(2), 107–122.
Munier-Jolain, N. M., Guyot, S. H. M. and Colbach, N. (2013). A 3D model for light interception in heterogeneous crop:weed canopies. Model structure and evaluation. *Ecological Modelling* 250, 101–110.
Munier-Jolain, N. M., Savois, V., Kubiak, P., Maillet-Mézeray, J., Jouy, L. and Quéré, L. (2005). DECID'Herb, a decision support system on the WEB, designed for sustainable weed management in cultivated fields. 13th International EWRS Symposium. Bari, Italy.
Murtaugh, P. A. (1966). The statistical evaluation of ecological indicators. *Ecological Applications* 6(1), 132–139.
Neyret, M., De Rouw, A., Colbach, N., Robain, H., Soulileuth, B. and Valentin, C. (2020). Year-to-year crop shifts promote weed diversity in tropical permanent rainfed cultivation. *Agriculture, Ecosystems and Environment* 301, 107023.
O'Donovan, J. T. (1996). Weed economic thresholds: useful agronomic tool or pipe dream? *Phytoprotection* 77, 13–28.
Oliver, L. R. (1988). Principles of weed threshold research. *Weed Technology* 2(4), 398–403.
Ould-Sidi, M. M. and Lescourret, F. (2011). Model-based design of integrated production systems: a review. *Agronomy for Sustainable Development* 31(3), 571–588.
Pannell, D. J., Stewart, V., Bennett, A., Monjardino, M., Schmidt, C. and Powles, S. B. (2004). RIM: a bioeconomic model for integrated weed management of *Lolium rigidum* in Western Australia. *Agricultural Systems* 79(3), 305–325.
Parsons, D. J., Benjamin, L. R., Clarke, J., Ginsburg, D., Mayes, A., Milne, A. E. and Wilkinson, D. J. (2009). Weed Manager–A model-based decision support system for weed management in arable crops. *Computers and Electronics in Agriculture* 65(2), 155–167.

Pelzer, E., Fortino, G., Bockstaller, C., Angevin, F., Lamine, C., Moonen, C., Vasileiadis, V., Guérin, D., Guichard, L., Reau, R. and Messéan, A. (2012). Assessing innovative cropping systems with DEXiPM, a qualitative multi-criteria assessment tool derived from DEXi. *Ecological Indicators* 18, 171–182.

Perthame, L., Petit, S. and Colbach, N. (2018). Cropping systems for driving biological regulation of weeds: a simulation study of seed predation by carabids. *Proceedings of the XVe ESA*. Geneva, Switzerland, 154.

Petit, S., Alignier, A., Colbach, N., Joannon, A., Le Coeur, D. and Thenail, C. (2013). Weed dispersal by farming at various spatial scale: a review. *Agronomy for Sustainable Development* 33(1), 205–217.

Peyrard, N., Sabbadin, R., Lô-Pelzer, E. and Aubertot, J. N. (2007). A graph-based Markov decision process framework for optimising collective management of diseases in agriculture: application to blackleg on canola. In: Oxley, L. and Kulasiri, D. (Eds) *MODSIM 2007 Internation Congress on Modelling and Simulation*. Modelling and Simulation Society of Australia and New Zealand, Christchurch, New-Zealand, 2175–2181.

Pointurier, O., Moreau, D., Pagès, L., Caneill, J. and Colbach, N. (2021). Individual-based 3D modelling of root systems in heterogeneous plant canopies at the multiannual scale: case study with a weed dynamics model. *Ecological Modelling* 440, 109376.

Press, W. H., Teukolsky, S. A., Vetterling, W. T. and Flannery, B. P. (2007). *Numerical Recipes: the Art of Scientific Computing* (3rd edn.). New York, NY: Cambridge University Press, 1235 p.

Primot, S., Valantin-Morison, M. and Makowski, D. (2006). Predicting the risk of weed infestation in winter oilseed rape crops. *Weed Research* 46(1), 22–33.

Renton, M. (2013). Shifting focus from the population to the individual as a way forward in understanding, predicting and managing the complexities of evolution of resistance to pesticides. *Pest Management Science* 69(2), 171–175.

Renton, M., Busi, R., Neve, P., Thornby, D. and Vila-Aiub, M. (2014). Herbicide resistance modelling: past, present and future. *Pest Management Science* 70(9), 1394–1404.

Ridier, A., Ghali, M. B. E., Nguyen, G. and Kephaliacos, C. (2013). The role of risk aversion and labor constraints in the adoption of low input practices supported by the CAP green payments in cash crop farms. *Revue d'Études en Agriculture et Environnement* 94(2), 195–219.

Röhrig, M., Stützel, H. and Alt, C. (1999). A three-dimensional approach to modeling light interception in heterogeneous canopies. *Agronomy Journal* 91(6), 1024–1032.

Royo-Esnal, A., Torra, J. and Chantre, G. R. (2020). Weed emergence models. In: Chantre, G. R. and González-Andújar, J. L. (Eds) *Decision Support Systems for Weed Management*. Cham: Springer International Publishing, 85–116.

Schoeny, A., Jeuffroy, M. H. and Lucas, P. (2001). Influence of take-all epidemics on winter wheat yield formation and yield loss. *Phytopathology* 91(7), 694–701.

Segal, M. and Xiao, Y. (2011). Multivariate random forests. *WIREs Data Mining and Knowledge Discovery* 1(1), 80–87.

Singh, M., Kaur, S. and Chauhan, B. S. (2020). Weed interference models. In: Chantre, G. R. and González-Andújar, J. L. (Eds) *Decision Support Systems for Weed Management*. Cham: Springer International Publishing, 117–142.

Swanton, C. J., Weaver, S., Cowan, P., Van Acker, R. C., Deen, W. and Shrestha, A. (1999). Weed thresholds: theory and applicability. *Journal of Crop Production* 2(1), 9–29.

Swets, J. A. (1988). Measuring the accuracy of diagnostic systems. *Science* 240(4857), 1285–1293.

Swinton, S. M., Buhler, D. D., Forcella, F., Gunsolus, J. L. and King, R. P. (1994). Estimation of crop yield loss due to interference by multiple weed species. *Weed Science* 42(1), 103-109.

Torra, J. and Monjardino, M. (2020). Ryegrass integrated management (RIM)-based decision support system. In: Chantre, G. R. and González-Andújar, J. L. (Eds) *Decision Support Systems for Weed Management*. Cham: Springer International Publishing, 249-278.

Van Himme, M. and Bulcke, R. (1975). Distribution, extension et importance d'*Alopecurus myosuroides* Huds. en Europe. *Proceedings of the European Weed Research Society Symposium Status, Biology and Control of Grassweeds in Europe*, 23-54.

Venter, G. (2010). Review of optimization techniques. In: Blockley, R. and Shyy, W. (Eds) *Encyclopedia of Aerospace Engineering*. Chichester, UK: John Wiley & Sons Ltd.

Vleeshouwers, L. M. and Bouwmeester, H. J. (2001). A simulation model for seasonal changes in dormancy and germination of weed seeds. *Seed Science Research* 11(1), 77-92.

Vos, J., Evers, J. B., Buck-Sorlin, G. H., Andrieu, B., Chelle, M. and De Visser, P. H. B. (2010). Functional-structural plant modelling: a new versatile tool in crop science. *Journal of Experimental Botany* 61(8), 2101-2115.

Wallach, D. (2006). Evaluating crop models. In: Wallach, D., Makowski, D. and Jones, J. W. (Eds) *Working with Dynamic Crop Models: Evaluating, Analyzing, Parameterizing and Using Them*. Paris: Éditions INRA, 11-53.

Wossink, G. A. A., De Buck, A. J., Van Niejenhuis, J. H. and Haverkamp, H. C. M. (1997). Farmer perceptions of weed control techniques in sugar beet. *Agricultural Systems* 55(3), 409-423.

Chapter 7

Developing decision support systems (DSS) for weed management

Panagiotis Kanatas, University of Patras, Greece; and Ilias Travlos, Ioannis Gazoulis and Alexandros Tataridas, Agricultural University of Athens, Greece

1 Introduction

Farmers are increasingly being encouraged to adopt more sustainable and environmentally friendly production systems. In the European Union (EU), this objective is set out in the recently adopted farm to fork strategy. Natural resources must be used wisely and biodiversity must be preserved whilst maintaining the volume and quality of agricultural production (Binfield et al., 2004). The contribution of precision agriculture, along with advances in sensors, computers and communication devices, is vital in achieving this transition to a more resource-efficient agriculture. Precision agriculture involves the use of information technology (IT) to target resources, maximize yields and produce high-quality products while minimizing impacts on the natural environment (Yost et al., 2017; Cisternas et al., 2020).

Computer-based tools known as decision support systems (DSS) for use in farm management have also been developed and integrated with precision agriculture. A DSS can be defined as an interactive, computer-based system intended to help farmers or other decision-makers in using data and models

http://dx.doi.org/10.19103/AS.2021.0098.08

to make better operational decisions (Zhai et al., 2020a). The introduction of DSSs in agriculture has shown great potential in various aspects of cultivation (Table 1). An example is the DSS model developed by Stanley et al. (2020) which uses data on spring wheat (*Triticum aestivum* L.) straw strength and tillering capacity data to give farmers recommendations on optimum seeding rates for maximizing yields. The LCIS DDS for irrigation management has been found to optimize irrigation in maize (*Zea mays* L.) whilst improving the final yield of the crop (Bonfante et al., 2019). The fertilization rates suggested by the DSS described by Meza-Palacios et al. (2020) optimized productivity of sugarcane (*Saccharum officinarum* Linn.) whilst reducing the amount of fertilization and its impact on the environment. The hand-held decision support tool (HH-DST) described by Pérez et al. (2020) suggested more targeted fungicide treatments for the management of potato late blight [*Phytophthora infestans* (Mont.) de Bary] disease in potato (*Solanum tuberosum* L.) and was, in some cases, more effective as well as more environmental friendly than existing spray regimes. Small et al. (2015) have highlighted the success of the BlightPro DSS in potato and tomato late blight management. The recommendations for the treatment of the olive fruit fly pest [*Bactrocera oleae* (Rossi, 1790)] in olive orchards (*Olea europaea* L.) by the ENPI DSS resulted in approximately 37% lower insecticide use while still effectively controlling adult insect populations (Miranda et al., 2019). The objectives of this chapter are to review what is achievable in DSS designed specifically for weed management, identify the obstacles to creating more effective DSS and discuss how to increase adoption by farmers.

2 Decision support systems for weed management: setting thresholds

The presence of weeds in the field is a major obstacle to agricultural production. Weed infestations are the cause of a 5%, 10% and 25% loss in agricultural production in the most-developed, less-developed and least-developed countries, respectively (Oerke and Dehne, 2004). Weeds are adaptable to all

Table 1 The objectives of DSSs introduced recently in agriculture

DSS names	Objectives	References
HH-DST	Potato late blight management	Pérez et al. (2020)
Not defined	Seeding rate selection in hard wheat	Stanley et al. (2020)
LCIS	Irrigation management in maize	Bonfante et al. (2019)
Not defined	NPK fertilization in sugarcane	Meza-Palacios et al. (2020)
ENPI	Olive fruit fly management	Miranda et al. (2019)
BlightPro	Potato and tomato late blight management	Small et al. (2015)

environments and, in the absence of any control methods, severe infestations can result in complete yield losses for some crops (Chauhan, 2020). Globally, weeds have been demonstrated as the most important pest group in wheat, rice, maize and soybean production systems (Oerke, 2006). Since they compete directly with the crop for natural resources, act as hosts for other pests, reduce crop yields and product quality, they increase the cost of cultivation (Boiteux et al., 2016; Zimdahl, 2018; Chauhan, 2020; Goss et al., 2020; Kanatas et al., 2020c; Nimu et al., 2020).

The use of DSSs for weed management has the potential to reduce herbicide inputs in modern agricultural systems, and, subsequently, the environmental risks associated with herbicide use whilst maintaining high yields and high-quality products (Montull et al., 2014). Such DSSs can perform tasks such as simulating weed abundance and diversity in the field and assessing potential weed impacts on crops and the effect of different management options on ecosystem structure and function (Blackshaw et al., 2006). Farmers can use DSS to assess different options for herbicide use and compare them with other control options. The final outcome may be optimization of herbicide efficacy or the ability to move beyond a short-term chemical-based to a more integrated approach to weed management (Colbach et al., 2017).

DSS have not so far been widely adopted in weed management. This is despite the fact that they have been reported to provide useful information on weed species abundance in agricultural fields, impact on crops and useful recommendations for weed control (Rydahl, 2003). Many DSS target the control of both grasses and broadleaf weeds with one herbicide application instead of two or three. DSSs can rank individual treatments and suggest potential combinations. When selecting a specific herbicidal treatment from the proposed list, information is provided on herbicide costs, application rates, weed densities either before or after treatment and expected yield losses for each case (Neeser et al., 2004; González-Andújar, 2020).

DSSs set thresholds for intervention according to yield loss estimations derived from models. Various parameters are usually considered in modelling crop yield losses from weed competition (Fig. 1). For example, the model used by Benjamin et al. (2010) in their weed manager DSS accounts for the following variables:

- weed seed germination and weed emergence;
- early growth;
- phenological development;
- herbicide and cultivation effects; and
- crop yield loss.

The model and the decision algorithm were validated by experts who confirmed that the predicted responses to herbicide application were sufficiently precise.

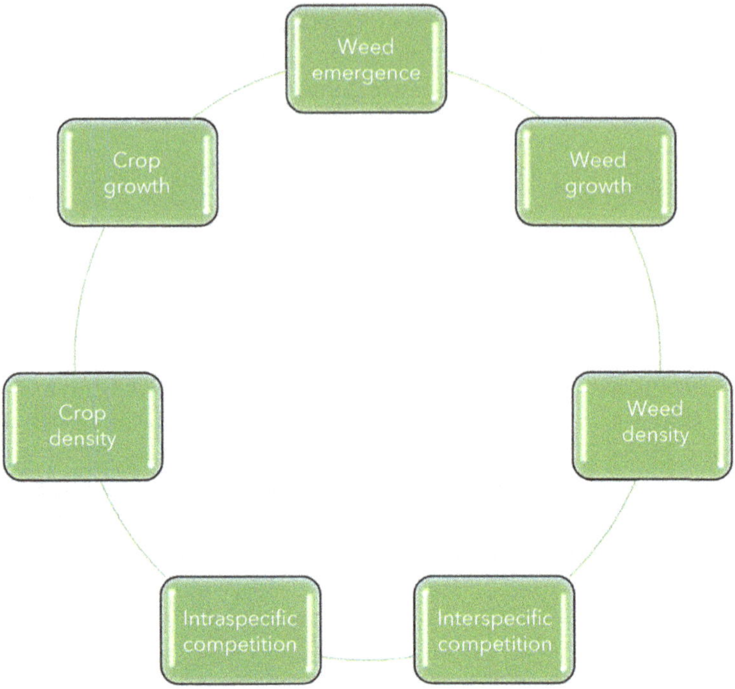

Figure 1 Parameters that are usually considered for the modelling of crop yield losses due to weed competition.

In the WeedSOFTt DSS described by Neeser et al. (2004), a competitive index was used to translate densities of different weed species into a common unit of weed pressure to determine potential crop yield losses. Field experiments showed that the DSS accurately estimated crop yield losses due to competition from non-controlled weeds (Neeser et al., 2004).

Hyperbolic, exponential and other mathematical models can be used to define the relationship between crop yield losses and the timing of weed emergence, density, vegetative growth and biomass accumulation (Cousens, 1985; Cousens et al., 1987; Lotz et al., 1990; Kropff and Spitters, 1991). DSS models have been reported to assess the yield losses of important crops such as winter cereals and soybean due to weed competition (Neeser et al., 2004; Pannell et al., 2004; González-Andújar, 2020; Molinari et al., 2020). All these DSSs-based models yield loss estimations both on interspecies competition between crops and weeds and on intraspecies competition between weed plants (Molinari et al., 2020). The competitiveness of a crop against a weed species depends on crop density and leaf area index (LAI) whose values can be calculated by decision support system for agrotechnology transfer (DSSAT) software (Jones et al., 2003). The negative impact of weed competition on crop yield can be expressed using the equation developed by Pannell et al. (2004). Molinari et al. (2020) have suggested using the total competition effects at the end of the season instead of the density of weeds that survive control measures to make the equation less complex.

3 The role of decision support systems in reducing herbicide use

One of the most important targets of DSS for weed management is to use no more herbicide than needed. Herbicide performance is predicted based on key variables such as weed flora and growth stage, crop competitiveness, climatic conditions, application technology, formulations and use of adjuvants and mixtures with other pesticides as summarized by Kudsk (2008). In determining optimum herbicide rates, the first step is to determine dose-response relationships for key weed species; the second step is validation of the DSS under real field conditions (Montull et al., 2020). Optimal herbicide combinations, the number of applications and application rates are based on herbicide dose-response curves, which are parameterized with experimental data from many years of herbicide testing under field conditions (Sønderskov et al., 2015).

A good example is the Danish crop protection online (CPO) DSS which provides information on when weed control is needed as well as optimal herbicide combinations and rates (Rydahl, 2003; Kudsk, 2008). The control level required for each weed species is determined according to the competitiveness of the weed species, seed dispersal rates as well as interference with harvestings. Target efficacy is adjusted according to season, as some weeds in autumn-sown crops are more important to control in autumn than in spring and vice versa. Using the DSS, Rydahl (2003) suggested supplementary herbicide applications for the control of late season weeds in winter cereal fields to reduce seed production from escapees.

There has been significant progress by DSS in optimizing herbicide use and efficacy (Kudsk, 2014; Montull et al., 2014; Sønderskov et al., 2015; Rydahl et al., 2018; González-Andújar, 2020; Torra and Monjardino, 2020). The Danish IPMwise DSS is able to make suggestions for either chemical or non-chemical weed control of more than 100 weed species (Rydahl et al., 2018). The Dutch minimum lethal herbicide dose DSS has been tested *in situ* and has provided useful information on weed flora and species abundance in the field (Haage et al., 2002; Kempenaar et al., 2002). The FlorSys model assesses the impact of weeds on both crop production and biodiversity within cropping systems (Colbach et al., 2014).

The CPO DSS was found to reduce herbicide use by 35% and 44% in spring and winter cereals, respectively, as compared to a standard treatment (Rydahl, 2003). The DSS also resulted in a 48% and a 80% less herbicide input in fodder beet (*Beta vulgaris* L. subsp. *vulgaris* var. *crassa*) and sugarbeet *Beta vulgaris* L.

subsp. *vulgaris* var. *altissima*) in comparison to a standard herbicide program. The DSS also resulted in approximately 30% less herbicide inputs in spring cereal crops (Netland et al., 2005). Reductions in herbicide inputs can reach 40–50% where herbicides are applied according to DSS recommendations instead of using the recommended rates on labelling of herbicide products (Jørgensen et al., 2007).

Other researchers have found potential savings in the amounts of herbicides used for broadleaf and grass weed species of 60% and 77%, respectively, using site-specific spraying (based on DSS recommendations) when compared to standard treatments (Gutjahr and Gerhards, 2010). Herbicide inputs can be decreased by up to 60% in cereals using precision-agriculture methods supported by DSS (Jensen et al., 2012). A 4-year study showed that precision agriculture site-specific techniques reduced herbicide use up to 41%, 78% and 90% in sugar beet, maize (*Zea mays* L.) and winter cereal fields, respectively, when compared to spraying the whole field (Gerhards and Christensen, 2003). The performance of CPOWeeds, a version of CPO adjusted for Spanish conditions, was tested in winter cereal field trials from 2010 to 2013 (Montull et al., 2014). Herbicide efficacy and grain yield were higher than expected with herbicide rates reduced by up to 30% in 9 out of the total of 17 field experiments (Montull et al., 2014). Similar findings were reported by Montull et al. (2020).

Sønderskov et al. (2014) have reported effective weed control with a 60% reduction in herbicide use in spring barley (*Hordeum vulgare* L.). According to Sønderskov et al. (2015), another DSS based on CPO, DSS herbicide, resulted in 20% and 40% less herbicide inputs in German and Polish fields of winter wheat (*T. aestivum*), respectively. AvenaNET and VallicoNET are other web-based DSSs developed for the management of *Lolium rigidum* Gaud. and *Avena sterilis* L. spp. *ludoviciana*, respectively, in Spain (González-Andújar, 2020). The PC versions

Table 2 Herbicide inputs' reductions in important crops as recorded when spraying was carried out according to DSSs' recommendations

DSS names	Herbicide reduction (%)	Crops	References
CPOWeeds	30	Winter cereals, maize	Montull et al. (2020)
DSSHerbicide	20-40	Winter wheat	Sønderskov et al. (2015)
CPOWeeds	30	Winter cereals, maize	Montull et al. (2014)
CPO	60	Spring barley	Sønderskov et al. (2014)
LOLIUM-PC	57	Winter wheat	González-Andújar et al. (2011)
AVENA-PC	65	Winter wheat	González-Andújar et al. (2010)
CPO	30	Spring cereals	Netland et al. (2005)
CPO	35-44	Winter/spring cereals	Rydahl (2003)
CPO	48-80	Fodder/sugar beet	Rydahl (2003)

of these DSS (LOLIUM-PC and AVENA-PC) have been reported to reduce herbicide inputs for the control of *L. rigidum* and *A. sterilis* spp. *ludoviciana* in Spanish winter wheat by 57% and 65%, respectively, as compared to conventional herbicide applications (González-Andújar et al., 2010; González-Andújar et al., 2011). Examples of the effectiveness of DSS in reducing herbicide use are summarized in Table 2. A more recent challenge for weed management DSS is the growing problem of herbicide resistance which will be discussed in the next section.

4 Decision support systems and preventing herbicide resistance

In most countries, herbicide application is still preferred over other weed management options for effective, low-cost weed control. Compared to mechanical methods of control, herbicide use has been associated with lower levels of soil erosion, fuel use and greenhouse gas emissions (Gianessi, 2013). However, the continued reliance on herbicides has resulted in a growing problem of herbicide resistance (Peterson et al., 2018). More than 500 different types of herbicide resistance have been reported globally (Heap, 2020).

Modern remote-sensing techniques have made significant advances in herbicide resistance detection and management. Chlorophyll fluorescence imaging sensors are able to identify the presence of herbicide-resistant weed populations in the field. These sensors have been used to detect resistant individuals of *Abutilon theophrasti* Medic., *Alopecurus myosuroides* Huds., *Apera spica-venti* L., and *Setaria faberi* Herrm. to different herbicide sites of action (Wang et al., 2017; Zhang et al., 2017; Wang et al., 2018). Computer simulation also allows rapid virtual experiments to investigate how weed control methods affect the rate of emergence of herbicide-resistant weeds (Renton et al., 2014). These advances have made it possible to develop DSS capable of providing recommendations to prevent and manage herbicide resistance (Montull et al., 2020; Torra and Monjardino, 2020). For example, a resistance prevention module has been developed in the Spanish version of CPO, able to estimate the appropriate weed growth stage for spraying, taking into account how susceptible each weed species is to a particular herbicide (Montull et al., 2014). Such findings may help limit the evolution of herbicide-resistant weeds.

However, a key problem is that using the lower application rates of herbicides recommended by DSS may result in more rapid herbicide resistance evolution since some weeds may survive lower doses (Neve and Powles, 2005; Manalil et al., 2011). The lower application rates of herbicides recommended by DSS are thus a controversial issue (Neve, 2007). The development of resistance seems to depend on species mode of reproduction since self-pollinated *Avena fatua* L. populations do not show the same response as cross-pollinated *L. rigidum* populations to low herbicide rates (Busi et al., 2016). An effective DSS should include recommendations on rotation of herbicides and mixtures of herbicides with different modes of action to prevent resistance from evolving (Beckie and Reboud, 2009). If resistance increases, weed management DSS need to recommend different tactics with alternative cultural methods of weed management.

The RIM DSS was developed for the management of *L. rigidum* in Australia, including recommendations for mixing herbicides and using cultural practices such as crop-pasture rotations to minimize resistance (Pannell et al., 2004). RIM has been used to develop other DSS for the management of specific herbicide-resistant weed species. Torra and Monjardino (2020) have summarized the following adaptations of RIM:

- Multispecies RIM for *L. rigidum* and *Raphanus raphanistrum* (L.) in Australia;
- Wild Radish RIM for *R. raphanistrum* in Australia;
- PIM for *Papaver rhoeas* (L.) in Spain;
- RIMPhil for *Echinochloa crus-galli* (L.) P. Beauv. in The Philippines;
- BYGUM for *Echinochloa colona* (L.) Link in Australia;
- PAM for *Amaranthus palmeri* S. Wats. in the United States of America;
- Brome RIM for *Bromus* spp. in Australia;
- Barley Grass RIM for *Hordeum glaucum* L. in Australia;
- SA-RIM for *L. rigidum* in South Africa; and
- DK-RIM for *Lolium multiflorum* Lam. in Denmark.

A new version of DK-RIM, DK-VIM, has been produced for *Vulpia myuros* (L.) C.C.Gmel in Denmark (Akhter, 2020). All these RIM-based DSS provide

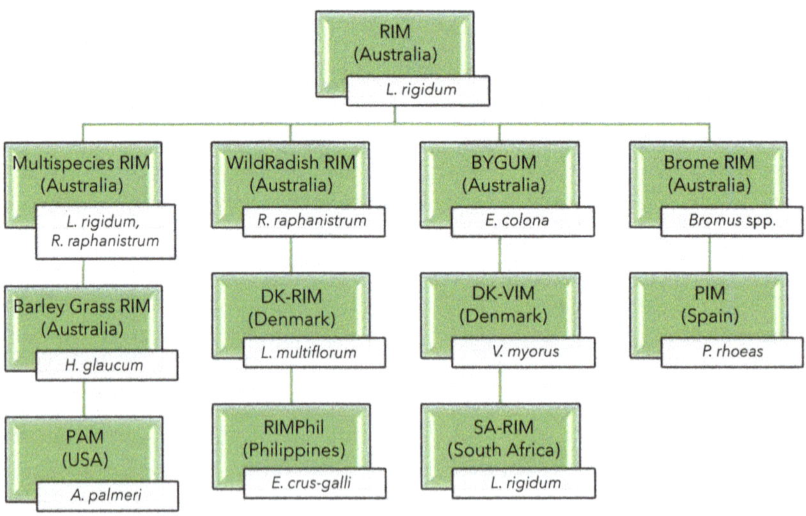

Figure 2 RIM-based DSSs designed for the management of herbicide-resistant weeds.

suggestions for the management of targeted species using both chemical and non-chemical methods to minimize herbicide resistance (Fig. 2).

However, RIM and its derivative DSSs have significant limitations. Suggestions are restricted to one location and a specific weed species, the best strategy is not always calculated immediately, whilst year-to-year variations in weather, potential yield losses and herbicide performance are not considered (Pannell et al., 2004). The complexity of herbicide resistance requires complex modelling. Large complex models with many parameters are more expensive to operate and less transparent, making it less clear how parameters interact with each other to produce reliable model outputs (Renton et al., 2014).

5 Decision support systems and long-term management of a broad spectrum of weed species

Another limitation of DSS is that, with some exceptions such as multispecies RIM, they have usually been developed only for specific weed species. AvenaNET and ValliconNET, for example, are focused on management of *A. sterilis* and *L. rigidum*, respectively (González-Andújar, 2020). The same is true for the RIM-based DSSs reviewed by Torra and Monjardino (2020). The focus on one or a few weed species does not address the real needs of farmers who need recommendations for management of a broad spectrum of weed species they may encounter. This is a complex challenge. DSS depends on modelling variables such as weed emergence, seed production, seedbank dynamics, inter- and intraspecific competition levels and yield loss estimations due to competition. This information is lacking for the majority of weed species (Chauhan, 2020).

This problem also affects the reliability of yield loss estimations made by DSS (Rydahl, 2003). In a study by Berti and Zanin (1997), GESTINF overestimated soybean (*Glycine max* L.) yield losses due to weed competition. The authors suggested this was due to different levels of competitiveness of weeds with different times of emergence. If yield losses are either over- or underestimated, recommendations will be unreliable. In addition, there are different equations to model yield losses and set thresholds for weed control operations (Cousens, 1985; Cousens et al., 1987; Lotz et al., 1990; Kropff and Spitters, 1991). Extended field trials may be needed to determine which equation fits best in each case and to simplify equations so they are easier to use (Singh et al., 2020). Another challenge is correct weed identification which can be difficult in crops such as cereals and soybean (*Glycine max* L.) given morphological similarities between the weed and crop plants during early stages of growth (Medlin et al., 2000; Gonzalez-Andujar et al., 2006). More broadly, there is a growing need to capture reliable data from the field as rapidly and precisely as possible, particularly given major advances in proximal and remote sensors capable of supplying a growing volume and range of real-time data on field conditions (Shaw, 2005; Jung et al., 2020).

To meet farmers' needs, DSSs should not only recommend strategies for a single growing season but also integrated weed management strategies across rotations (Kanatas et al., 2020b). DSS should also be able to recommend cultural practices for weed control, based on experience of real field conditions over a significant period to ensure consistent results (Munier-Jolain et al., 2002). It is important to consider parameters such as soil and climatic conditions, weed seedbank dynamics and weed emergence for long-term cultural weed management (Kanatas et al., 2020a,c). The weed manager DSS developed by Parsons et al. (2009) was based on weed seedbank dynamics in shallow and deep soil layers but only used data from trials in 2005–2006 and 2006–2007. More recently, Molinari et al. (2020) modelled parameters such as inter- and intraspecific competition, seedbank dynamics, weed seed production and growth to develop a more effective DSS for long-term weed management.

Crop rotation is also an important cultural practice for long-term weed management. The Fruchtfolge DSS uses big data and explicit spatial modelling to recommend an optimal rotation plan for each field (Pahmeyer et al., 2020). However, this DSS was not developed for weed management. There is a need to design a DSS capable of planning a crop rotation system in line with weed control targets for each field (Macé et al., 2007). As noted earlier, this requires detailed understanding of weed biology and ecology, accounting for variability between species (Chauhan and Johnson, 2010).

Another factor to consider is variability in soil and climatic conditions in different geographical locations that can limit the potential of a DSS to give realistic recommendations for weed control, especially in the case of cultural practices. For example, the DSS described by González-Andújar (2020) target short and long-term management of *A. sterilis* and *L. rigidum*. They suggest that the best long-term management strategy for controlling these weeds is to combine application of the herbicide Diclofop-methyl with delayed crop sowing (González-Andújar, 2020). However, this recommendation is impractical if delayed sowing or the establishment of a false/stale seedbed are prevented by adverse weather conditions in a particular location (Rasmussen, 2004). Such limitations highlight the fact that many models underpinning DSS are still best suited to research and need significant further development to meet farmers' needs in complex, real environments (González-Andújar, 2020; Molinari et al., 2020; Torra and Monjardino, 2020).

These problems highlight the challenge of accounting for a wide range of variables. Weed populations are heterogeneous in composition and distribution at field level, and any suggested control practice that does not address this heterogeneity will not be fully effective (Brown and Noble, 2005). As mentioned, other factors include soil and climatic conditions and cultural practices such as

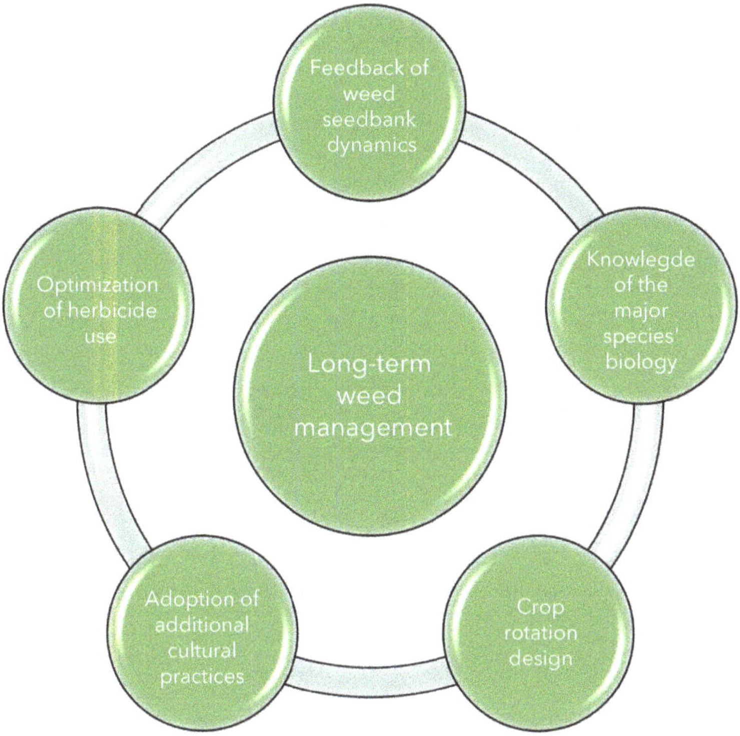

Figure 3 Necessary ingredients of integrated weed management strategies in the long-term period.

tillage and fertilization which are key factors affecting weed seed dormancy and germination, timing of weed emergence, weed abundance and ecology, weed flora composition and finally, the level of competition between crops and weeds (Travlos et al., 2018, 2020). To implement effective weed management strategies in the long-term period, a deep knowledge of weed biology and ecology is required to optimize herbicide use and integrate it with cultural practices (Fig. 3).

6 Increasing adoption of weed management decision support systems by farmers

The impact of weed management DSS is also limited by user constraints. One problem is the relatively *ad hoc* way in which many DSS have been developed which limits their widespread use by farmers (Walling and Vaneeckhaute, 2020). Zhai et al. (2020a) have listed some major constraints which limit the usefulness of DSSs. These include:

- a focus on solving particular weed problems in specific conditions (with limited relevance to other weed problems or locations); and

- poorly designed interfaces which make DSS difficult for non-specialists to understand and use.

These problems mean some farmers deviate from DSS recommendations due to a lack of understanding or trust (Möhring et al., 2020). Persuading farmers to use DSS in their crop production processes is a complex issue. This is partly because DSSs have made an important but still limited contribution to solving practical problems in crop protection under real field conditions (Parker et al., 1997).

Many farmers remain unfamiliar with new technologies (Kernecker et al., 2020). They are also confused by the pace of change in technology and consequently wary of investing in it (Johannsen et al., 2000; Jørgensen et al., 2007; Kanatas et al., 2020b). DSS often do not fully target farmers' needs or reflect real-field conditions (Magarey et al., 2002). From a user perspective, there are three general limitations to DSS (Rossi et al., 2012):

- the lack of accessibility of models which can be hard to understand and use (Donnelly et al., 2002, Welch et al., 2002);
- the need for adequate knowledge and expertise and continuous updating of information (Kerr, 2004); and
- delays in data processing and lack of relevance of information to users.

To deal with these problems, developers of DSS need to carefully identify potential stakeholders, establish good relationships with them and then prioritize their needs (Ingram and Gaskell, 2019). It is vital to carry out surveys, host workshops and group meetings in order to interact with farmers, design the right DSS and improve adoption. A crucial issue is to persuade farmers that pesticide volumes can be reduced or even eliminated while maintaining crop productivity at an adequate level (Kristensen and Rasmussen, 2002). The value of DSSs could also be strengthened by integration of risk assessments and assessments of the environmental impact of pesticides (Zhan and Zhang, 2012). A major challenge for DSS developers is to keep stakeholders informed at all stages of development, from model design to the delivery of specific recommendations, while maintaining transparency and rigour in the development process (Walling and Vaneeckhaute, 2020).

Modelling inputs and outputs need extensive research and development by both DSS developers and farmers (Kipling et al., 2016; Colas et al., 2020). Decision-making needs a participatory approach and a transparent, open structure to ensure the right technologies are developed and data are properly validated at different spatial and temporal scales (Booltink et al., 2001). There is a need to translate complex modelling processes into simple, clear and usable information (Travlos et al., 2018). Colas et al. (2020) have suggested the introduction of:

- more familiar vocabulary for describing agricultural practices; and
- a broad range of weed impact indices so that farmers can select the most relevant control practice for them.

One suggestion is to use decision trees with numerical weed impact indices in workshops with farmers, in order to define the impacts of multiple cultural practices combinations (Colas et al., 2020). In one case, a repeated cross-validation of data was used to fit ten decision tree models in a DSS to guide seeding rate selection for hard red spring wheat cultivars (Stanley et al., 2020).

Given that current DSS recommendations might not be sufficiently robust to be widely adopted by stakeholders, a key issue is to improve reliability in decision-making (Li et al., 2020; Starke and Baber, 2020). Robustness can be improved through measures such as use of expert knowledge and utilization of historical data (Zhai et al., 2020a). One problem is that most algorithms created for DSS are mostly case-specific. Zhai et al. (2020b) have proposed use of case-based reasoning systems (CBR) using a case retrieval algorithm to retrieve the most relevant past cases when a new situation arises. The thorough evaluation of similarities and differences between past and new entries into the DSS leads to more efficient algorithms capable of providing more accurate decision support to users. Jones et al. (2017) highlighted the need for comprehensive input of data to assist model design and predict variables for a range of situations, from small holder farming systems to intensive production systems.

The enhancement and adoption of real-time decision support systems depend on developments in other technologies such as sensors, drones and machine learning algorithms (Chauhan et al., 2020). Jørgensen et al. (2007) have also emphasized the need to give economic incentives to farmers to persuade them to use new technologies instead of their usual agronomic practices. The growing utilization by a new generation of farmers of IT and mobile applications in general may facilitate the take-up of DSS, especially if properly supported by experts on using DSS to guide their site-specific weed management strategies (Ogunti et al., 2018). Key issues in adoption are summarized in Fig. 4.

7 Conclusions

Significant research has been carried out during the last two decades on designing effective DSS for weed management. Modelling crop yield losses due to weed competition and setting thresholds for intervention is an achievable goal for current DSSs. One of the most important advantages of DSS for weed management is to reduce herbicide use to the minimum required, reducing herbicide inputs significantly. Looking to the future, developing DSSs to reduce herbicide resistance and manage herbicide-resistant weeds is a complex challenge. Significant effort is needed to create DSSs capable

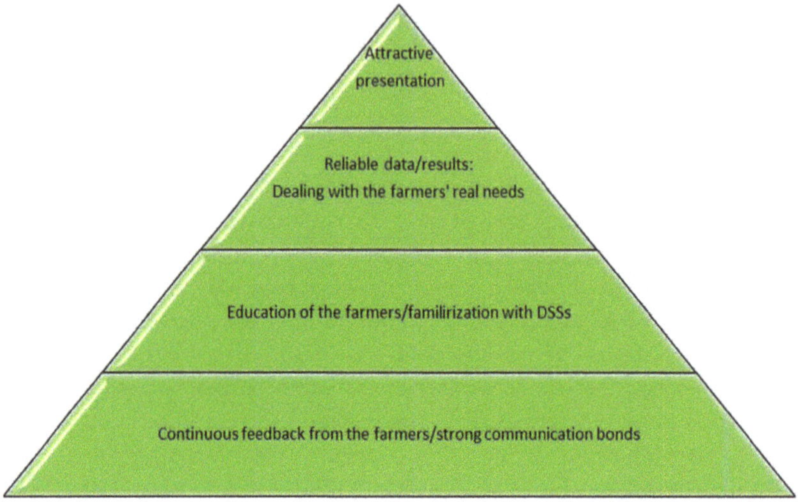

Figure 4 Crucial points for increasing the adoption of weed management DSSs from the farmers.

of giving farmers recommendations for long-term weed management based on the integration of herbicide use and cultural practices. There is a need for DSS to be able to provide feedback on management of a wide range of weed species. Another challenge is to get farmers familiar with using DSS as a useful tool for developing effective weed management strategies. Validated and reliable results, as well as continuous communication between experts and the farmers, are key factors which will help increase DSS adoption in weed management. Although there are many difficulties, rapid progress has been made. Further research is needed to create more effective and complete DSSs to provide all farmers with a customized tool to handle their individual needs for weed management.

8 Where to look for further information

Readers unfamiliar with the general concepts of DSS methodologies are encouraged to collect information from the following book by Armstrong (2020) before gathering further information on Decision Support Systems (DSS) developed for weed management purposes.

Armstrong, L. (2020). *Improving data management and decision support systems in agriculture* (1st edn.), Burleigh Dodds Science Publishing. https://doi.org/10.1201/9781003047872.

The above book was recently published by Burleigh Dodds Science Publishing and provides a standard introduction to understanding how DSS can be used in various aspects of crop production. It contains several chapters that cover data collection, data analysis, and how to share recommendations with farmers. The following link also serves as a valuable source of information on decision support tools developed in the European Union in a more popular version.

https://ec.europa.eu/eip/agriculture/en/digitising-agriculture/developing-digital-technologies/decision-support-tools.

For specialists in weed science, the following book by Chantre and González-Andújar (2020) is one of the first attempts to provide a novel guide to the state of the art of DSS in weed and future prospects, which will hopefully be of interest to students, academics and professionals in related fields.

Chantre, G. R., and González-Andújar, J. L. (Eds.). (2020). *Decision Support Systems for Weed Management.* Springer International Publishing. https://doi.org/10.1007/978-3-030-44402-0.

Several chapters in this book address mathematical modelling as an excellent framework for weed management, present integrated weed management tactics from both an operational and strategic planning perspective, and evaluate the environmental impacts of various weed management methods.

In addition, the fact sheet of the DSS Crop Protection Online (CPO) provided by the Faculty of Agricultural Sciences of the University of Aarhus explains in simple language the general concepts and website tools of this commercialized Danish DSS. Techniques are explained for determining weed control needs and quantifying optimal herbicide rates. The same applies to the Additive Dose Model (ADM), which is used to automatically optimize tank mixtures of herbicides consisting of more than two components in herbicide mixtures. The fact sheet can be found at the following link:

https://plantevaernonline.dlbr.dk/cp/documents/InfoFactSheet2.pdf.

As for other sources, the website IPM Decisions Network (https://www.ipmdecisions.net/) has valuable information on a platform that will soon go online to give farmers and advisors access to a wide range of existing DSS for their regional conditions. Such DSS are expected to play a central role in the development of effective and sustainable Integrated Pest Management (IPM) and Integrated Weed Management (IWM) systems in Europe in the coming years. Farmers, researchers and also readers from other scientific fields can visit the website and learn more about this project. It should be noted that this pan-European project is an important part of the Horizon 2020 research and innovation program of European Union.

Information on the development of web-based software applications that allow the retrieval of data and the creation/observation of different scenarios using environmental models and user-driven analysis in the areas of weed management, crop protection, and agriculture can also be found on the websites of the project's leading partners, in particular, the following:

- RSK ADAS Ltd (https://adas.co.uk/news/).
- Norwegian Institute of Bioeconomy Research (https://www.nibio.no/en).
- Aarhus University (https://www.au.dk/).
- Burgundy School of Business (https://www.bsb-education.com/?lang=en).
- Delphy B.V. (https://delphy.nl/en/).
- Agricultural University of Athens (https://www2.aua.gr/en).
- Swedish University of Agricultural Sciences (https://www.slu.se/en/).

9 References

Akhter, M. J. (2020). Biology and management of Vulpia myuros in arable farming. PhD thesis, Department of Agroecology, Aarhus University, Denmark, p. 143.

Beckie, H. J. and Reboud, X. (2009). Selecting for weed resistance: herbicide rotation and mixture. *Weed Technol.* 23(3), 363-370.

Benjamin, L. R., Milne, A. E., Parsons, D. J. and Lutman, P. J. W. (2010). A model to simulate yield losses in winter wheat caused by weeds, for use in a weed management decision support system. *Crop Prot.* 29(11), 1264-1273.

Berti, A. and Zanin, G. (1997). GESTINF: a decision model for post-emergence weed management in soybean (Glycine max (L.) Merr.). Crop Prot. 16(2), 109-116.

Binfield, J., Donnellan, T., Hanrahan, K. F., Hart, C. E. and Westhoff, P. C. (2004). CAP reform and the WTO: potential impacts on EU agriculture. *Selected Paper prepared for presentation at the American Agricultural Economics Association Annual Meeting*, Denver, CO (accessed July 1-4, 2004). Available at: https://mospace.umsystem.edu/xmlui/bitstream/handle/10355/3066/CAPReformWTO.pdf?sequence=1&isAllowed=y (accessed 8 November 2020).

Blackshaw, R. E., O'Donovan, J. T., Harker, K. N., Clayton, G. W. and Stougaard, R. N. (2006). Reduced herbicide doses in field crops: a review. *Weed Biol. Manag.* 6(1), 10-17.

Boiteux, L. S., Fonseca, M. E. N., Reis, A., Costa, A. F., Fontes, M. G. and González-Arcos, M. (2016). Wild radish (*Raphanus* spp.) and garden rocket (*Eruca sativa*) as new Brassicaceae hosts of Tomato chlorosis virus in South America. *Plant Dis.* 100(5), 1027-1027.

Bonfante, A., Monaco, E., Manna, P., De Mascellis, R., Basile, A., Buonanno, M., Cantilena, G., Esposito, A., Tedeschi, A., De Michele, C., Belfiore, O., Capatano, I., Ludeno, G., Salinas, K. and Brook, A. (2019). LCIS DSS–an irrigation supporting system for water use efficiency improvement in precision agriculture: a maize case study. *Agric. Syst.* 176, 102646.

Booltink, H. W. G., Van Alphen, B. J., Batchelor, W. D., Paz, J. O., Stoorvogel, J. J. and Vargas, R. (2001). Tools for optimizing management of spatially-variable fields. *Agric. Syst.* 70(2-3), 445-476.

Brown, R. B. and Noble, S. D. (2005). Site-specific weed management: sensing requirements–what do we need to see? *Weed Sci.* 53(2), 252-258.

Busi, R., Girotto, M. and Powles, S. B. (2016). Response to low-dose herbicide selection in self-pollinated *Avena fatua. Pest Manag. Sci.* 72(3), 603-608.

Chauhan, B. S. (2020). Grand challenges in weed management. *Front. Agron.* 1, 3.

Chauhan, B. S. and Johnson, D. E. (2010). The role of seed ecology in improving weed management strategies in the tropics. *Adv. Agron.* 105, 221-262.

Chauhan, S. S., Singh, U. P. and Jain, S. (2020). Applications of computer vision in plant pathology: a survey. *Arch. Comput. Methods Eng.* 27(2), 611-632.

Cisternas, I., Velásquez, I., Caro, A. and Rodríguez, A. (2020). Systematic literature review of implementations of precision agriculture. *Comput. Electron. Agric.* 176, 105626.

Colas, F., Cordeau, S., Granger, S., Jeuffroy, M.-H., Pointurier, O., Queyrel, W., Rodriguez, A., Villerd, J. and Colbach, N. (2020). Co-development of a decision support system for integrated weed management: contribution from future users. *Eur. J. Agron.* 114, 126010.

Colbach, N., Busset, H., Roger-Estrade, J. and Caneill, J. (2014). Predictive modelling of weed seed movement in response to superficial tillage tools. *Soil Till. Res.* 138, 1-8.

Colbach, N., Colas, F., Pointurier, O., Queyrel, W. and Villerd, J. (2017). A methodology for multi-objective cropping system design based on simulations: application to weed management. *Eur. J. Agron.* 87, 59-73.

Cousens, R. (1985). A simple model relating yield loss to weed density. *Ann. Appl. Biol.* 107(2), 239-252.

Cousens, R., Brain, P., O'Donovan, J. T. and O'Sullivan, P. A. (1987). The use of biologically realistic equations to describe the effects of weed density and relative time of emergence on crop yield. *Weed Sci.* 35(5), 720-725.

Donnelly, J. R., Freer, M., Salmon, L., Moore, A. D., Simpson, R. J., Dove, H. and Bolger, T. P. (2002). Evolution of the GRAZPLAN decision support tools and adoption by the grazing industry in temperate Australia. *Agric. Syst.* 74(1), 115-139.

Gerhards, R. and Christensen, S. (2003). Real-time weed detection, decision making and patch spraying in maize, sugarbeet, winter wheat and winter barley. *Weed Res.* 43(6), 385-392.

Gianessi, L. P. (2013). The increasing importance of herbicides in worldwide crop production. *Pest Manag. Sci.* 69(10), 1099-1105.

González-Andújar, J. L. (2020). AvenaNET and VallicoNET: DSS for *Avena sterilis* and *Lolium rigidum* control in Spanish dryland cereal crops. In: Chantre, G. and González-Andújar, J. (Eds), *Decision Support Systems for Weed Management*, pp. 299-309, Springer Nature Switzerland AG.

González-Andújar, J. L., Fernández-Quintanilla, C., Bastida, F., Calvo, R., González-Díaz, L., Izquierdo, J., Lezaún, J. A., Perea, F., Sánchez Del Arco, M. J. and Urbano, J. M. (2010). Field evaluation of a decision support system for herbicidal control of *Avena sterilis* ssp. *ludoviciana* in winter wheat. *Weed Res.* 50(1), 83-88.

González-Andújar, J. L., Fernández-Quintanilla, C., Bastida, F., Calvo, R., Izquierdo, J. and Lezaún, J. A. (2011). Assessment of a decision support system for chemical control of annual ryegrass (*Lolium rigidum*) in winter cereals. *Weed Res.* 51(3), 304-309.

Gonzalez-Andujar, J. L., Fernandez-Quintanilla, C., Izquierdo, J. and Urbano, J. M. (2006). SIMCE: an expert system for seedling weed identification in cereals. *Comput. Electron. Agric.* 54(2), 115-123.

Goss, E. M., Kendig, A. E., Adhikari, A., Lane, B., Kortessis, N., Holt, R. D., Clay, K., Harnon, P. F. and Luke Flory, S. L. (2020). Disease in invasive plant populations. *Annu. Rev. Phytopathol.* 58(1), 97-117.

Gutjahr, C. and Gerhards, R. (2010). Decision rules for site-specific weed management. In: Oerke, E. C., Gerhards, R., Menz, G. and Sikora, R. A. (Eds), *Precision Crop Protection–The Challenge and Use of Heterogeneity*, pp. 223-239, Springer Science+Business Media B.V.

Haage, I., Bastiaans, L. and Kempenaar, C. (2002). Exploring options for improved low dose application based on the MLHD-technology. In: Van Laar, H. H., Bastiaans, L., Baumann, D. T., Christensen, S., Hatcher, P. E., Kudsk, P., Grundy, A. C., Marshall, E. J. P., Streibig, J. C. and Tei, F. (Eds), *Proceedings of the 12th EWRS Symposium Wageningen*, pp. 200–201, Papendal, Arnhem, The Netherlands, 24–27 June 2002.

Heap, I. (2020). International survey of herbicide resistant weeds. Available at: http://www.weedscience.org/Pages/RecentCases.aspx (accessed 8 November 2020).

Ingram, J. and Gaskell, P. (2019). Searching for meaning: co-constructing ontologies with stakeholders for smarter search engines in agriculture. *NJAS - Wagen J. Life Sci.* 90–91, 100300.

Jensen, H. G., Jacobsen, L. B., Pedersen, S. M. and Tavella, E. (2012). Socioeconomic impact of widespread adoption of precision farming and controlled traffic systems in Denmark. *Precis. Agric.* 13(6), 661–677.

Johannsen, C. J., Carter, P. G., Morris, D. K., Ross, K. and Erickson, B. (2000). The real applications of remote sensing to agriculture. In: *Proceedings of the Second International Conference on Geospatial Information in Agriculture and Forestry* (vol. 1), pp. 1–5, Lake Buena Vista, FL, 10–12 January 2000.

Jones, J. W., Antle, J. M., Basso, B., Boote, K. J., Conant, R. T., Foster, I., Godfray, H. C. J., Herrero, M., Howitt, R. E., Janssen, S., Keating, B. A., Munoz-Carpena, R., Portera, C. H., Rosenzweig, C. and Wheelerk, T. R. (2017). Toward a new generation of agricultural system data, models, and knowledge products: state of agricultural systems science. *Agric. Syst.* 155, 269–288.

Jones, J. W., Hoogenboom, G., Porter, C. H., Boote, K. J., Batchelor, W. D., Hunt, L. A., Wilkens, P. W., Singh, U., Gijsman, A. J. and Ritchie, J. T. (2003). The DSSAT cropping system model. *Eur. J. Agron.* 18(3–4), 235–265.

Jørgensen, L. N., Noe, E., Langvad, A. M., Jensen, J. E., Ørum, J. E. and Rydahl, P. (2007). Decision support systems: barriers and farmers' need for support. *Bull. OEPP* 37(2), 374–377.

Jung, J., Maeda, M., Chang, A., Bhandari, M., Ashapure, A. and Landivar-Bowles, J. (2020). The potential of remote sensing and artificial intelligence as tools to improve the resilience of agriculture production systems. *Curr. Opin. Biotechnol.* 70, 15–22.

Kanatas, P. J., Travlos, I. S., Gazoulis, J., Antonopoulos, N., Tsekoura, A., Tataridas, A. and Zannopoulos, S. (2020a). The combined effects of false seedbed technique, post-emergence chemical control and cultivar on weed management and yield of barley in Greece. *Phytoparasitica* 48(1), 131–143.

Kanatas, P., Travlos, I. S., Gazoulis, I., Tataridas, A., Tsekoura, A. and Antonopoulos, N. (2020b). Benefits and limitations of Decision Support Systems (DSS) with a special emphasis on weeds. *Agronomy* 10(4), 548.

Kanatas, P., Travlos, I., Papastylianou, P., Gazoulis, I., Kakabouki, I. and Tsekoura, A. (2020c). Yield, quality and weed control in soybean crop as affected by several cultural and weed management practices. *Not. Bot. Horti Agrobo.* 48(1), 329–341.

Kempenaar, C., Groeneveld, R. M. W., Uffing, A. J. M., Van Der Weide, R. Y. and Wevers, J. (2002). New insights and developments in the MLHD-concept of weed control. In: Van Laar, H. H., Bastiaans, L., Baumann, D. T., Christensen, S., Hatcher, P. E., Kudsk, P., Grundy, A. C., Marshall, E. J. P., Streibig, J. C. and Tei, F. (Eds), *Proceedings of the 12th EWRS Symposium, Wageningen*, pp. 98–99, Papendal, Arnhem, The Netherlands, 24–27 June 2002.

Kernecker, M., Knierim, A., Wurbs, A., Kraus, T. and Borges, F. (2020). Experience versus expectation: farmers' perceptions of smart farming technologies for cropping systems across Europe. *Precis. Agric.* 21(1), 34-50.
Kerr, D. (2004). Factors influencing the development and adoption of knowledge-based decision support systems for small, owner-operated rural business. *Artif. Intell. Rev.* 22(2), 127-147.
Kipling, R. P., Bannink, A., Bellocchi, G., Dalgaard, T., Fox, N. J., Hutchings, N. J., Kjeldsen, C., Lactera, N., Sinabell, F., Topp, C. F. E., van Oijen, M., Virkajarvi, P. and Scollan, N. D. (2016). Modeling European ruminant production systems: facing the challenges of climate change. *Agric. Syst.* 147, 24-37.
Kristensen, K. and Rasmussen, I. A. (2002). The use of a Bayesian network in the design of a decision support system for growing malting barley without use of pesticides. *Comput. Electron. Agric.* 33(3), 197-217.
Kropff, M. J. and Spitters, C. J. T. (1991). A simple model of crop loss by weed competition from early observations on relative leaf area of weeds. *Weed Res.* 31(2), 97-105.
Kudsk, P. (2008). Optimising herbicide dose: a straightforward approach to reduce the risk of side effects of herbicides. *Environmentalist* 28(1), 49-55.
Kudsk, P. (2014). Reduced herbicide rates: present and future. *Julius Kühn Archiv.* 443, 37-44.
Li, H., Zhao, Y. and Zheng, F. (2020). The framework of an agricultural land-use decision support system based on ecological environmental constraints. *Sci. Total Environ.* 717, 137149.
Lotz, L. A. P., Kropff, M. J. and Groeneveld, R. M. W. (1990). Modelling weed competition and yield losses to study the effect of omission of herbicides in winter wheat. *NJAS - Wagen J. Life Sci.* 38(4), 711-718.
Macé, K., Morlon, P., Munier-Jolain, N.. and Quéré, L. (2007). Time scales as a fac-tor in decision-making by French farmers on weed management in annual crops. *Agric. Syst.* 93(1-3), 115-142.
Magarey, R. D., Travis, J. W., Russo, J. M., Seem, R. C. and Magarey, P. A. (2002). Decision support systems: quenching the thirst. *Plant Dis.* 86(1), 4-14.
Manalil, S., Busi, R., Renton, M. and Powles, S. B. (2011). Rapid evolution of herbicide resistance by low herbicide dosages. *Weed Sci.* 59(2), 210-217.
Medlin, C. R., Shaw, D. R., Gerard, P. D. and LaMastus, F. E. (2000). Using remote sensing to detect weed infestations in *Glycine max*. *Weed Sci.* 48(3), 393-398.
Meza-Palacios, R., Aguilar-Lasserre, A. A., Morales-Mendoza, L. F., Rico-Contreras, J. O., Sánchez-Medel, L. H. and Fernández-Lambert, G. (2020). Decision support system for NPK fertilization: a solution method for minimizing the impact on human health, climate change, ecosystem quality and resources. *J. Environ. Sci. Health A* 55(11), 1267-1282.
Miranda, M. Á., Barceló, C., Valdés, F., Feliu, J. F., Nestel, D., Papadopoulos, N., Sciarretta, A., Ruiz, M. and Alorda, B. (2019). Developing and implementation of Decision Support System (DSS) for the control of olive fruit fly, *Bactrocera Oleae*, in Mediterranean olive orchards. *Agronomy* 9(10), 620.
Möhring, N., Wuepper, D., Musa, T. and Finger, R. (2020). Why farmers deviate from recommended pesticide timing: the role of uncertainty and information. *Pest Manag. Sci.* 76(8), 2787-2798.
Molinari, F. A., Blanco, A. M., Vigna, M. R. and Chantre, G. R. (2020). A simulation model as the core for Integrated Weed Management Decision Support Systems: the case

of *Avena fatua*-Winter Wheat in the semiarid Pampean Region of Argentina. In: Chantre, G. and González-Andújar, J. (Eds), *Decision Support Systems for Weed Management*, pp. 311-333, Springer Nature Switzerland AG.

Montull, J. M., Sønderskov, M., Rydahl, P., Bøjer, O. M. and Taberner, A. (2014). Four years validation of decision support optimising herbicide dose in cereals under Spanish conditions. *Crop Prot.* 64, 110-114.

Montull, J. M., Taberner, A., Bøjer, O. and Rydahl, P. (2020). IPMwise: a decision support system for multispecies weed control in cereal crops. In: Chantre, G. and González-Andújar, J. (Eds), *Decision Support Systems for Weed Management*, pp. 279-298, Springer Nature Switzerland AG.

Munier-Jolain, N. M., Chavvel, B. and Gasquez, J. (2002). Long-term modelling of weed control strategies: analysis of threshold-based options for weed species with contrasted competitive abilities. *Weed Res.* 42(2), 107-122.

Neeser, C., Dille, J. A., Krishnan, G., Mortensen, D. A., Rawlinson, J. T., Martin, A. R. and Bills, L. B. (2004). WeedSOFT(R): a weed management decision support system. *Weed Sci.* 52(1), 115-122.

Netland, J., Torresen, K. S. and Rydahl, P. (2005). Evaluation of the weed module in the Danish decision support system "Crop Protection Online" adapted to Norwegian conditions. In: *Proceedings of the 13th EWRS Symposium*, pp. 19-23, Bari, Italy, 19-23 June 2005.

Neve, P. (2007). Challenges for herbicide resistance evolution and management: 50 years after Harper. *Weed Res.* 47(5), 365-369.

Neve, P. and Powles, S. (2005). High survival frequencies at low herbicide use rates in populations of *Lolium rigidum* result in rapid evolution of herbicide resistance. *Heredity* 95(6), 485-492.

Nimu, S. F., Paul, S. K., Salam, M. A. and Sarkar, S. K. (2020). Influence of weed free periods on the growth, yield and quality of soybean (*Glycine max* L.). *Fundam. Appl. Agric.* 5(1), 99-107.

Oerke, E.-C. (2006). Crop losses to pests. *J. Agric. Sci.* 144(1), 31-43.

Oerke, E.-C. and Dehne, H.-W. (2004). Safeguarding production-losses in major crops and the role of crop protection. *Crop Prot.* 23(4), 275-285.

Ogunti, E. O., Akingbade, F. K., Segun, A. and Oladimeji, O. (2018). Decision support system using mobile applications in the provision of day to day information about farm status to improve crop yield. *Period. Eng. Nat. Sci.* 6(2), 89-99.

Pahmeyer, C., Kuhn, T. and Britz, W. (2020). 'Fruchtfolge': a crop rotation decision support system for optimizing cropping choices with big data and spatially explicit modeling. No. 305287, Discussion Papers, University of Bonn, Institute for Food and Resource Economics. Available at: https://ageconsearch.umn.edu/record/305287/ (accessed 9 November 2020).

Pannell, D. J., Stewart, V., Bennett, A., Monjardino, M., Schmidt, C. and Powles, S. B. (2004), RIM: a bioeconomic model for integrated weed management of Lolium rigidum in Western Australia. *Agric. Syst.* 79(3), 305-325.

Parker, C. G., Campion, S. and Kure, H. (1997). Improving the uptake of decision support systems in agriculture. In: Kure, H., Thysen, I. and Kristensen, A. R. (Eds), *Proceedings of the First European Conference for Information Technology in Agriculture*, pp. 129-134, Denmark, Copenhagen.

Parsons, D. J., Benjamin, L. R., Clarke, J., Ginsburg, D., Mayes, A., Milne, A. E. and Wilkinson, D. J. (2009). Weed Manager–a model-based decision support system for weed management in arable crops. Comput Electron. Agric. 65(2), 155-167.

Pérez, W., Arias, R., Taipe, A., Ortiz, O., Forbes, G. A., Andrade-Piedra, J. and Kromann, P. (2020). A simple, hand-held decision support designed tool to help resource-poor farmers improve potato late blight management. *Crop Prot.* 134, 105186.

Peterson, M. A., Collavo, A., Ovejero, R., Shivrain, V. and Walsh, M. J. (2018). The challenge of herbicide resistance around the world: a current summary. *Pest Manag. Sci.* 74(10), 2246-2259.

Rasmussen, I. A. (2004). The effect of sowing date, stale seedbed, row width and mechanical weed control on weeds and yields of organic winter wheat. *Weed Res.* 44(1), 12-20.

Renton, M., Busi, R., Neve, P., Thornby, D. and Vila-Aiub, M. (2014). Herbicide resistance modelling: past, present and future. *Pest Manag. Sci.* 70(9), 1394-1404.

Rossi, V., Caffi, T. and Salinari, F. (2012). Helping farmers face the increasing complexity of decision-making for crop protection. *Phytopathol. Mediterr.* 51(3), 457-479.

Rydahl, P. (2003). A web-based decision support system for integrated management of weeds in cereals and sugarbeet. *Bull. OPPE* 33(3), 455-460.

Rydahl, P., Bøjer, O. M., Jørgensen, R. N., Dyrmann, M., Andersen, P., Jensen, N. P. and Sorensen, M. (2018). Spatial variability of optimized herbicide mixtures and dosages. In: *Proceedings of the 14th International Conference on Precision Agriculture*, Montreal, Quebec, Canada, 24-27 June 2018.

Shaw, D. R. (2005). Translation of remote sensing data into weed management decisions. *Weed Sci.* 53(2), 264-273.

Singh, M., Kaur, S. and Chauhan, B. S. (2020). Weed interference models. In: Chantre, G. and González-Andújar, J. (Eds), *Decision Support Systems for Weed Management*, pp. 117-142, Springer Nature Switzerland AG.

Small, I. M., Joseph, L. and Fry, W. E. (2015). Development and implementation of the BlightPro decision support system for potato and tomato late blight management. Comput. Electron. Agric. 115, 57-65.

Sønderskov, M., Fritzsche, R., de Mol, F., Gerowitt, B., Goltermann, S., Kierzek, R., Krawczyk, R., Bøjer, O. M. and Rydahl, P. (2015). DSSHerbicide: weed control in winter wheat with a decision support system in three South Baltic regions - field experimental results. *Crop Prot.* 76, 15-23.

Sønderskov, M., Kudsk, P., Mathiassen, S. K., Bøjer, O. M. and Rydahl, P. (2014). Decision support system for optimized herbicide dose in spring barley. *Weed Technol.* 28(1), 19-27.

Stanley, J. D., Mehring, G. H., Wiersma, J. J. and Ransom, J. K. (2020). A decision support system to guide grower selection of optimal seeding rates of wheat cultivars in diverse environments. *Front. Plant Sci.* 11, 779.

Starke, S. D. and Baber, C. (2020). The effect of known decision support reliability on outcome quality and visual information foraging in joint decision making. *Appl. Ergon.* 86, 103102.

Torra, J. and Monjardino, M. (2020). Ryegrass integrated management (RIM)-based decision support system. In: Chantre, G. and González-Andújar, J. (Eds), *Decision Support Systems for Weed Management*, pp. 249-278, Springer Nature Switzerland AG.

Travlos, I. S., Cheimona, N., Roussis, I. and Bilalis, D. J. (2018). Weed-species abundance and diversity indices in relation to tillage systems and fertilization. *Front. Environ. Sci.* 6, 11.

Travlos, I., Gazoulis, I., Kanatas, P., Tsekoura, A., Zannopoulos, S. and Papastylianou, P. (2020). Key factors affecting weed seeds' germination, weed emergence, and their possible role for the efficacy of false seedbed technique as weed management practice. *Front. Agron.* 2, 1.

Walling, E. and Vaneeckhaute, C. (2020). Developing successful environmental decision support systems: challenges and best practices. *J. Environ. Manag.* 264, 110513.

Wang, P., Peteinatos, G., Li, H., Brändle, F., Pfündel, E., Drobny, H. G. and Gerhards, R. (2018). Rapid monitoring of herbicide-resistant *Alopecurus myosuroides* Huds. using chlorophyll fluorescence imaging technology. *J. Plant Dis. Prot.* 125(2), 187–195.

Wang, P., Peteinatos, G. G. and Gerhards, R. (2017). In field identification of herbicide resistant *Apera spica-venti* using chlorophyll fluorescence. *Adv. Anim. Biosci.* 8(2), 283–287.

Welch, S. M., Jones, J. W., Brennan, M. W., Reeder, G. and Jacobson, B. M. (2002). PCYield: model-based decision support for soybean production. *Agric. Syst.* 74(1), 79–98.

Yost, M. A., Kitchen, N. R., Sudduth, K. A., Sadler, E. J., Drummond, S. T. and Volkmann, M. R. (2017). Long-term impact of a precision agriculture system on grain crop production. *Precis. Agric.* 18(5), 823–842.

Zhai, Z., Martínez, J. F., Beltran, V. and Martínez, N. L. (2020a). Decision support systems for agriculture 4.0: survey and challenges. *Comput. Electron. Agric.* 170, 105256.

Zhai, Z., Ortega, J. M., Martínez, N. L. and Xu, H. (2020b). An efficient case retrieval algorithm for agricultural case-based reasoning systems, with consideration of case base maintenance. *Agriculture* 10(9), 387.

Zhan, Y. and Zhang, M. (2012). PURE: a web-based decision support system to evaluate pesticide environmental risk for sustainable pest management practices in California. *Ecotoxicol. Environ. Saf.* 82, 104–113.

Zhang, J., Jäck, O., Menegat, A., Li, G. and Wang, X. (2017). Chlorophyll fluorescence measurement: a new method to test the effect of two adjuvants on the efficacy of topramezone on weeds. In: *IFIP Advances in Information and Communication Technology. Proceedings of the International Conference on Computer and Computing Technologies in Agriculture*, pp. 206–216, Springer.

Zimdahl, R. L. (2018). *Fundamentals of Weed Science* (5th edn.), Academic Press, 125 London Wall, London EC2Y 5AS, UK.

Chapter 8

Advanced detection technologies for weed scouting

C. Fernandez-Quintanilla, J. Dorado and J. M. Peña, Instituto de Ciencias Agrarias (CSIC), Spain; and D. Andújar, Centro de Automatica y Robótica (CSIC), Spain

1 Introduction

Crop scouting is an essential part of integrated pest management (IPM). Through a regular and systematic field-sampling program, scouting aims to provide accurate and timely field-specific information on pest pressure. This information is essential for appropriate selection and application of pest management programs (Fishel et al., 2009). Although scouting is most often used for monitoring insect damages and diseases, it is also used for assessing weed infestations.

Traditionally, weed scouting has been conducted by growers or consultants who regularly visit the fields and record problems caused by weeds, making use of their expertise and a series of diagnostic procedures. New tools (e.g. specialized handheld instruments with global positioning system [GPS] receiver) are now available to increase the effectiveness of this 'manual' scouting task, enabling frequent and accurate gathering and geotagging of weed data. In recent years, a rich arsenal of technologies has been developed for this purpose: differential GPS (DGPS) and other global navigation satellite systems (GNSS); geographic information systems (GIS); sensors; information and communications technology (ICT) networks; unmanned aerial vehicles (UAV);

http://dx.doi.org/10.19103/AS.2021.0098.09

very-high-resolution (VHR) satellite platforms; and artificial intelligence (AI). Vast amounts of data can be collected with these tools. These datasets could be analyzed and interpreted later, helping farmers to make more informed and appropriate decisions (van Evert et al., 2017). Next, these decisions could be implemented through advanced machinery and robots equipped with smart actuators (Aravind et al., 2017). Finally, farmers could get feedback on the impact of their actions.

All these technological advances should be integrated within a given production system. Nowadays, diverse approaches are being proposed as alternatives for conventional agriculture: biodynamic agriculture, organic farming, permaculture, low-input agriculture, conservation agriculture and precision farming. All these systems try to increase sustainability, promoting a diverse range of alternative practices designed to reduce dependence on

Table 1 Potential uses of weed scouting for integrated weed management (IWM) (SSWM: site-specific weed management)

Type of IWM practice	Examples of application	Technology	References
	EFFICIENCY		
Management and application of herbicides	SSWM in wheat	Late season scouting from satellite and map-based zone spraying	(Castillejo-González et al., 2019)
	SSWM in maize	Early season on-ground scouting and map-based spot spraying	(Andújar et al., 2011b)
	SUBSTITUTION		
Alternative management practices	Breeding competitive wheat cultivars	UAV phenotyping for germplasm screening	(Ostos-Garrido et al., 2019)
	Improved tillage in wheat	Early season on-ground scouting and map-based harrowing	(Rueda-Ayala et al., 2013)
	Improved rotations	Late season pluri-annual landscape satellite scouting and identification of interactions rotation-weeds	
	REDESIGN		
Agroecological designs	On-farm research	SSWM experiments in farmer's fields and on-ground scouting	(Luschei et al., 2001)
	Weed competition studies	Experiments in farmer's fields and UAV scouting	(Rasmussen and Nielsen, 2020)
	Presence and spread of herbicide resistance	Ground- and UAV-based hyperspectral imagery	(Scherrer et al., 2019)

fertilizers and pesticides, protect the soil and biodiversity, cut production costs and diminish adverse environmental consequences (Altieri et al., 2017).

The transition from conventional to a more sustainable agriculture has been constrained by inadequate frameworks for assessing the best strategies to pursue (Hill, 1998; Meynard et al., 2012). Four decades ago, Hill (1985) proposed the efficiency, substitution, redesign (ESR) conceptual framework for assessing strategies to support this transition. In this chapter, we will use Hill's ESR paradigm to assess how new weed scouting technologies can contribute to integrated weed management (IWM) (Table 1). This contribution can be through more targeted and optimized herbicide treatments (higher *efficiency*), the replacement of high impact herbicide interventions by more benign ones (*substitution*) and the design and adoption of research and management approaches to meet sustainable agriculture goals (*redesign*) (Hill, 1998; Padel et al., 2020; Pretty, 2018).

2 Efficiency: optimizing herbicide treatments

Currently, the transition from conventional agriculture toward a more sustainable strategy for weed management is mainly focused on its first stage: improving herbicide efficiency. This goal can be achieved by three major pathways: applying herbicides only when needed (timely application at specific weed thresholds), applying the optimum dose (variable-rate spraying) and spraying only in weed-infested areas (site-specific weed management - SSWM). An essential component of SSWM is weed detection, which can be addressed with two major approaches: 'map-based' and 'real-time' (Christensen et al., 2009; Fernández-Quintanilla et al., 2018; López-Granados, 2011). The former proposes to map the weeds in one operation and spray herbicides in a posterior field operation, while the latter proposes to conduct weed detection and spraying in a single operation. In map-based weed management (Fig. 1), spatial information generated in advance is used to select the most suitable herbicides to be sprayed, locate the zones to be treated and more accurately fit the volumes required for spraying. Although weed maps can be generated with either discrete or continuous sampling techniques, discrete sampling is costly and labor-intensive. Continuous sampling is more appropriate for SSWM, thanks to the current availability of the above-mentioned digital technologies.

2.1 On-ground weed detection

Weed scouting can be done at different crop stages and from diverse ground vehicles. All-terrain vehicles (ATV) mounted with a DGPS receiver and a computer were used to create continuous *Avena fatua* (wild oat) seedling maps

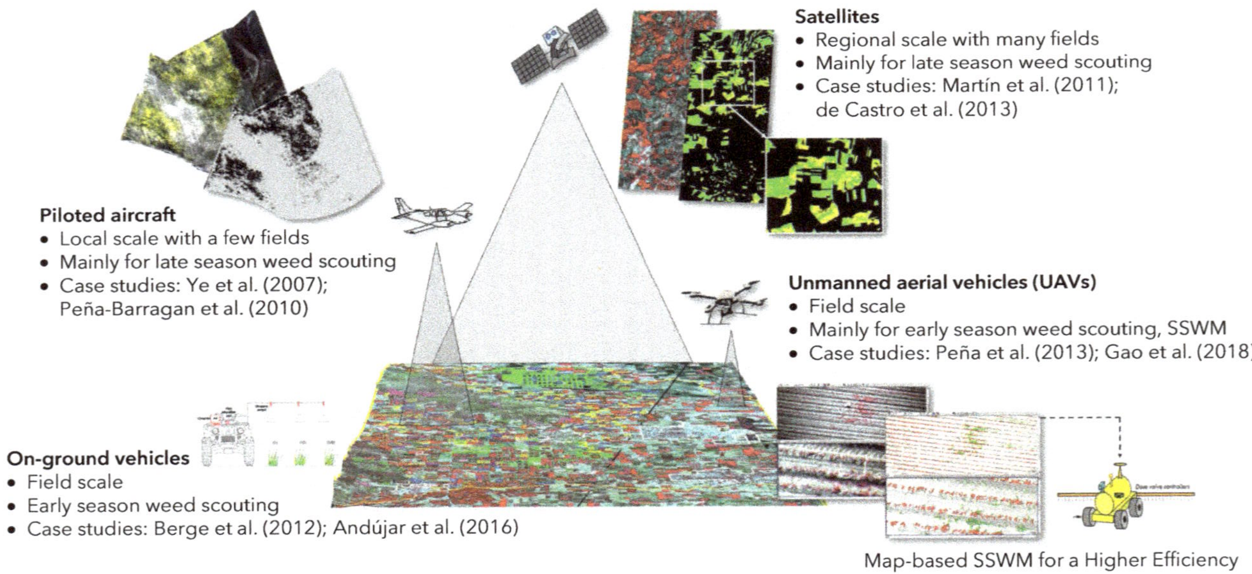

Figure 1 Different approaches for map-based weed scouting to generate continuous spatial information, from remote sensing (satellites, aircrafts, UAVs) to on-ground sensing.

in small grain crops in Montana (USA), in which visual data were recorded by expert crop consultants (Van Wychen et al., 2002). The accuracy of these maps, determined from georeferenced quadrats of wild oat densities, ranged from 48% to 87%. Although this system generated excellent maps, it required an additional field operation. A more practical approach would be to map wild oats during the harvest operation. The driver of the harvester can switch a logger on and off manually when the header passes through a patch of wild oats (Colliver et al., 1996). The accuracy of these panicle maps ranged from 66% to 91%. Economic assessment of various management options indicated that a site-specific herbicide application to areas mapped as wild oat presence generated higher net returns than a herbicide application over the entire field in most sites. Ruiz et al. (2006) surveyed 31 *Avena sterilis* (winter wild oat) infested barley fields at harvest time using a visual approach. These authors reported that site-specific adjusted-dose herbicide application offered 61–74% potential herbicide savings. Site-specific treatments were economically advantageous in sites with high barley yields and high returns. However, in areas with low yields and low infestation levels, the most profitable strategy was, generally, no herbicide application. Andújar et al. (2011b) used the same visual to assess *Sorghum halepense* (johnsongras) infestations in 37 maize fields at harvest time. Simulation results showed that SSWM was the most profitable strategy when johnsongras-infested areas ranged between 7% and 19% (22% of the surveyed fields). In fields with less than 7% infestation, yield losses were slight and the most profitable strategy was using no herbicide. Actual herbicide savings obtained with the various strategies depended on the proportion of the field infested.

Visual assessment of weed infestations has two major drawbacks: (1) high cost associated with manual labor and (2) low accuracy and consistency of human perception. Automatic assessment using cameras or sensors allows overcoming these two limitations. Gerhards and Oebel (2006) assessed weed infestations in various crops using digital bi-spectral cameras. Images were analyzed with a special software to identify crops, grass weeds and up to three groups of broad-leaved weeds. Herbicide application maps were created based on interpolated maps of weed distribution, and control thresholds for the various classes of weed species were estimated. A decision algorithm for patch spraying (DAPS) (Christensen et al., 2003) was used to decide on herbicides and doses. Herbicide use with this map-based approach was reduced in winter cereals by 6% to 81% for herbicides against broad-leaved weeds and 20% to 79% for grass weed herbicides.

Berge et al. (2008) used a near-ground red-green-blue (RGB) sensor and an object-oriented algorithm to estimate the total density and cover of broad-leaved weed seedlings in cereal fields. The ability of this algorithm to predict 'spray'/'no spray' decisions according to a previously suggested spray decision model was tested with images from two wheat fields. This system provided correct decisions for 65–85% of the test images. Three additional map-based trials showed that estimations obtained from the system were generally adequate (Berge et al., 2012).

Andújar et al. (2011a) used ultrasonic sensors mounted on the front of a tractor, pointing straight downwards toward the ground in the inter-row area, in combination with a DGPS receiver and a computer, to automatically detect and discriminate various grass and broad-leaved weeds in a maize field. The sensor readings permitted the discrimination of pure stands of grasses (up to 81% success) and pure stands of broad-leaved weeds (up to 99% success). Most of the errors were related to low-density areas (with less than 15% of weed coverage) defined as the threshold. Using LIDAR sensors, instead of ultrasonic sensors, these researchers were able to create 3D vegetation models that allowed the identification of positions of maize rows, vegetation-free areas and weed-infested areas. Since weed height was generally lower than maize height, this method resulted in good discrimination of both vegetation types. The major weed present, *S. halepense*, was identified in almost 80% of the cases (Andújar et al., 2013). This procedure was later improved by using RGB-D cameras that combined RGB images with depth perception. This system allowed a clear discrimination between maize plants, grasses and broad-leaved weeds, even when they were combined. Also, this method allowed quantifying the number of weed classes in mixed patches (Andújar et al., 2016).

Weed scouting does not necessarily require a specific operation. Barroso et al. (2017) used the optical sensor installed in the combine harvester for the measurement of grain protein concentration to detect the presence of green plant matter in flowing grains. Based on this information, they were able to map *Kochia scoparia* (kochia), *Salsola tragus* (Russian thistle) and *Lactuca serriola* (prickly lettuce) infesting dryland wheat. Maps of the chlorophyll signal showed a 78% agreement with visual evaluations of the three weed species conducted by experts prior to harvest and at harvest time.

2.2 Remotely sensed weed detection

Remote sensing is being used for weed scouting at appropriate times to locate weed-infested areas and to get information on the occurrence, coverage and growth stage of major weeds present in the monitored area. The images from satellite platforms operating in low earth orbits (400–700 km altitude) are a good option when the aim is to assess medium or large weed patches in large areas including many individual fields, while piloted aircrafts flying at medium altitudes (1–2 km altitude) generally have been used at the scale of a few farms (Singh et al., 2020). As mentioned earlier, the images obtained with these platforms have the advantage of covering large extensions. However, their moderate spatial resolution (0.5–3 m of ground sample distance) only enables the detection of weed patches of several meters in size. This technology basically relies on detecting the spectral difference (i.e. color) between the targeted species due to certain phenology changes observed in specific time windows (Müllerová et al., 2017). Under these circumstances, it is possible to monitor invasive plants and arable weeds at late-season (Huang and Asner, 2009; López-Granados, 2011). For example, Castillejo-González et al. (2014) and de Castro et al. (2012a) mapped late-season winter wild oat and cruciferous weed patches, respectively, in numerous wheat fields at the regional scale with multispectral QuickBird satellite scenes, reporting classification accuracies in the range of 89–91% in all the fields. Similarly, Martín et al. (2011) studied winter wild oat in barley crops and established a density of ten plants/m^2 as a minimum threshold to reach high accuracies (86–94%) with QuickBird images. Using piloted aircraft, other investigators also reported good results in mapping weed patches in citrus (Ye et al., 2007), sunflower (Peña-Barragán et al., 2010), wheat, broad bean and pea (de Castro et al., 2012b; López-Granados et al., 2006) and invasive plant species in complex landscapes (Dorigo et al., 2012; Müllerová et al., 2005).

Since these weed maps were generated at the final growth stages of crops and weeds, the information obtained could not be used to plan site-specific herbicide treatments in the current season. This leads to exploiting late-season maps for implementing SSWM measures in the following years, which is possible if the weed patches are persistent over time, as in the case of winter wild oat (Barroso et al., 2004; Castillejo-González et al., 2019). Despite the constraint to apply in-season weed treatments with satellite-based maps, de Castro et al. (2012a) concluded that the average reduction in herbicide costs would be approximately 96% if herbicides were only applied to cruciferous weed-infested areas. Similarly, a simulation study conducted with data from consecutive periods of 3 years (wheat in the first and third years and sunflower in the second year) showed that although the economic profit achieved with SSWM treatments was modest, any of the site-specific treatments tested were preferred to herbicide broadcast over the entire field to reduce herbicide and environmental pollution (Castillejo-González et al., 2019).

Both satellite- and aircraft-based procedures have a series of limitations, such as moderate spatial resolutions, coverage problems due to cloudy days and restricted flying schedules or revisit times, interference of soil background (Christensen et al., 2009; Fernández-Quintanilla et al., 2018; López-Granados, 2011); however, the potential reduction in herbicide consumption at the scale of many fields, could more than justify the implementation of this technology whenever possible despite their mentioned limitations.

Currently, innovative UAV platforms have allowed overcoming some of these limitations. Their potential lies in their flexibility to work on demand according to the weeding goal (timely temporal window) and their capacity to collect aerial images with the proper spectral wavelengths and the ultra-high spatial resolution (in the range of a few centimeters or even millimeters) needed for discriminating small weeds at early crop stages (Peña et al., 2015). However, strong expertise is required to optimize the UAV flight mission according to the size of the smaller object to be discriminated (i.e. weed seedlings in early season or weed patches in late season) and in agreement with the covered area, flight length/altitude and sensor specifications (Torres-Sánchez et al., 2013).

Numerous studies have proved the reliability of this tool to perform a detailed weed scouting in maize (López-Granados et al., 2016a; Peña et al., 2013), sunflower (de Castro et al., 2018; Pérez-Ortiz et al., 2015, 2016), cotton (de Castro et al., 2018), wheat and barley (Jurado-Expósito et al., 2019; Rasmussen et al., 2019), oats (Gašparović et al., 2020), rice (Barrero and Perdomo, 2018; Huang et al., 2018a,b), vineyards (de Castro et al., 2020; Jiménez-Brenes et al., 2019), sugarcane (Girolamo-Neto et al., 2019), sugar beet (Mink et al., 2018) and many other cropping scenarios. Compared to late-season weed maps created from satellites or airplanes, the great benefit of UAV technology is the generation of prescription maps at the moment recommended for weed control (Peña et al., 2013). These maps can provide site-specific information of weed occurrence (e.g. position and coverage), which would enable planning in-season weed treatments prior to the operation and help to choose herbicides and dose, leading to increased application efficiency in terms of relevant herbicide savings, optimizing machinery routes and environmental benefits (Castaldi et al., 2017; Huang et al., 2018a; López-Granados et al., 2016b).

The retrieval of geo-referenced and detailed weed information is the main bottleneck for the implementation of the aforementioned sensors and platforms in real cropping scenarios, although a wide variety of image-processing procedures is available to perform this task. In the past, most weed detection studies were focused on exploring the specific spectral signature of the crop and weed species (Gómez-Casero et al., 2010; Gray et al., 2009), using various vegetation indices (e.g. normalized difference vegetation index [NDVI], as the basic metric for weed discrimination (Barrero and Perdomo, 2018; Meyer and Neto, 2008). In this approach, efficiency can be improved by applying automatic threshold methods, for example, the Otsu method (Otsu, 1979), which allow the determination of vegetation cover fraction (VCF) as a preliminary phase to mask bare soil from vegetation (Montalvo et al., 2012; Torres-Sánchez et al., 2015). Next, spectral similarities between crop and weed species are resolved with object-based image analysis (OBIA) methods (Blaschke et al., 2015), a new paradigm that combines pixel data with additional contextual features for successful weed discrimination (Gao et al., 2018; Peña et al., 2013).

In recent years, machine learning (ML) and deep learning (DL) have emerged to create new opportunities for analyzing image data (Behmann et al., 2015; Rehman et al., 2019; Su, 2020). In this regard, projects such as Deepweed (Olsen et al., 2019) have provided a multiclass image dataset of weed species that intends to be the baseline for the creation of robust classification methods. The use of these advanced algorithms has also overcome the classical challenge of weed classification within the crop rows (Pérez-Ortiz et al., 2015, 2016), reaching an average level of correct classifications higher than 85% in sunflower and cotton crops (de Castro et al., 2018). However, there are still many issues that must be improved regarding the use of ML and DL techniques for robust weed identification in diverse agricultural scenarios with changing crop and weed phenology. These techniques should face real-field conditions with uncontrolled illumination, occlusion and leave overlapping or weed mixtures at differential growth stages.

3 Substitution: replacement of herbicide treatments for other control tactics

The overdependence on chemical weed control and the various problems created by this overuse (herbicide resistance, water pollution) have led to a movement of substituting herbicides with less environmentally impacting interventions, such as competitive crop varieties, mechanical tools, cover crops, crop rotations and improved tillage operations (Liebman and Davis, 2000; Liebman et al., 2016; Liebman and Dyck, 1993).

A desirable alternative or supplement to herbicides would be to breed and select for weed-suppressive crop genotypes. Cultivars and accessions of various crops have been screened for weed-suppressive ability in numerous field trials. Competitive ability is generally conferred by a combination of morphological traits that allow the crop to access more limited resources than neighboring weeds. Understanding what traits are most strongly associated with competitive advantages enables breeders to indirectly select for weed suppressive lines in weed-free environments, allowing them to screen their entire breeding nurseries (Worthington and Reberg-Horton, 2013). These traits could be monitored using proximal or remote sensing and imaging techniques designed for field phenotyping (Araus and Cairns, 2014; Ostos-Garrido et al., 2019).

Mechanical weed control provides another good alternative to reduce weed pressure. The efficacy of mechanical weeders can be optimized if site-specific conditions are taken into account. Rueda-Ayala et al. (2013) used various sensors to estimate the field variability of crop leaf cover, weed density and soil density. Crop leaf cover and weed density were assessed using bispectral cameras through differential image analysis. These researchers developed and tested an algorithm to automatically adjust the harrowing intensity by varying the tine angle and number of passes. The system developed was able to take into account the variability of crop-weed-soil conditions present in the field. Its application for automatic harrowing controlled satisfactory high weed densities and did not reduce crop yield under low weed levels. Although various types of smart mechanical weeding machines have been developed and commercialized in recent years (Fennimore et al., 2016), they are generally based on real-time sensors, not on weed maps.

Weed population density and biomass production may be markedly reduced using crop rotation (temporal diversification) (Liebman and Davis, 2000; Liebman and Dyck, 1993). However, assessing experimentally the impact of various crop rotations is a difficult challenge. Taking into consideration the high complexity of analyzing the effects of different crop sequences, initial conditions and effects of the sites and the climatic years, this task may require conducting long-term experiments in a variety of sites. The use of remote sensing at the landscape scale may provide an alternative methodology to assess these impacts.

The mapping of crops using remote sensor data has shown good potential for characterizing the extent, distribution and condition of croplands (Thenkabail et al., 2009). Lunetta et al. (2010) have used the moderate resolution imaging spectrometer (MODIS), NDVI, to characterize the crop area distributions and changes in crop rotations during 3 consecutive years. A similar approach, using higher resolution aerial platforms and sensors, could be used to monitor the population of a given weed species over an entire landscape with individual fields and with different crop rotations.

No-till, high-residue crop management practices are key components of conservation agriculture and are crucial to promote sustainable cropping systems (Hobbs et al., 2008). Hively et al. (2018) used WorldView-3 satellite images to assess crop residue and tillage intensity at a regional level. Maps were produced depicting the percentage of crop residue on fields with minimal green vegetation cover throughout the agricultural landscape. This type of map may be useful to farmers for monitoring temporal trends and spatial variability in conservation tillage implementation. In addition, we hypothesize that systematic scouting of numerous nearby fields under conventional and no-tillage regimes may provide useful information on the interactions between tillage, residue cover and weed populations.

4 Redesign: contributing to agroecosystems sustainability

Although various academic studies have focused on redesigning farming systems (Hill, 1985, 1998; Meynard et al., 2012; Padel et al., 2020; Pretty, 2018), achieving this goal is not simple. Several complex scientific, agronomic, economic, social and political factors should be taken into account.

In order to reach sustainable agriculture goals, it is necessary to design and implement research programs that are more collaborative, provide a deeper agroecological knowledge, a holistic understanding of cropping systems and a better assessment of the impacts of control actions (Fig. 2). Although the use of weed scouting technologies in all these tasks has been very limited, in the following sections we will suggest that they may offer opportunities in this regard.

4.1 Collaborative research

The linear model of innovation (science to farmer) is not valid for the redesign of agricultural systems. In recent years, this model has progressively been replaced by a participatory or 'side-by-side' approach, in which innovation is 'co-produced,' thanks to the interaction between farmers, researchers and various intermediate actors (advisors, input providers, experts) (European Commission, 2016). In this approach, farmers can support their decision-making by knowledge exchange among themselves so that learning emerges from a shared interest in a problem or challenge through experience and mutual support (Moschitz et al., 2014). In this case, the cost of collaboration can be reduced, thus facilitating cooperation and giving farmers the confidence to invest in collective activities (others will also do so).

There are some good examples of the use of this approach. As early as 1999, Norton et al. (1999) described a case study of participatory IPM research in the United Kingdom. Although this study emphasized on-farm research, there was an extrapolation domain beyond the single farm and in some cases, beyond the local region or country. Using this approach required farmers to get involved in experimentation.

Digital technologies can be powerful tools for engaging individual farmers in on-farm research and in other ways that improve information accessibility. Farmers using these technologies can digitize their fields and receive satellite remote-sensing imagery, weather information, and localized agronomic models that can provide crop-specific local information relevant for decision-making (Huang and Brown, 2018). Luschei et al. (2001) provide a good example of on-farm experiments that utilize producer equipment to test SSWM practices. By using this approach, it was possible to consider all of the uncertainty faced in real production situations. Ultimately, advances in digital technology will require proof of concept, approval and training. In this sense, social aspects

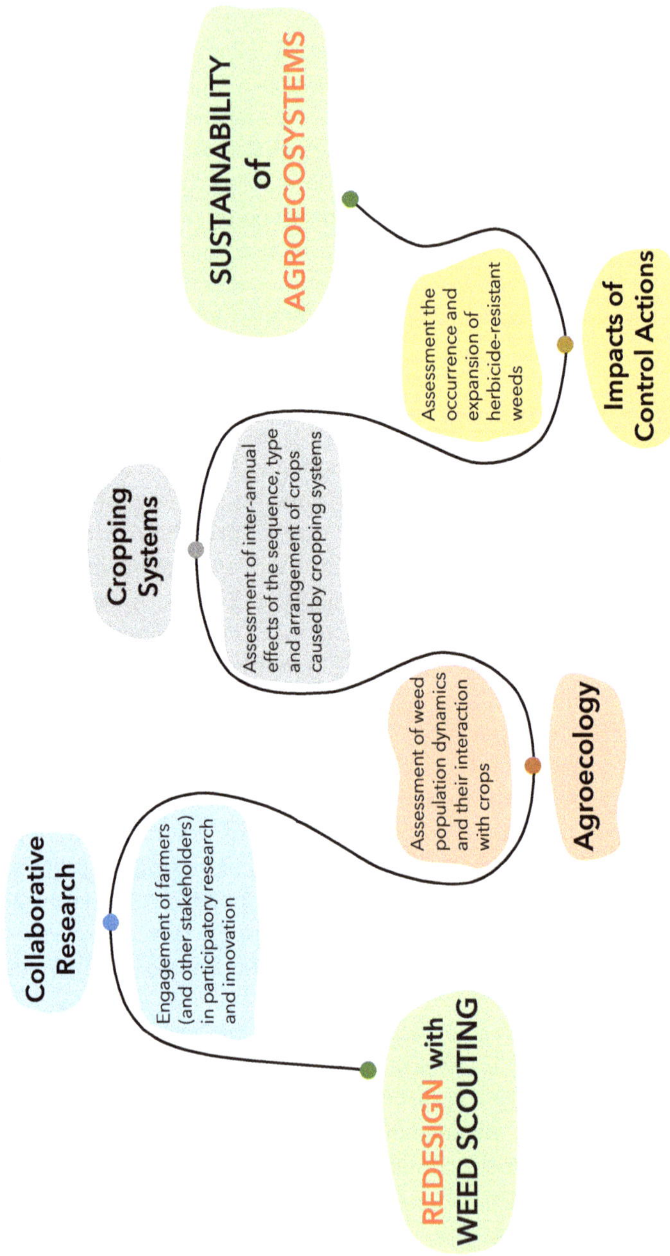

Figure 2 Weed scouting applications to farming redesign aimed at achieving agroecosystems sustainability.

play an essential role in understanding and facilitating schemes on the use of this new diversity of tools so that it would be possible to combine both the human and the digital sides (Leveau et al., 2019).

4.2 Agroecology

Agroecology is an approach emphasizing ecological principles and practices in the design and management of agroecosystems. It integrates the long-term protection of natural resources as an element of food, fuel and fiber production (Lampkin et al., 2015). The redesign of weed management in the context of agroecology should be based on a thorough understanding of weed populations and their interaction with crops.

Knowing the population dynamics of weeds is crucial for the design and implementation of integrated weed management (IWM) programs. Methods used to assess weed dynamics are time-consuming, and the population sampled is usually restricted to one part of a single field. Aerial images can provide estimates of areas covered by certain weed populations and have been used as a tool for quantitative assessment of plant infestations (McCormick, 1999; Stow et al., 2000) and their dispersal dynamics (Mast et al., 1997). Müllerová et al. (2005) used aerial photographs as a tool for assessing the regional dynamics of *Heracleum mantegazzianum* (giant hogweed). Ten sites invaded by this weed were monitored on 11 sampling dates between 1947 and 2000. The information gathered from these images allowed managers to identify dispersal foci and to focus control efforts on landscape structures with developing populations. Knowledge of the rate of spread and habitat vulnerability to invasion facilitated the identification of areas at the highest risk of immediate invasion.

The use of reliable weed thresholds is one of the cornerstones of integrated weed management. The traditional methods of establishing control thresholds involve small-scale trials on research farms. However, considering the existence of numerous sources of variation (soils, rainfall, temperatures and weed distribution patterns), the notion that we can extrapolate small-scale studies over hundreds of kilometers becomes tenuous (Maxwell, 1999). The availability of very precise tools for weed detection may allow new approaches for these studies. By using these tools, weed researchers could move from traditional plot studies in a very limited number of experimental stations to farm-scale studies in numerous sites (Luschei et al., 2001). A good example of this approach is the study conducted by Rasmussen and Nielsen (2020) in Denmark. These researchers used UAV images to estimate the competitive ability of *Cirsium arvense* (creeping thistle) in spring barley in eight commercial fields. Similar results were obtained by field scoring *C. arvense* patches (an uncomfortable and slow procedure when done manually) or by using aerial vehicle imagery.

4.3 Cropping systems

A cropping system refers to the crop type, sequence, and arrangement and management practices used on a particular field over years. The design of IWM strategies requires the assessment of cropping systems considering inter-annual data (e.g. the impacts of succeeding crops) from a holistic approach (Lechenet et al., 2017).

Methods that can be used to study cropping systems include field experiments and farm surveys. Field experiments provide appropriate support to assess the impact of specific agricultural practices and/or compare various production paradigms. Farm surveys may provide a better understanding of the links between diverse cropping systems and their associated impacts on various agroecosystem components, contributing to identify the drivers of cropping system performance in real farming conditions. However, both field experiments and farm surveys are costly, labor-intensive and time-consuming (Lechenet et al., 2017).

Remote sensing has proven to be an effective tool for monitoring cropping practices. Due to a large variety of onboard sensors on an increasing number of civilian satellites, the spectral characterization of the land surface resulting from human practices can be achieved and monitored at different spatial and temporal scales (Bégué et al., 2018). The combination and proper analysis of images obtained over a series of years in a wide area (landscape approach), together with specific measurements and data from the history of the fields present in that area, can result in a robust evaluation of crop and weed management in those fields. In order, to obtain reliable knowledge on the impact of cropping systems, farmers should be involved in the process of image analysis, interpretation and validation (Zhang and Kovacs, 2012).

4.4 Impacts of control actions

An important component of weed management programs is the assessment of the efficacy of the control actions. This objective is particularly relevant in the case of herbicide resistance. Assessing the occurrence and expansion of patches of resistant weed biotypes is of paramount importance for research and management purposes. Current methods of identifying herbicide resistance are laborious and require a relatively long time to produce results. To achieve more real-time classification and detection capabilities, Nugent et al. (2018) developed ground-based hyperspectral imaging and machine learning algorithms to accurately discriminate glyphosate-resistant and dicamba-resistant biotypes from susceptible biotypes of *Kochia scoparia*. Their results, obtained from mature plants grown in the greenhouse, showed classification accuracies ranging from 67% to 80%. Based on these results, they used a similar procedure to collect ground- and UAV-based hyperspectral imagery of three weeds: *K. scoparia*, *Conyza canadensis* and *Chenopodium album* at various growth stages and with multiple levels and types of resistance. The images were classified using a neural network machine learning algorithm (Scherrer et al., 2019). Depending on the species, growth stage and lighting conditions, the classification accuracies ranged from 77% to 99% for on-ground images and from 25% to 79% for UAV images. An interesting observation from this research was that the younger plants showed greater accuracy, which would be very useful for the detection of resistant weed biotypes at an early stage. This research results open new opportunities to detect the presence of patches of resistant weed biotypes at the field level and to monitor their spread.

5 Conclusion and future trends

We have a relatively large arsenal of platforms (satellites, airplanes, UAVs, combines and ATVs), sensors and image-processing procedures available to detect weeds at various spatial and temporal scales. Up to now, these technologies have been mainly used for map-based targeted herbicide spraying, improving the efficiency of this operation. In addition, they have been used for map-based mechanical weeding tools, facilitating the replacement of herbicide interventions. In the future, weed scouting may be used in the development of other environmentally benign interventions (competitive varieties, improved rotations ant tillage practices). In addition, they may be used in the redesign of cropping and weed management systems, facilitating collaborative research, providing new agroecological knowledge and improved cropping systems and allowing an automatic assessment of the impact of control activities.

6 Where to look for further information

6.1 Further reading

- Christensen, S., Sogaard, H. T., Kudsk, P., Nørremark, M., Lund, I., Nadimi, E. S. and Jørgensen, R. (2009). Site specific weed control technologies. *Weed Res.* 49, 233–241.
- Fernández-Quintanilla, C., Peña, J. M., Andújar, D., Dorado, J., Ribeiro, A. and López-Granados, F. (2018). Is the current state of the art of weed monitoring suitable for site-specific weed management in arable crops? *Weed Res.* 58, 259–272.
- López-Granados, F. (2011). Weed detection for site-specific weed management: mapping and real-time approaches. *Weed Res.* 51, 1–11.

6.2 Key journals and conferences

- WR, WS, and WT (*Weed Research, Weed Science, and Weed Technology*) are weed journals that regularly publish articles related to spatial weed detection and management.
- COMPAG (*Computers and Electronics in Agriculture*) provides coverage of advances in computer hardware, software, electronic instrumentation and control systems for precision agriculture.
- PA (*Precision Agriculture*) is an international journal on advances in precision agriculture and regularly publishes articles related to site-specific weed management.
- RS (*Remote Sensing*) is an open access international journal about the science and application of remote sensing technology and often runs special issues to create collections on specific topics, such as those related to weed detection, mapping and management based on RS.
- SE (*Sensors*) is the leading international open-access journal on the science and technology of sensors and often publishes articles on sensors and smart systems for agriculture, in which some focus is given to weed science applications.
- The International Conferences on Precision Agriculture (ICPAs), which are celebrated every 2 years in North America (ISPA), Latin America (CLAP), Europe (ECPA), Asian-Australasian (ACPA), and in the year 2020 for the first time in Africa (AfCPA), aim at supporting the PA community within every continent and showing the main advances in this discipline worldwide, which logically includes everything related to weed sensing and technology in the context of PA.

6.3 Major international research projects

- Flourish (http://flourish-project.eu/) 2015–2018. By combining the aerial survey capabilities of a small autonomous multi-copter UAV with a multi-purpose agricultural unmanned ground vehicle (UGV), the system will be able to survey a field from the air, perform a targeted intervention on the ground and provide detailed information for decision support, all with minimal user intervention.
- GALIRUMI (https://cordis.europa.eu/project/id/870258/es) (2019–2022). The objective of this project is to develop and demonstrate a Galileo-assisted robot to tackle the weed Rumex obtusifolius. A number of innovative technologies in weed detection, weed degradation, autonomous vehicles and robot-as-a-service for precision dairy farming will be based on precise navigation provided by European Global Navigation Satellite System.

- IWMPRAISE (https://iwmpraise.eu/) 2017–2022. The project will review current socioeconomic and agronomic barriers to the uptake of IWM in Europe and develop and optimize novel alternative weed control methods. On this basis, the project will create a toolbox of validated IWM tools. The objective of one of its work packages is to investigate the potential of sensors and satellite images for the detection of specific weed problems to support the planning of IWM strategies.

7 Acknowledgments

To the Projects AGL2017-83325-C4-1-R and AGL2017-83325-C4-3-R, funded by AEI/FEDER, UE.

8 References

Altieri, M. A., Nicholls, C. I. and Montalba, R. 2017. Technological approaches to sustainable agriculture at a crossroads: an agroecological perspective. *Sustainability (Basel)* 9(3), 349.

Andújar, D., Escolà, A., Dorado, J. and Fernández-Quintanilla, C. 2011a. Weed discrimination using ultrasonic sensors. *Weed Res.* 51(6), 543–547.

Andújar, D., Ribeiro, Á., Fernández-Quintanilla, C. and Dorado, J. 2011b. Reliability of a visual recognition system for detection of Johnsongrass (*Sorghum halepense*) in corn. *Weed Technol.* 25(4), 645–651.

Andújar, D., Escolà, A., Rosell-Polo, J. R., Fernández-Quintanilla, C. and Dorado, J. 2013. Potential of a terrestrial LiDAR-based system to characterise weed vegetation in maize crops. *Comput. Electron. Agr.* 92, 11–15.

Andújar, D., Dorado, J., Fernández-Quintanilla, C. and Ribeiro, A. 2016. An approach to the use of depth cameras for weed volume estimation. *Sensors (Basel)* 16(7), 972.

Araus, J. L. and Cairns, J. E. 2014. Field high-throughput phenotyping: the new crop breeding frontier. *Trends Plant Sci.* 19(1), 52–61.

Aravind, K. R., Raja, P. and Pérez-Ruiz, M. 2017. Task-based agricultural mobile robots in arable farming: a review. *Span. J. Agric. Res.* 15(1), 1–16.

Barrero, O. and Perdomo, S. A. 2018. RGB and multispectral UAV image fusion for Gramineae weed detection in rice fields. *Precis. Agric.* 19(5), 809–822.

Barroso, J., Fernandez-Quintanilla, C., Ruiz, D., Hernaiz, P. and Rew, L. J. 2004. Spatial stability of *Avena sterilis* ssp. *ludoviciana* populations under annual applications of low rates of imazamethabenz. *Weed Res.* 44(3), 178–186.

Barroso, J., McCallum, J. and Long, D. 2017. Optical sensing of weed infestations at harvest. *Sensors (Basel)* 17(10), 2381.

Bégué, A., Arvor, D., Bellon, B., Betbeder, J., De Abelleyra, D., Ferraz, R. P. D., Lebourgeois, V., Lelong, C., Simões, M. and Verón, S. R. 2018. Remote sensing and cropping practices: a review. *Remote Sens. (Basel)* 10, 99.

Behmann, J., Mahlein, A. K., Rumpf, T., Römer, C. and Plümer, L. 2015. A review of advanced machine learning methods for the detection of biotic stress in precision crop protection. *Precis. Agric.* 16(3), 239–260.

Berge, T. W., Aastveit, A. H. and Fykse, H. 2008. Evaluation of an algorithm for automatic detection of broad-leaved weeds in spring cereals. *Precis. Agric.* 9(6), 391–405.

Berge, T. W., Goldberg, S., Kaspersen, K. and Netland, J. 2012. Towards machine vision based site-specific weed management in cereals. *Comput. Electron. Agr.* 81, 79–86.

Blaschke, T., Kelly, M. and Merschdorf, H. 2015. Object-based image analysis: evolution, history, state of the art, and future vision. In: *Remotely Sensed Data Characterization, Classification, and Accuracies*. UC Berkeley. http://dx.doi.org/10.1201/b19294.

Castaldi, F., Pelosi, F., Pascucci, S. and Casa, R. 2017. Assessing the potential of images from unmanned aerial vehicles (UAV) to support herbicide patch spraying in maize. *Precis. Agric.* 18(1), 76–94.

Castillejo-González, I. L., Peña-Barragán, J. M., Jurado-Expósito, M., Mesas-Carrascosa, F. J. and López-Granados, F. 2014. Evaluation of pixel- and object-based approaches for mapping wild oat (*Avena sterilis*) weed patches in wheat fields using QuickBird imagery for site-specific management. *Eur. J. Agron.* 59, 57–66.

Castillejo-González, I. L., De Castro, A. I., Jurado-Expósito, M., Peña, J.-M., García-Ferrer, A. and López-Granados, F. 2019. Assessment of the persistence of *Avena sterilis* L. patches in wheat fields for site-specific sustainable management. *Agronomy* 9(1), 30.

Christensen, S., Heisel, T., Walter, A. M. and Graglia, E. 2003. A decision algorithm for patch spraying. *Weed Res.* 43(4), 276–284.

Christensen, S., Søgaard, H. T., Kudsk, P., Nørremark, M., Lund, I., Nadimi, E. S. and Jørgensen, R. 2009. Site-specific weed control technologies. *Weed Res.* 49(3), 233–241.

Colliver, C. T., Maxwell, B. D., Tyler, D. A., Roberts, D. W. and Long, D. S. 1996. Georeferencing wild oat infestations in small grains: accuracy and efficiency of three weed survey techniques. In: Proceedings of the Third International Conference on Precision Agriculture. John Wiley & Sons, Ltd, Minneapolis, MN, pp. 453–463.

de Castro, A. I., Jurado-Expósito, M., Gómez-Casero, M. T. and López-Granados, F. 2012a. Applying neural networks to hyperspectral and multispectral field data for discrimination of cruciferous weeds in winter crops. *Sci. World J.* 2012, 11 pages.

de Castro, A. I., Jurado-Expósito, M., Peña-Barragán, J. M. and López-Granados, F. 2012b. Airborne multi-spectral imagery for mapping cruciferous weeds in cereal and legume crops. *Precis. Agric.* 13(3), 302–321.

de Castro, A. I., Torres-Sánchez, J., Peña, J. M., Jiménez-Brenes, F. M., Csillik, O. and López-Granados, F. 2018. An automatic random forest-OBIA algorithm for early weed mapping between and within crop rows using UAV imagery. *Remote Sens. (Basel)* 10(3), 285.

de Castro, A. I., Peña, J. M., Torres-Sánchez, J., Jiménez-Brenes, F. M., Valencia-Gredilla, F., Recasens, J. and López-Granados, F. 2020. Mapping *Cynodon dactylon* infesting cover crops with an automatic decision tree-OBIA procedure and UAV imagery for precision viticulture. *Remote Sens. (Basel)* 12(1), 56.

Dorigo, W., Lucieer, A., Podobnikar, T. and Čarni, A. 2012. Mapping invasive *Fallopia japonica* by combined spectral, spatial, and temporal analysis of digital orthophotos. *Int. J. Appl. Earth Obs.* 19, 185–195.

European Commission (Ed.) 2016. *Agricultural Knowledge and Innovation Systems towards the Future: a Foresight Paper, EUR*. Publications Office of the European Union, Luxembourg.

Fennimore, S. A., Slaughter, D. C., Siemens, M. C., Leon, R. G. and Saber, M. N. 2016. Technology for automation of weed control in specialty crops. *Weed Technol.* 30(4), 823-837.

Fernández-Quintanilla, C., Peña, J. M., Andújar, D., Dorado, J., Ribeiro, A. and López-Granados, F. 2018. Is the current state of the art of weed monitoring suitable for site-specific weed management in arable crops? *Weed Res.* 58(4), 259-272.

Fishel, F., Bailey, W. C., Boyd, M. L., Johnson, W. G., William, G., O'Day, M. H., Sweets, L., Wiebold, W. J. and William, J. 2009. *Integrated Pest Management. Introduction to Crop Scouting*. University of Misouri Extension. IPM 1006. 24 pp.

Gao, J., Liao, W., Nuyttens, D., Lootens, P., Vangeyte, J., Pižurica, A., He, Y. and Pieters, J. G. 2018. Fusion of pixel and object-based features for weed mapping using unmanned aerial vehicle imagery. *Int. J. Appl. Earth Obs.* 67, 43-53.

Gašparović, M., Zrinjski, M., Barković, Đ. and Radočaj, D. 2020. An automatic method for weed mapping in oat fields based on UAV imagery. *Comput. Electron. Agr.* 173, 105385.

Gerhards, R. and Oebel, H. 2006. Practical experiences with a system for site-specific weed control in arable crops using real-time image analysis and GPS-controlled patch spraying. *Weed Res.* 46(3), 185-193.

Girolamo-Neto, C. D., Sanches, I. D., Neves, A. K., Prudente, V. H. R., Körting, T. S., Picoli, M. C. A. and de Aragão, L. E. O. e C. 2019. Assessment of texture features for Bermudagrass (*Cynodon dactylon*) detection in sugarcane plantations. *Drones* 3(2), 36.

Gómez-Casero, M. T., Castillejo-González, I. L., García-Ferrer, A., Peña-Barragán, J. M., Jurado-Expósito, M., García-Torres, L. and López-Granados, F. 2010. Spectral discrimination of wild oat and canary grass in wheat fields for less herbicide application. *Agron. Sustain. Dev.* 30(3), 689-699.

Gray, C. J., Shaw, D. R. and Bruce, L. M. 2009. Utility of hyperspectral reflectance for differentiating soybean (*Glycine max*) and six weed species. *Weed Technol.* 23(1), 108-119.

Hill, S. 1985. Redesigning the food system for sustainability. *Alternatives* 12, 32-36.

Hill, S. B. 1998. Redesigning agroecosystems for environmental sustainability: a deep systems approach. *Syst. Res. Behav. Sci.* 15(5), 391-402.

Hively, W. D., Lamb, B. T., Daughtry, C. S. T., Shermeyer, J., McCarty, G. W. and Quemada, M. 2018. Mapping crop residue and tillage intensity using WorldView-3 satellite shortwave infrared residue indices. *Remote Sens. (Basel)* 10(10), 1657.

Hobbs, P. R., Sayre, K. and Gupta, R. 2008. The role of conservation agriculture in sustainable agriculture. *Philos. Trans. R. Soc. Lond. B Biol. Sci.* 363(1491), 543-555.

Huang, C. Y. and Asner, G. P. 2009. Applications of remote sensing to alien invasive plant studies. *Sensors (Basel)* 9(6), 4869-4889.

Huang, H., Deng, J., Lan, Y., Yang, A., Deng, X., Wen, S., Zhang, H. and Zhang, Y. 2018a. Accurate weed mapping and prescription map generation based on fully convolutional networks using UAV imagery. *Sensors (Basel)* 18(10), 3299.

Huang, H., Deng, J., Lan, Y., Yang, A., Deng, X. and Zhang, L. 2018b. A fully convolutional network for weed mapping of unmanned aerial vehicle (UAV) imagery. *PLoS ONE* 13(4), e0196302.

Huang, Y. and Brown, M. E. 2018. Advancing to the next generation of precision agriculture. In: *Agriculture & Food Systems to 2050, World Scientific Series in Grand Public Policy Challenges of the 21st Century*. World Scientific, Singapore, pp. 285-314.

Jiménez-Brenes, F. M., López-Granados, F., Torres-Sánchez, J., Peña, J. M., Ramírez, P., Castillejo-González, I. L. and de Castro, A. I. 2019. Automatic UAV-based detection of *Cynodon dactylon* for site-specific vineyard management. *PLoS ONE* 14(6), e0218132.

Jurado-Expósito, M., de Castro, A. I., Torres-Sánchez, J., Jiménez-Brenes, F. M. and López-Granados, F. 2019. *Papaver rhoeas* L. mapping with cokriging using UAV imagery. *Precis. Agric.* 20(5), 1045–1067.

Lampkin, N., Pearce, B., Leake, A., Creissen, H., Gerrard, C., Girling, R., Lloyd, S., Padel, S., Smith, J., Smith, L., Vieweger, A. and Wolfe, M. 2015. *The Role of Agroecology in Sustainable Intensification*. Report for the Land Use Policy Group. Organic Research Centre, Elm Farm and Game & Wildlife Conservation Trust.

Lechenet, M., Deytieux, V., Antichi, D., Aubertot, J.-N., Bàrberi, P., Bertrand, M., Cellier, V., Charles, R., Colnenne-David, C., Dachbrodt-Saaydeh, S., Debaeke, P., Doré, T., Farcy, P., Fernandez-Quintanilla, C., Grandeau, G., Hawes, C., Jouy, L., Justes, E., Kierzek, R., Kudsk, P., Lamichhane, J. R., Lescourret, F., Mazzoncini, M., Melander, B., Messéan, A., Moonen, A.-C., Newton, A. C., Nolot, J.-M., Panozzo, S., Retaureau, P., Sattin, M., Schwarz, J., Toqué, C., Vasileiadis, V. P. and Munier-Jolain, N. 2017. Diversity of methodologies to experiment Integrated Pest Management in arable cropping systems: analysis and reflections based on a European network. *Eur. J. Agron.* 83, 86–99.

Leveau, L., Bénel, A., Cahier, J.-P., Pinet, F., Salembier, P., Soulignac, V. and Bergez, J.-E. 2019. Information and communication technology (ICT) and the agroecological transition. In: Bergez, J.-E., Audouin, E. and Therond, O. (Eds), *Agroecological Transitions: From Theory to Practice in Local Participatory Design*. Springer International Publishing, Cham, pp. 263–287.

Liebman, M. and Davis, A. S. 2000. Integration of soil, crop and weed management in low-external-input farming systems. *Weed Res.* 40(1), 27–47.

Liebman, M., Baraibar, B., Buckley, Y., Childs, D., Christensen, S., Cousens, R., Eizenberg, H., Heijting, S., Loddo, D., Merotto, A., Renton, M. and Riemens, M. 2016. Ecologically sustainable weed management: how do we get from proof-of-concept to adoption? *Ecol. Appl.* 26(5), 1352–1369.

Liebman, M. and Dyck, E. 1993. Crop rotation and intercropping strategies for weed management. *Ecol. Appl.* 3(1), 92–122.

López-Granados, F. 2011. Weed detection for site-specific weed management: mapping and real-time approaches. *Weed Res.* 51(1), 1–11.

López-Granados, F., Jurado-Expósito, M., Peña-Barragán, J. M. and García-Torres, L. 2006. Using remote sensing for identification of late-season grass weed patches in wheat. *Weed Sci.* 54(2), 346–353.

López-Granados, F., Torres-Sánchez, J., De Castro, A.-I., Serrano-Pérez, A., Mesas-Carrascosa, F.-J. and Peña, J.-M. 2016a. Object-based early monitoring of a grass weed in a grass crop using high resolution UAV imagery. *Agron. Sustain. Dev.* 36(4), 67.

López-Granados, F., Torres-Sánchez, J., Serrano-Pérez, A., de Castro, A. I., Mesas-Carrascosa, F.-J. and Peña, J. M. 2016b. Early season weed mapping in sunflower using UAV technology: variability of herbicide treatment maps against weed thresholds. *Precis. Agric.* 17(2), 183–199.

Lunetta, R. S., Shao, Y., Ediriwickrema, J. and Lyon, J. G. 2010. Monitoring agricultural cropping patterns across the Laurentian Great Lakes Basin using MODIS-NDVI data. *Int. J. Appl. Earth Obs.* 12(2), 81–88.

Luschei, E. C., Van Wychen, L. R., Maxwell, B. D., Bussan, A. J., Buschena, D. and Goodman, D. 2001. Implementing and conducting on-farm weed research with the use of GPS. *Weed Sci.* 49(4), 536-542.

Martín, M. P., Barreto, L. and Fernández-Quintanilla, C. 2011. Discrimination of sterile oat (*Avena sterilis*) in winter barley (*Hordeum vulgare*) using QuickBird satellite images. *Crop Prot.* 30(10), 1363-1369.

Mast, J. N., Veblen, T. T. and Hodgson, M. E. 1997. Tree invasion within a pine/grassland ecotone: an approach with historic aerial photography and GIS modeling. *Forest Ecol. Manag.* 93(3), 181-194.

Maxwell, B. D. 1999. My view. *Weed Sci.* 47(2), 129-129.

McCormick, C. M. 1999. Mapping exotic vegetation in the Everglades from large-scale aerial photographs - sciencebase-catalog. *Photogramm. Eng. Rem. S.* 65, 179-184.

Meyer, G. E. and Neto, J. C. 2008. Verification of color vegetation indices for automated crop imaging applications. *Comput. Electron. Agr.* 63(2), 282-293.

Meynard, J. M., Dedieu, B. and Bos, A. P. (Bram) 2012. Re-design and co-design of farming systems: an overview of methods and practices. In: Darnhofer, I., Gibbon, D. and Dedieu, B. (Eds), *Farming Systems Research into the 21st Century: the New Dynamic*. Springer Netherlands, Dordrecht, pp. 405-429.

Mink, R., Dutta, A., Peteinatos, G. G., Sökefeld, M., Engels, J. J., Hahn, M. and Gerhards, R. 2018. Multi-temporal site-specific weed control of *Cirsium arvense* (L.) Scop. and *Rumex crispus* L. in maize and sugar beet using unmanned aerial vehicle based mapping. *Agriculture* 8(5), 65.

Montalvo, M., Pajares, G., Guerrero, J. M., Romeo, J., Guijarro, M., Ribeiro, A., Ruz, J. J. and Cruz, J. M. 2012. Automatic detection of crop rows in maize fields with high weeds pressure. *Expert Syst. Appl.* 39(15), 11889-11897.

Moschitz, H., Tisenkopfs, T., Brunori, G., Home, R., Kunda, I. and Sumane, S. 2014. *Final Report of the SOLINSA Project*. Project Deliverable D8.2 of the SOLINSA project (GA 266306) (Report). Research Institute of Organic Agriculture (FiBL), Frick.

Müllerová, J., Pyšek, P., Jarošík, V. and Pergl, J. 2005. Aerial photographs as a tool for assessing the regional dynamics of the invasive plant species *Heracleum mantegazzianum*. *J. Appl. Ecol.* 42(6), 1042-1053.

Müllerová, J., Brůna, J., Bartaloš, T., Dvořák, P., Vítková, M. and Pyšek, P. 2017. Timing is important: unmanned aircraft vs. satellite imagery in plant invasion monitoring. *Front. Plant Sci.* 8, 887.

Norton, G. W., Rajotte, E. G. and Gapud, V. 1999. Participatory research in integrated pest management: lessons from the IPM CRSP. *Agr. Hum. Values* 16(4), 431-439.

Nugent, P. W., Shaw, J. A., Jha, P., Scherrer, B., Donelick, A. and Kumar, V. 2018. Discrimination of herbicide-resistant kochia with hyperspectral imaging. *J. Appl. Remote Sens.* 12(1), 016037.

Olsen, A., Konovalov, D. A., Philippa, B., Ridd, P., Wood, J. C., Johns, J., Banks, W., Girgenti, B., Kenny, O., Whinney, J., Calvert, B., Azghadi, M. R. and White, R. D. 2019. DeepWeeds: A multiclass weed species image dataset for deep learning. *Sci. Rep.* 9(1), 2058.

Ostos-Garrido, F. J., de Castro, A. I., Torres-Sánchez, J., Pistón, F. and Peña, J. M. 2019. High-throughput phenotyping of bioethanol potential in cereals using UAV-based multi-spectral imagery. *Front. Plant Sci.* 10, 948.

Otsu, N. 1979. A threshold selection method from gray-level histograms. *IEEE Trans. Syst. Man Cybern.* 9(1), 62-66.

Padel, S., Levidow, L. and Pearce, B. 2020. UK farmers' transition pathways towards agroecological farm redesign: evaluating explanatory models. *Agroecol. Sustain. Food Syst.* 44(2), 139-163.

Peña, J. M., Torres-Sánchez, J., de Castro, A. I., Kelly, M. and López-Granados, F. 2013. Weed mapping in early-season maize fields using object-based analysis of unmanned aerial vehicle (UAV) images. *PLoS ONE* 8(10), e77151.

Peña, J. M., Torres-Sánchez, J., Serrano-Pérez, A., De Castro, A. I. and López-Granados, F. 2015. Quantifying efficacy and limits of unmanned aerial vehicle (UAV) technology for weed seedling detection as affected by sensor resolution. *Sensors (Basel)* 15(3), 5609-5626.

Peña-Barragán, J. M., López-Granados, F., Jurado-Expósito, M. and García-Torres, L. 2010. Sunflower yield related to multi-temporal aerial photography, land elevation and weed infestation. *Precis. Agric.* 11(5), 568-585.

Pérez-Ortiz, M., Peña, J. M., Gutiérrez, P. A., Torres-Sánchez, J., Hervás-Martínez, C. and López-Granados, F. 2015. A semi-supervised system for weed mapping in sunflower crops using unmanned aerial vehicles and a crop row detection method. *Appl. Soft Comput.* 37, 533-544.

Pérez-Ortiz, M., Peña, J. M., Gutiérrez, P. A., Torres-Sánchez, J., Hervás-Martínez, C. and López-Granados, F. 2016. Selecting patterns and features for between- and within-crop-row weed mapping using UAV-imagery. *Expert Syst. Appl.* 47, 85-94.

Pretty, J. 2018. Intensification for redesigned and sustainable agricultural systems. *Science* 362(6417), eaav0294.

Rasmussen, J. and Nielsen, J. 2020. A novel approach to estimating the competitive ability of *Cirsium arvense* in cereals using unmanned aerial vehicle imagery. *Weed Res.* 60(2), 150-160.

Rasmussen, J., Nielsen, J., Streibig, J. C., Jensen, J. E., Pedersen, K. S. and Olsen, S. I. 2019. Pre-harvest weed mapping of Cirsium arvense in wheat and barley with off-the-shelf UAVs. *Precis. Agric.* 20(5), 983-999.

Rehman, T. U., Mahmud, Md. S., Chang, Y. K., Jin, J. and Shin, J. 2019. Current and future applications of statistical machine learning algorithms for agricultural machine vision systems. *Comput. Electron. Agr.* 156, 585-605.

Rueda-Ayala, V., Weis, M., Keller, M., Andújar, D. and Gerhards, R. 2013. Development and testing of a decision making based method to adjust automatically the harrowing intensity. *Sensors (Basel)* 13(5), 6254-6271.

Ruiz, D., Escribano, C. and Fernández-Quintanilla, C. 2006. Assessing the opportunity for site-specific management of *Avena sterilis* in winter barley fields in Spain. *Weed Res.* 46(5), 379-387.

Scherrer, B., Sheppard, J., Jha, P. and Shaw, J. A. 2019. Hyperspectral imaging and neural networks to classify herbicide-resistant weeds. *J. Appl. Remote Sens.* 13(4), 044516.

Singh, V., Rana, A., Bishop, M., Filippi, A. M., Cope, D., Rajan, N. and Bagavathiannan, M. 2020. Chapter Three. Unmanned aircraft systems for precision weed detection and management: prospects and challenges. In: Sparks, D. L. (Ed.), *Advances in Agronomy*. Academic Press, pp. 93-134.

Stow, D., Hope, A., Richardson, D., Chen, D., Garrison, C. and Service, D. 2000. Potential of colour-infrared digital camera imagery for inventory and mapping of alien plant invasions in South African shrublands. *Int. J. Remote Sens.* 21(15), 2965-2970.

Su, W. H. 2020. Advanced machine learning in point spectroscopy, RGB- and hyperspectral-imaging for automatic discriminations of crops and weeds: a review. *Smart Cities* 3(3), 767–792.

Thenkabail, P., Lyon, J. G., Turral, H., Biradar, C., Lyon, J. G., Turral, H. and Biradar, C. 2009. *Remote Sensing of Global Croplands for Food Security*. CRC Press, Boca Raton, FL.

Torres-Sánchez, J., López-Granados, F., De Castro, A. I. and Peña-Barragán, J. M. 2013. Configuration and specifications of an Unmanned Aerial Vehicle (UAV) for early site specific weed management. *PLoS ONE* 8(3), e58210.

Torres-Sánchez, J., López-Granados, F. and Peña, J. M. 2015. An automatic object-based method for optimal thresholding in UAV images: application for vegetation detection in herbaceous crops. *Comput. Electron. Agr.* 114, 43–52.

van Evert, F. K., Fountas, S., Jakovetic, D., Crnojevic, V., Travlos, I. and Kempenaar, C. 2017. Big Data for weed control and crop protection. *Weed Res.* 57(4), 218–233.

Van Wychen, L. R., Luschei, E. C., Bussan, A. J. and Maxwell, B. D. 2002. Accuracy and cost effectiveness of GPS-assisted wild oat mapping in spring cereal crops. *Weed Sci.* 50(1), 120–129.

Worthington, M. and Reberg-Horton, C. 2013. Breeding cereal crops for enhanced weed suppression: optimizing allelopathy and competitive ability. *J. Chem. Ecol.* 39(2), 213–231.

Ye, X., Sakai, K., Asada, S.-I. and Sasao, A. 2007. Use of airborne multispectral imagery to discriminate and map weed infestations in a citrus orchard. *Weed Biol. Manag.* 7(1), 23–30.

Zhang, C. and Kovacs, J. M. 2012. The application of small unmanned aerial systems for precision agriculture: a review. *Precis. Agric.* 13(6), 693–712.

Chapter 9

Advances in precision application technologies for weed management

Ran N. Lati, Newe Ya'ar Research Center, Agricultural Research Organization, Israel; Roland Gerhards, University of Hohenheim, Germany; Hanan Eizenberg and Maor Matzrafi, Newe Ya'ar Research Center, Agricultural Research Organization, Israel; Lior Blank, Agricultural Research Organization – Volcani Center, Israel; and Svend Christensen, University of Copenhagen, Denmark

1 Introduction

As we experience the current rapid advancement in technology, image-driven data have become a major part of many aspects of everyday life, including agriculture. Remote and proximal sensing methods are now playing a key role in the development of new advanced precision agriculture methodologies and, more specifically, of precise weed management (PWM) protocols (Rasmussen et al., 2019). Most PWM-related research activity (at least as reflected in scientific papers) is focused on different aspects of weed scouting and mapping, aiming to optimize the weed detection performances of various sensors and imaging platforms (e.g., Franco et al., 2017; Louargant et al., 2017; Kounalakis et al., 2019; Sharpe et al., 2020). It appears that much less attention is being devoted to other aspects of the PWM pipeline, which is unfortunate, since the development and implementation of complementary tools and methodologies would contribute enormously to the success of the overall objective of PWM of minimizing herbicide application. Thus, this chapter aims

http://dx.doi.org/10.19103/AS.2021.0098.10

to highlight some of the recent findings in the field of precise application technologies.

Economic criteria, in terms of yield losses, determine the basis for many decisions regarding weed control. Image-driven data on the spatial and temporal characteristics of weed communities in the field can improve our decision-making and prevent sub-optimum estimation of potential damage. Any additional data would further improve precision application and increase the accuracy of spot spraying (Mink et al., 2018). The information available will obviously also affect decisions about application tactics and the tool of choice. Such decisions will also be influenced by the increasing stringency of restrictions on herbicide application and by the development of herbicide-resistance populations, and these considerations may indeed lead to the adoption of alternative control tactics (Kunz et al., 2015).

Improvements in herbicide application results can be achieved by simple adaptations of conventional methods, such as precise inter-row cultivation in close proximity to the crops using off-the-shelf RGB cameras. It can also be improved by the implementation of state-of-the-art technologies, such as autonomous vehicles that scout and control the weeds with no human intervention. In concert, other cutting-edge technologies, such as genome editing (e.g. CRISPR/Cas) and nanosensing, can be harnessed to improve herbicide application results and facilitate PWM (Giraldo et al., 2019). These complementary approaches can improve the results of image-driven weed detection methods and hence the accuracy and the overall robustness of current integrated weed management (IWM) over a wide range of field conditions. Lastly, integration of herbigation into current IWM protocols can promote the optimized application in terms of the nature of the active ingredient and the rate and timing of the application. Adding temporal aspects to the weed population dynamics in PWM methods can support the precise application of pre-emerged weeds, a concept that, so far, did not get enough attention. As IWM is about combing several control methods for optimal results, within the field of PWM, new application technologies should also be adopted and integrated for effective and robust results.

2 Advances in precision weed control systems

2.1 Precise herbicide application

The concept of economic weed thresholds (ETs) has enabled farmers to estimate yield losses caused by weed competition at the time that weed control decisions are made. Therefore, ETs provide simple decision rules for herbicide application. Until the 1990s, ETs were determined at the field level to decide whether herbicide application across the whole field was required or not (Coble and Mortensen, 1992). Threshold levels exist for weed groups such as grass weeds, broadleaved weeds or single weed species such as *Galium aparine* (Gerowitt and Heitefuss, 1990). Usually, empirical models were used to estimate crop yield loss by weed competition based on early observations of weed density (Cousens, 1985) and relative weed cover (Kropff and Spitters, 1991; Lotz et al., 1996). The models fit better to relative weed cover than to weed density (Ali et al., 2013), because relative weed cover accounts for the size of the crop and the weeds and the relative time of emergence (Cousens et al., 1987). However, it still remains problematic to quantify the competitive effects of mixed weed populations.

Weed control according to ETs is a useful strategy to avoid herbicide application in fields with low weed infestation and to reduce selection pressure on weed populations towards herbicide resistance. However, ETs do not take into consideration the spatial variation of weed populations within a field. The use of the average weed density for fields with spatially heterogeneous weed populations results in yield loss predictions that are too low in locations where weed density is high and predictions that are too high in parts of the field where weed densities are low or weeds are absent (Gerhards et al., 1997). Spatial variation in weed density must therefore be considered using ETs.

ETs regained interest in the development of precision farming technologies in weed management. Weed mapping and precise herbicide spraying resulted in 23-89% herbicide savings in 58 field experiments in cereals, maize, sugar beet and peas (Gerhards and Oebel, 2006; Christensen et al., 2009; Berge et al., 2012). PWMs contain three important steps that can be completed in three consecutive passes (offline) or simultaneously (online): 1) weed mapping, 2) decision-making and 3) patch spraying. We discuss these three interconnected issues below.

Offline patch spraying requires a georeferenced weed distribution map. This can be created on the basis of grid sampling (Rew and Cousens, 2001; Fernández-Quintanilla et al., 2018), aerial imaging or ground-based image analysis. The accuracy and efficiency of ground-based weed/crop identification have improved significantly in the past two decades. Classical image analysis combined with machine learning has resulted in 80-90% correct classification of weed species/groups based on selected features of shape, texture and colour of weed species (Gerhards and Oebel, 2006; Tillett et al., 2008; Weis et al., 2008). With the introduction of convolutional neural networks (CNN), the accuracy of classification rates has increased to more than 90% (Dyrmann et al., 2016; Peteinatos et al., 2020).

Online patch spraying was introduced by Longchamps et al. (2014), who assessed weed coverage between maize rows in 19 fields in Canada, in which an RGB camera was used to control a single spray nozzle of a spraying system. Crop consultants had decided that weed control thresholds should be 0.06%

for farmers predicting a low risk of weed competition and 0.312% for farmers predicting a higher risk. At the low threshold, 20% herbicide savings were achieved, while at the high threshold, a 50% reduction in herbicide application was recorded.

Gerhards and Kühbauch (1993) defined ETs in winter cereals based on weed coverage. Image-based model was used to measure weed coverage at different growth stages of the crop (Table 1) and showed that weed cover measurements provided accurate predictors of yield losses in cereals. Testing of the model on 22 farms showed that it provided correct recommendations on herbicide use in 65 of 72 experiments (92%). The best correlation with yield loss was found for relative weed cover (%) measured at the time of regular weed control. This parameter can serve for the on/off control of a camera-based patch sprayer or spray robot.

However, patch spraying is not just a simple on/off decision (Gutjahr and Gerhards, 2010). If the ET is exceeded for a weed patch, then decision rules for patch spraying are used to select the right herbicide and the herbicide application rate as a function of the weed species composition and the weed growth stages (Fig. 1). Long-term effects of the particular weed species may be taken into consideration by using population dynamics models to prevent increases in the size and density of the weed patch (Gutjahr and Gerhards, 2010).

Site-specific weed control implies adjustments of application technology for herbicides. If standard boom sprayers with one tank containing the herbicide mixture were used for site-specific weed control, the full potential for herbicide savings could not be exploited. This is shown in Fig. 2.

Modern artificial-intelligence-based technologies, such as CNN, broaden the spectrum of applications of PWM. They increase the accuracy of weed

Table 1 Economic weed thresholds based on weed cover (%) measurements in winter cereals (Gerhards and Kühbauch, 1993).

BBCH-stage crop	Winter wheat	Winter rye	Winter barley
11	0.15	0.25	0.15
12	0.2	0.35	0.2
13	0.35	0.5	0.25
15	0.4	0.65	0.35
17	0.6	0.9	0.5
19	0.85	1.4	0.6
21	1.25	1.8	0.7
23	2.0	3.5	1.2
25	3.3	6.5	1.9
27	4.3	8.0	2.7
29	5.4	8.9	5.5

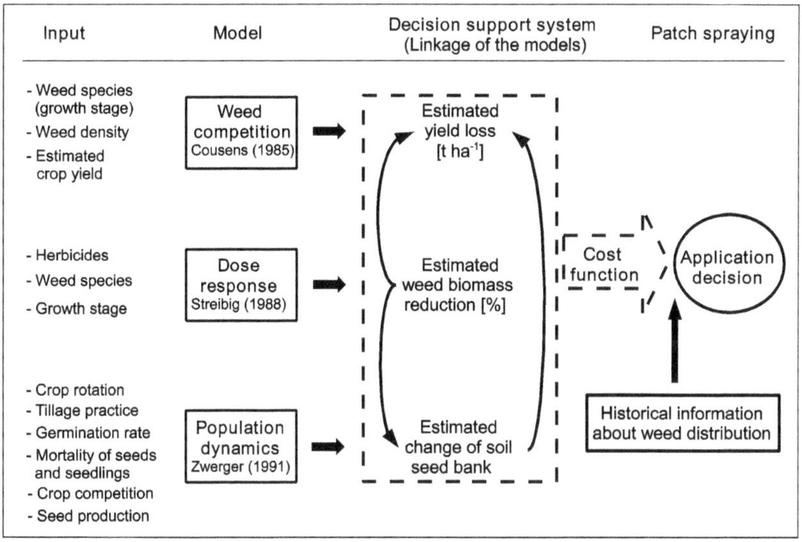

Figure 1 Architecture for a decision support system for weed control, including yield loss, dose response, population dynamics and cost functions (Gutjahr and Gerhards, 2010).

identification and allow differentiation between many different plant species. Therefore, they can be used for scouting, mapping and patch spraying. The improved computing capacities of controllers and on-board computers now allow the control of single nozzles and single hoeing blades. With better communication between sensors, actors and users, it has become easier to include decision support systems (DSSs) on patch sprayers. Such DSS can generate precise decisions for controlling problematic weed species, such as herbicide-resistant blackgrass (*Alopecurus myosuroides*), and for protecting rare weed species or species with ecological benefits.

2.2 Mechanical weed control

Restrictions on herbicide use, the negative environmental side-effects of herbicides, the problem of herbicide residues in the food chain and the spread of herbicide-resistant weed populations have all contributed to driving precision farming technologies for physical weed control. Camera-guided inter-row hoeing with automatic side-shift control has significantly improved the efficacy of mechanical weed control (Tillett et al., 2002; Kunz et al., 2015, 2018). Kunz et al. (2015) tested the efficacy of a Kult-Robocrop® hoe and an OEM CLAAS stereo camera in combination with an Einböck Row-Guard hoe with duck-foot blades in weeding between the rows in sugar beet and soybean. This setup increased weed control efficacy by 12% compared to machine hoeing with manual guidance. The authors explained that the improved weed

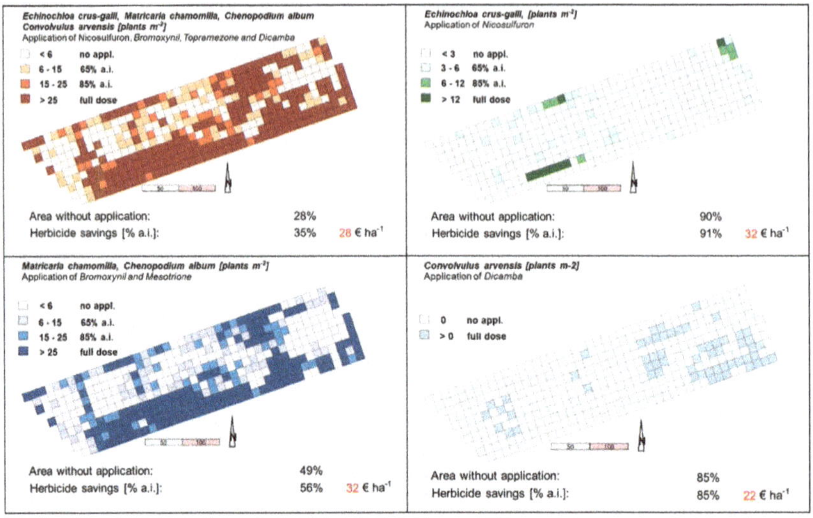

Figure 2 Precise herbicide application in maize based on a map of all weed species present (one tank sprayer) (top left) and based on weed distributions of weed species groups separately (multiple tank sprayer/direct injection system) (Gutjahr et al., 2012).

control efficacy resulted from guiding the duck-foot blades closer to the crop rows and from the better burial of the intra-row weeds into the soil due to the higher driving speed. In this setup, the K.U.L.T. Vision Control System provides high accuracy of row guiding. Its RGB camera allows robust row detection even in partly overlapping crop rows, such as cereals with a 12.5 cm row distance. The camera faces diagonally forward and scans four to six crop rows at a time. Images are segmented into green plants and background consisting of soil and mulch. In the regions of the highest green pixel densities, the tracking of crop rows is achieved by an extended Kalman filter (Tillett et al., 2002). Artificial light improves the quality of row detection. At present, several manufacturers offer camera-based automatic side-shift control systems (Table 2).

2.3 Robotic weeding

Robots have been – and continue to be – developed for chemical and physical weed control. They target single weeds and differentiate between crop plants and weeds in the intra-row space (Rasmussen et al., 2012). Ruckelshausen et al. (2006), for example, developed and applied a hoeing robot for maize. A sensor system identified maize plants based on plant height and shape parameters. When a maize plant was detected, the intra-row hoeing blade was moved out of the crop row. Gobor et al. (2013) designed a rotary hoe to be positioned over the crop row, but their system was not combined with a sensor to differentiate between weeds and crop plants. Several studies have

Table 2 Inter-row camera guidance systems for mechanical hoes

Commercial product	Accuracy (cm)	Minimum row distance (cm)	Camera technology	Maximum speed (km/h)
K.U.L.T. Vision Control	± 2	12.5	1D bi-spectral (NIR + red)	15
Garford Robocrop	± 4	20	2D visible spectrum	14
Steketee IC	± 4	25	2D bi-spectral (NIR + red)	10
OEM Claas Row-Guard, Einböck, Hatzenbichler, Schmotzer	± 4	25	3D stereo	14

been conducted to record the position of crop seeds during seeding using a real-time kinematic positioning-global navigation satellite system (RTK-GNSS), followed by pre- and post-emergence hoeing around the crop plants directed by an RTK-GNSS-controlled guidance system (Nørremark et al., 2012). Other examples for robotic weeders are the two commercial sensor-based intra-row hoeing robots that have been introduced by Garford (*Garford Robocrop InRow Weeder*) and Steketee IC for transplanted crops. The systems work precisely if weeds are relatively small compared to the crop; that is, the systems controlled approximately 70% of the intra-row weeds. However, their working speeds are low, being a maximum of 1.8 km h^{-1}, which makes them relevant mainly for cash crops like lettuce and broccoli (Tillett et al., 2002, 2008).

3 New technologies for optimizing application precision

PWM protocols aim to target control tactics to weed-infested areas exclusively, while avoiding any treatment to the crop plants (Thorp and Tian, 2004; Burgos-Artizzu et al., 2011; López-Granados, 2011; Fernández-Quintanilla et al., 2018). An essential stage in the application of PWM protocols is thus the autonomous detection and classification of weed plants from crop plants (Wang et al., 2019). Over the past two decades, numerous studies have proposed various methodologies to detect weeds for real-time or mapping applications, using RGB (Lambert et al., 2018), multispectral (Sa et al., 2018), hyperspectral (Pantazi et al., 2016), time-of-flight (Piron et al., 2011) and light detection and ranging (LiDAR) (Peteinatos et al., 2014) sensors, among others. In addition, advanced image processing and data analysis algorithms have been employed to improve the robustness of weed/crop detection and classification and to handle the large amounts of data generated by these advanced sensors (Raja et al., 2020a). Despite the promising results demonstrated with the different sensors and image processing approaches, classification accuracy and robustness are still affected by a variety of factors (Kennedy et al., 2019; Raja et al., 2020c; Su, 2020a). For example, outdoor illumination can affect the reflectance pattern of any plant species (weed or crop), thereby leading to misclassification or over-estimation of the weed-related area, which may substantially reduce the overall accuracy of these image-driven classification models (Lati et al., 2019). Misclassification of weed-related area may also occur at high weed infestation levels with overlapping weed and crop canopies and shading/occlusion of the weed leaves (Bakhshipour et al., 2017) and complex backgrounds that affect classification accuracies (Bakhshipour and Jafari, 2018). In addition, studies have shown that several plant species have very similar reflectance characteristics and canopy shapes, making their classification extremely difficult (Slaughter et al., 2008). Handling these limitations calls for novel approaches that can provide robust classification results under challenging imaging conditions.

3.1 Genetic modifications

It has been suggested that genetic engineering of the crop phenotype may be used as an alternative approach to improve crop detection and differentiation from the weeds and hence to provide more robust detection results (Slaughter et al., 2008; Lati et al., 2014). The focus here is the identification of crop plants rather than weeds. This idea has been put into practice for tomato and tobacco, as model plants. For tomato, genetically modified plants (i.e. germplasm AN-113) that constitutively overexpress the *ANTHOCYANIN1* (*ANT1*) gene, an MYB transcriptional regulator of anthocyanin biosynthesis, were produced (Mathews et al., 2003). These plants have purple leaves, which are dramatically different from the green leaves of the surrounding weeds. Slaughter et al. (2008) exploited the sharp hue difference to facilitate classification using low-cost cameras (off-the-shelf RGB cameras) and a non-complex algorithm that can be easily applied in real time. They emphasized that the contributions of the modified purple tomato plants for real-time application lay in reducing the pre-processing time, by saving several stages and reducing the impact of occluded leaves on the detection accuracy (Slaughter et al., 2008). Similarly, Lati et al. (2014) demonstrated high weed detection levels (~95%) that were not affected by different imaging and field conditions, including high weed infestation levels with overlapping weed and crop canopies (Fig. 3). In those experiments, a hue-invariant segmentation algorithm was used to ensure robust classification results under varying illumination conditions that included extreme light and shading (Lati et al., 2011). Misclassifications were attributed mainly to the small size of weed at early stages (10 days after seeding), which were omitted by the size-based filter that aimed to handle noise during processing (Lati et al., 2014). It is likely that for a crop of purple-leaved tomatoes, the drastic differences in hue between the crop plants and the weeds will also

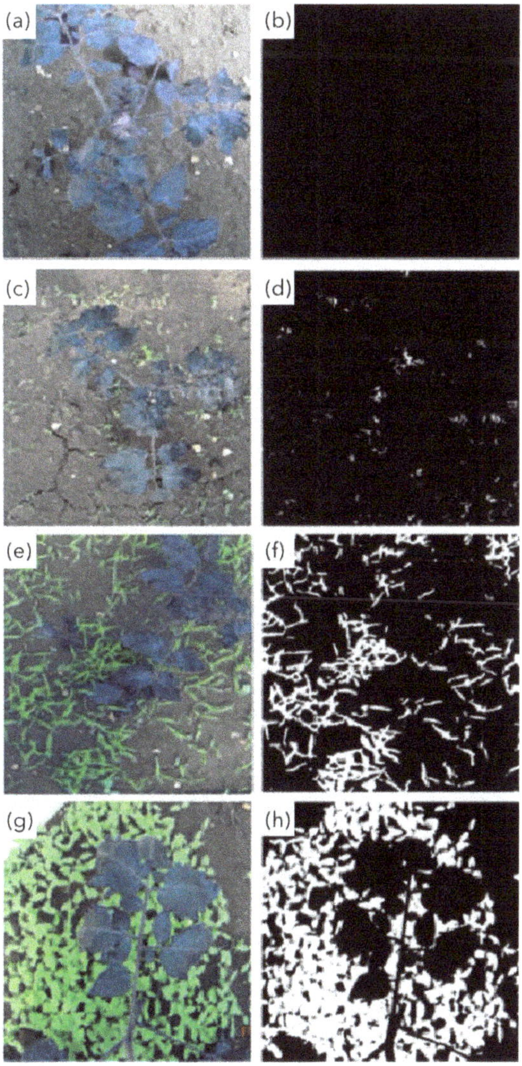

Figure 3 Mixture of AN-113 genetically modified tomato plants and plants of the weed species *Solanum nigrum* and *Alopecurus utriculatus*, and their respective weed/crop segmented images captured before weed emergence (a, b), and 10 (c, d), 20 (e, f) and 30 (g, h) days after weed seeding (weeds are shown in white).

be identifiable from remote-sensing platforms, and thus this technique could be suitable for weed mapping from unmanned aerial vehicle (UAVs) and other aerial vehicles.

A follow-up study on tobacco [*Nicotiana tabacum* (Xanthi)] used a gene-silencing technique [virus-induced gene silencing (VIGS)] to prevent any potential fitness and crop yield cost that could result from the overexpression of anthocyanin in the plant leaves (Aly et al., 2019). The tobacco plants were thus subjected to gene silencing that significantly reduced the overexpression of anthocyanin pigments 40 days after they had been induced. Although the purple hue in the tobacco leaves faded gradually, the tobacco leaves still had a sufficient purple hue for accurate classification from the green weeds within the first few weeks (Fig. 4). In addition, there were no fitness costs in terms of plant height or leaf number in the silenced *vs.* non-silenced tobacco transgenes. For many crops, the critical timing for weed control is the early period of the growing season; hence, a purple phenotype in the first month can be sufficient for the implementation of PWMs.

Improved crop detection using genetic modifications can also be achieved by the insertion of fluorescent proteins into the crop plants. These proteins have been used in genetic and biological research for many years with the aim of detecting successful transformation and protein synthesis in transgenic plants (Cormack et al., 1996). For example, green fluorescent protein (GFP) emits a bright green fluorescence when exposed to light in the range of blue to UV (van Thor et al., 2002). Pray (2008) tested this fluorescent-protein approach by inserting GFP into tobacco plants. When the plants were excited with 470-nm (blue) radiation, a bright green reflectance was detected in the canopy regions that were infiltrated with the fluorochrome. More recently, Rigoulot et al. (2019) tested mixtures of multiple fluorescent proteins (out of 20 various proteins)

Figure 4 Overexpression and suppression of anthocyanin gene *VlmybA1-2* in *Nicotiana tabacum* (Xanthi) plants. Wild-type (a) and transgenic (b) *N. tabacum* (Xanthi) seedlings overexpressing anthocyanin pigments before (upper row) and 40 days after (lower row) VIGS treatment.

for phenomics purposes. Combinations of different fluorescent proteins were imaged by using a fluorescence-inducing laser projector (FILP) platform with a low-noise camera. Four of the fluorescent proteins were detected in most of the plant canopy, suggesting a potential contribution for weed detection. The authors mentioned that despite the fact that the FLIP was tested under controlled lab conditions, adaptations of the system may be relevant in the future for crop detection, even from drones (Rigoulot et al., 2019). This statement suggests the potential applicability of the technology for close-range herbicide application and for remote weed mapping.

Despite their potential applicability in PWM, genetic modifications have not been widely used for that purpose. It seems that sociological aspects and public opinion regarding the consumption of genetically modified food are the main barriers. Furthermore, genetic engineering methodologies are expensive and time-consuming, as they involve complicated production stages before commercialization. Nonetheless, we cannot ignore the technological leaps that have taken place in the fields of genotyping and genome editing, since these fields carry the potential for crop improvement. The CRISPR/Cas-based genome-editing technique has indeed become widely used in the agricultural arena, since its high efficiency, low cost and simplicity make it a suitable technology for accelerating crop improvement (Chen et al., 2019): genome editing provides scientists with the ability to precisely and quickly insert or remove a desired/undesirable trait (Bari et al., 2019). Thus, fundamental breakthroughs in crop improvement, disease resistance and other aspects of agriculture, such as weed detection, could be developed on the basis of genome editing. Alternative genes, expressing pigments as traceable markers that differentiate transgenes from the surrounding weeds, are available, which could promote precise application using CRISPR\Cas-based genome-editing techniques. For example, knockout and repair of the phytoene desaturase (PDS) gene involved in the carotenoid biosynthesis pathway in crop plants might impact the reflectance pattern and facilitate improved classification between the crop plants from the surrounding weeds. It seems that the public is more open to CRISPR/Cas-modified crops than the genetically modified ones; however, regulation aspects must be resolved to accelerate research.

3.2 Crop signalling

Recently, crop signalling was developed and tested as another methodology to improve crop detection and classification from weeds (Kennedy et al., 2019; Su, 2020b,c; Su et al., 2020). As was the case for the genetic modifications, the focus here is on the crop and not on the weeds. Since no genetic manipulations are employed, this novel method may thus be better accepted and used by farmers. The main concept is to expose the crop plants to a signalling compound (i.e.

fluorescent compound) before or at planting (Su, 2020b). Subsequent excitation of the plants at specific wavelengths will cause the compound to generate very strong fluorescent emissions from the canopy or the stem, which can then be easily detected by a sensor. The methodology allows robust differentiation between the crop and the weeds and, more importantly, enables accurate setting of the crop location to facilitate precision application of herbicide or alternative mechanical means. To be suitable for real-time detection and agricultural purposes, the signalling compounds should meet several demands. From the crop perspective, the compounds should be safe for food crops with no toxic effects. They should also have no effect on plant development or cause any reduction in the yield quality or quantity. From the sensor perspective, these compounds should be photostable for a sufficient period of time, allowing accurate detection at the relevant growth stages for weeding. They should also produce strong and singular signals to ensure detection under field conditions without any linkages to the surrounding weeds and the soil.

Three main application approaches have been suggested for the crop signalling methodology: the physical, the topical and the systemic. The concept of the physical approach is the most basic one, in which plant labels painted with a fluorescent compound are used to determine crop location. The plant labels can be made from various materials in different shapes and sizes, but the labels should be large enough so that the fluorescent compound that covers them provides sufficiently strong signals that will ensure adequate identification levels by moving platforms under high weed densities (Raja et al., 2019). Low cost and biodegradability are additional desirable characteristics, and therefore, plastic straws made of polylactic acid (PLA) or of maize-based plastic are the materials of choice for plant labelling. The plant labels are placed next to the crop plant, at planting pots or even in the transplant flat at seeding time. To identify the labels, an imaging chamber equipped with a top-view camera, six UV lights (to excite the fluorescent compound) and six mirrors located on both sides of the crop plant were developed. The top-view camera coupled with the six view angles, provided by the mirrors, allowed to set the location of the crop plants according to the geometric appearance. The imaging chamber is connected to pneumatic knives that perform the weeding operation according to the detection, with small 2 cm safety zone left around the stem of the crop plant. The physical labelling approach was evaluated on lettuce and tomato, representing crops with different morphologies - rosette vs. vertical growth. Results showed 90% and 66% more weed removal compared to the standard cultivator in tomato and lettuce, respectively, in fields with various weed species such as purslane [*Portulaca oleracea*], lambsquarter [*Chenopodium album*], pigweed [*Amaranthus retroflexus*] and black nightshade [*Solanum nigrum*] without causing any damage to the crop plants or affecting their final yield. The multi-view mirror system allowed real-time detection with the platform moving at 3.2 km h^{-1}. It also provided robust detection under high infestation levels when the crop was shaded by weeds from one of the viewing sides. The plant labels approach offers a general solution that can be applied to any crop without the need to pre-define crop-specific spectral or morphological features. It also minimizes false-negative misdetections, as none of the weeds contains fluorescent compounds. There can, however, be misclassifications resulting mainly from the removal of labels by irrigation, wind or other field-related factors, or because the labels did not remain in good proximity to the crop. There are two other limitations associated with the approach: the manual marking is time-consuming and costly, and the approach, which was developed for robotic weeding where imaging is performed at close range with a unique imaging chamber, may not be relevant for remote-sensing or other sensor technologies.

3.3 Topical markers

The topical markers approach employs water-based latex fluorescent paints, which are applied to the seedling canopy or steam, simultaneously with the transplanting step. By doing so, the tedious, time-consuming and costly marking stage, which involves manual attaching of the labels to the crop seedlings/transplants, is avoided. The topical markers are automatically applied from a moving platform, that is, a standard row crop transplanter that has been modified to include a spraying system for precise real-time application of these compounds (Vuong et al., 2017). The topical marker approach is quite general and can be applied to a wide variety of crops. However, a delicate application balance is required: the area coved by the fluorescent compound should be optimized to ensure a strong signal that can be accurately identified by a sensor but should not cover the leaves or other plant parts to the extent that photosynthesis is reduced or growth is inhibited. They must stay on the plant for several weeks period, until weeding timing, under various field conditions that include extreme light and irrigation. The topical marker approach was tested on lettuce and tomato. Here, the detection system was coupled to a micro-jet herbicide-spraying device (Raja et al., 2020b). For tomato, the stem parts, optimally at least 25 cm, were sprayed with the signalling compound and herbicide application was performed 3 weeks after transplanting. Based on the unique signal from the paint, the vision system showed ~99% accuracy in three different field trials with no false positives. The main reasons for misdetections were occlusion, preventing the signal from reaching the sensor and misapplication of the signalling compound. For lettuce, the rosette leaves were sprayed with the topical marker and two sets of imaging systems were employed. The first one, with white light, aimed to detect all vegetation, while the second one, with UV light, aimed to detect the crop plants. Then, the two images were combined to identify the location of the crop plants in real time and the technique showed 98% accuracy. At crop harvesting, the topical markers were left only on the sprayed leaves that had dried out with no residual paint on the crop.

3.4 Systemic markers

Unlike topical markers, systemic signalling compounds are applied to the seed coat or to the root system of the seedling in transplant trays. The compound is absorbed by the seedling and systemically translocated to the stem and the canopy during plant development. The compounds located in these organs allow the induction of an optical signal that is detected by a sensor under a specific combination of exciting light and an optical filter. The lipophilicity of the compound is an essential characteristic for its functionality, as it must be able to move easily through the plant xylem. It should also be photostable with no negative impacts on the environment. Finally, a low concentration of the compound should be sufficient to give very strong fluorescence from the plants. One such compound is rhodamine B (RhB), which emits a unique signal under UV light. Experiments with this compound have been performed in snap beans, lettuce and celery. Snap bean seeds were soaked in an RhB solution and then planted in pots. When the seedlings developed, the compound absorbed from the seeds was sufficient for 100% detection among three different weed species under controlled conditions and allowed accurate setting of the stem location (Su et al., 2019). The experiment revealed that RhB could be detected in the bean stems, with limited transport of the compound into the leaves. In the lettuce experiment, the root system of the seedlings was treated with RhB and the seedlings were then transplanted into pots. The plants were excited with UV light, which resulted in the emission of bright red-orange light that was detected by the sensor. The phenotypic reflectance of the treated lettuce plants differed significantly from that of non-treated plants, which appeared darker. For celery, preliminary field trials showed photostability of RhB under full sunlight for 5 weeks. However, the concentration and treatment duration of RhB were found to affect the later development of the celery plants (Su et al., 2020), indicating that the impact of the RhB marker on plant health and vigour has not been yet fully elucidated. This aspect needs further study, as the cytotoxicity of this fluorescent compound in high doses is well known.

3.5 Nanotechnology

The concept of systemic markers can be taken one step further with the use of nanomaterials and nanotechnology methodologies. This novel and evolving research field manipulates and uses materials with dimensions smaller than 100 nm for different applications in medicine, environmental science, agriculture and food processing (Shang et al., 2019). Generally, nanomaterials can be harnessed to promote plant growth and protection by optimizing the application of various agricultural inputs, such as fertilizers and pesticides (Ghormade et al., 2011). Nanocarriers offering controlled release that delivers these inputs at the right time and to the right place can improve crop productivity while minimizing negative environmental effects (Hofmann et al., 2020). More specifically, nanosensors can contribute to PWM by the detection of invisible plant stress-related chemical signals into unique fluorescence signals (Wilson et al., 2015). These optical signals can be recorded in real time by currently available sensing equipment (e.g., multispectral camera) to provide indications as to the plant health status (Giraldo et al., 2019). There are a number of reports of the monitoring of plant signalling molecules by nanosensors; these signalling molecules include reactive oxygen species (ROS) (Giraldo et al., 2015), calcium (Krebs et al., 2012) and plant hormones, such as jasmonic acid (Larrieu et al., 2015). More recently, late blight (caused by *Phytophthora infestans*) was successfully detected by a low-cost smart-phone-based volatile organic compound monitoring device (Li et al., 2019). This handheld device contains a disposable colorimetric sensor array consisting of plasmonic nanocolorants with chemically responsive organic dyes that rapidly detect low levels of key plant volatiles that are associated with late blight. An excellent example of how the technology can be harnessed is for the detection of ROS: since an early physiological response of crop plants that develop in the close vicinity of weeds is the accumulation of ROS (H_2O_2), even at early stages before resources become limited (Afifi and Swanton, 2012; McKenzie-Gopsill et al., 2019), nanosensors can also be used for online monitoring of the presence of weeds. In summary, nanotechnology may offer potential solutions for sustainable agriculture in general, and for PWM specifically. Smart sensors can be used to translate chemical signals associated with weed-related stress into wireless, electrical and optical signals, and transfer the data through existing agricultural electronic devices. However, several barriers must be overcome before commercial application of these technologies becomes feasible, including field-level efficacy, regulatory and safety concerns and above all, consumer acceptance.

4 Herbigation

In 1997–1999, irrigated land represented about 20% of the total arable area but cereal production constituted 40–60% of this area and is expected to increase further. In developing countries, the irrigated area is expected to expand from 202 million hectares in 1997–1999 to 242 million hectares by 2030 (http://www.fao.org/land-water/en/). Expanding the area of irrigated crops worldwide may require similarly increased use of methods such as chemigation for pest control, where chemigation is defined as the application of agricultural chemicals injected into the water flowing through an irrigation system

The application of herbicides through an irrigation system – herbigation – offers many advantages, including economic use of existing equipment, reduction in soil compaction, activation using the irrigation water and incorporation of chemicals to the soil (Myers, 1985). Drip application may also provide farmers with a timely and cost-effective approach for applying pre-emergence herbicides. In comparison with spraying, the application of herbicides via drip irrigation may offer additional benefits, including the better movement of the chemical into the target root zone, increased safety for field workers and reduced crop damage due to herbicide drift (Thomas et al., 2003; Wang et al., 2009). The application protocol comprises an initial wetting of the soil, herbicide application and then an irrigation flush to wash herbicide residues from the irrigation system. Herbigation can be applied to a variety of fields, trees and vine crops and ornamentals.

4.1 Herbigation for weed control in various crops

Several studies have illustrated the potential of herbigation for weed control. For weed control in container-grown ornamentals, such as Azalea (*Rhododendron*), liriope (*Liriope muscari*) and Japanese holly (*Ilex crenata*), *S*-metolachlor, oryzalin and napropam applied via the irrigation system were found to be more effective in controlling large crabgrass (*Digitaria sanguinalis*), than applied by foliar spray (Caviness et al., 1988).

In the case of field-grown vegetables, the application of halosulfuron to an eggplant crop was efficient for the control of nutsedges (*Cyperus* spp.) but an inverse linear relationship was observed between halosulfuron rate and eggplant growth and final yield (Webster and Culpepper, 2005). For tomato, drip-applied fomesafen, halosulfuron and *S*-metolachlor provided good control of yellow nutsedge (*Cyperus esculentus*) but yellow nutsedge control diminished with the distance from the drip emitters (Adcock et al., 2008). Herbigation has also been used for the control of parasitic weed species in different vegetable crops, such as tomato, carrot and parsley. For tomato, integrating chlorsulfuron and triasulfuron into drip irrigation or overhead sprinkler irrigation successfully controlled *Phelipanche aegyptiaca* (Eizenberg et al., 2012).

A series of field trials conducted at the Aberdeen Research and Extension Center, University of Idaho evaluated the efficacy of site-specific weed management using differential herbigation in zones of a field with variable infestation levels (Eberlein et al., 2000). In an experiment conducted in a potato field, a mixture of *S*-metolachlor and metribuzin was herbigated at different rates according to the infestation levels of *Brassica juncea* and *Setaria italica* in each zone. Overall, weed control was very good for all the tested zones, showing the potential of this in adjusting herbicide doses according to weed infestation levels.

4.2 Limitations of herbigation

Herbigation can be conducted by sprinklers and drippers, but sprinklers can be used only under optimal environmental conditions with no wind. For dripper herbigation, the situation is technically more complicated, as distribution may be affected by the chemical and physical parameters of both the soil and the applied herbicide. The outcome could be a reduction in the homogeneity of herbicide distribution through the soil and hence increased herbicide leaching and reduced availability at the plant roots. For both sprinklers and drippers, herbicide doses must be carefully calibrated for crop safety and for meeting regulatory requirements. In addition, some plant species, such as ornamental plants, are highly sensitive to herbicides and may not survive herbigation. Thus, it is necessary to conduct trials for specific species prior to commercial application.

5 Spatial distribution patterns of weeds: the need for precise pre-emergence management

Most PWM studies have shown that different weed species in different crop systems tend to cluster spatially, and thus spot spraying can reduce herbicide application while ensuring adequate weed control levels (Dieleman and Mortensen, 1999; Heijtimg et al., 2007; Andujar et al., 2012; Martín et al., 2015; Blank et al., 2019; Rozenberg et al., 2021). Today, PWM represents a relatively well-studied discipline that utilizes spatial aspects of weeds in agricultural fields to facilitate weed control. However, the temporal aspect of PWM, which requires data for a period of years, has generally been overlooked. Several studies focusing on the stability of weed patches indicated that these patches remain stable over time (Wilson and Brain, 1991; Blanco-Moreno et al. 2006; Heijtimg et al., 2007; Blank et al., 2019). Heijtimg et al. (2007) attributed patch temporal instability to both the dispersion mechanism of the species, being greater for wind-dispersed seeds, and for species with sparser populations. Other field studies showed that pre-harvest dispersal was important for patch stability of annual weed species, since it will result in compact and dense seed patches (Wilson and Brain, 1991; Dieleman and Mortensen, 1999; Gerhards et al., 1997). Another potential characteristic affecting patch stability is seed weight. Heavy seeds are likely to generate dense and stable patches (Heijtimg et al., 2007). When studying patch stability, all of the above mechanisms and characteristics, along with their interaction, should be considered. *Ecballium elaterium* seeds, for example, are relatively heavy and thus are expected to generate stable patches. However, as this species has a unique dispersal mechanism and seeds are actively dispersed and could disperse to almost all parts of the plot, it was expected that temporal stability will be limited. However, Blank et al. (2019) showed that the size and location of the *E. elaterium* infested areas remained similar for 3 years, supporting the notion that in this case the spatial aggregation was not the result of dispersal limitation but was rather based on niche limitation.

A lack of temporal stability (Izquierdo et al., 2009) or stability over short time spans (Colbach et al., 2000) has also been reported for some weed species. Thus, further research is needed to better understand which species aggregate and produce temporally stable patches and which species exhibit low temporal stability. Such information can be used to direct pre-emergence herbicide treatments when visual information is not available for a field. In this regard, the information regarding the spatial distribution of weeds in a particular year could serve as the basis for making spraying decisions in the following year.

Pre-emergence application is an important tool for an effective weed management programme targeting weeds that have not yet emerged from the soil. The most effective weed control programmes include pre-emergence herbicides that offer extended control of germination (known as 'residual herbicides'). The advantages of pre-emergence herbicides are that they mitigate yield losses by reducing weed competition (Sarangi and Jhala, 2017), provide an extended time for the crop to grow and establish (Eizenberg and Goldwasser, 2018), reduce the selection pressure for resistance to post-emergence herbicides (Moss et al., 2019) and effectively reduce or even eliminate weed seed banks. In addition, the inclusion of pre-emergence herbicides into a PWM programme can reduce the use of post-emergence herbicides (Kaur et al., 2020). The stability of weeds patches favours precise PRE treatments. Farmers can estimate the quantity of herbicides needed in advance, optimize their purchase of herbicides and reduce costs and space for storage. Understanding the spatio-temporal dynamics of weed distribution can facilitate more effective and precise PRE herbicide management. This, in turn, has obvious advantages for cost savings and reduces the environmental impact of weeds control.

6 Conclusion and future trends

New developments in PWM should not be limited to innovations in the fields of imaging techniques and sensing platforms: Complementary methodologies and disciplines should also be integrated into PWM. For example, we can significantly improve the performances of the currently used weeding machines and robots, in terms of detection and robustness, even without revolutionized sensors or detection algorithms. In the future, biological innovations should take a greater share in the development of new PWM. Additionally, temporal information should be taken into consideration to optimize the pre-emergence application. This can also be achieved by new herbicide application methods that optimize the time and rate of application. There is still room for improvement in the basic components of herbicide application, and to achieve the overall goal of minimizing herbicide amounts, all aspects of the pipeline should be optimized.

7 Where to look for further information

7.1 Further reading

- Kunz, C., Weber, J. F. and Gerhards, R. (2015). Benefits of precision farming technologies for mechanical weed control in soybean and sugar beet - comparison of precision hoeing with conventional mechanical weed control. *Agronomy* 5, 130-142.
- Raja, R., Slaughter, D. C., Fennimore, S. A., et al. (2019). Crop signalling: A novel crop recognition technique for robotic weed control. *Biosystems Engineering* 187, 278-291.

7.2 Key journals and conferences

WR, *WS*, *WT* (*Weed Research*, *Weed Science*, *Weed Technology*) are the main weed-oriented journals that publish articles about advanced methodologies and innovations in weed control and management.

- *PMS* (*Pest Management Science*) this journal provide coverage about advanced methodologies and novel concept to control weeds. This interdisciplinary journal focus on wide aspects of pest control, thus, regularly reports about non-conventional new weed control methods.
- *YBENG* (*Biosystems Engendering*) is an international journal that reports about the development of new robotics and vision-based machinery for weed detection and control.
- *PA* (*Precision Agriculture*) is an international journal on advances in precision
- agriculture and publish articles related to weed sensing and detection.
- *COMPAG* (*Computers and Electronics in Agriculture*) provides coverage of advances in computer hardware, software, electronic instrumentation and control systems for precision agriculture.

8 References

Adcock, C. W., Foshee, W. G., Wehtje, G. R. and Gilliam, C. H. (2008). Herbicide combinations in tomato to prevent nutsedge (*Cyperus esculentus*) punctures in plastic mulch for multicropping systems. *Weed Technology* 22(1), 136-141.

Afifi, M. and Swanton, C. (2012). Early physiological mechanisms of weed competition. *Weed Science* 60(4), 542-551.

Ali, A., Streibig, J. C. and Andreasen, C. (2013). Yield loss prediction models based on early estimation of weed pressure. *Crop Protection* 53, 125-131.

Aly, R., Lati, R., Bari, V. K., Abu-Nassar, J. and Eizenberg, H. (2019). Use of a visible reporter marker- myb-related gene in crop plants to minimize herbicide usage against weeds. *Plant Signaling and Behavior* 14(4), e1581558.

Andujar, D., Baroso, J., Fernandez-Quintanilla, C. and Dorado, J. (2012). Spatial and temporal dynamics of Sorghum halepense patches in maize crops. *Weed Research* 52(5), 411-420.

Bakhshipour, A. and Jafari, A. (2018). Evaluation of support vector machine and artificial neural networks in weed detection using shape features. *Computers and Electronics in Agriculture* 145, 153-160.

Bakhshipour, A., Jafari, A., Nassiri, S. M. and Zare, D. (2017). Weed segmentation using texture features extracted from wavelet sub-images. *Biosystems Engineering* 157, 1-12.

Bari, V. K., Nassar, J. A., Kheredin, S. M., Gal-On, A., Ron, M., Britt, A., Steele, D., Yoder, J. and Aly, R. (2019). CRISPR/Cas9-mediated mutagenesis of CAROTENOID CLEAVAGE DIOXYGENASE 8 in tomato provides resistance against the parasitic weed Phelipanche aegyptiaca. *Scientific Reports* 9(1), 11438.

Berge, T. W., Goldberg, S., Kaspersen, K. and Netland, J. (2012). Towards machine vision based site-specific weed management in cereals. *Computers and Electronics in Agriculture* 81, 79-86.

Blanco-Moreno, J. M., Chamorro, L. and Sans, F. X. (2006). Spatial and temporal patterns of Lolium rigidum-Avena sterilis mixed populations in a cereal field. *Weed Research* 46(3), 207-218.

Blank, L., Birger, N. and Eizenberg, H. (2019). Spatial and temporal distribution of Ecballium elaterium in almond orchards. *Agronomy* 9(11) .

Burgos-Artizzu, X. P., Ribeiro, A., Guijarro, M. and Pajares, G. (2011). Real-time image processing for crop/weed discrimination in maize fields. *Computers and Electronics in Agriculture* 75(2), 337-346.

Caviness, D. M., Talbert, R. E. and Klingaman, G. L. (1988). Chemigation and spray application of herbicides on container-grown ornamentals. *Weed Technology* 2(4), 418-422.

Chen, K., Wang, Y., Zhang, R., Zhang, H. and Gao, C. (2019). CRISPR/Cas genome editing and precision plant breeding in agriculture. *Annual Review of Plant Biology* 70, 667-697.

Christensen, S., Sogaard, H. T., Kudsk, P., Nørremark, M., Lund, I., Nadimi, E. S. and Jørgensen, R. (2009). Site-specific weed control technologies. *Weed Research* 49(3), 233-241.

Coble, H. D. and Mortensen, D. A. (1992). The threshold concept and its application to weed science. *Weed Technology* 6(1), 191-195.

Colbach, N., Forcella, F. and Johnson, G. A. (2000). Spatial and temporal stability of weed populations over five years. *Weed Science* 48(3), 366-377.

Cormack, B. P., Valdivia, R. H. and Falkow, S. (1996). FACS-optimized mutants of the green fluorescent protein (GFP). *Gene* 173(1), 33-38.

Cousens, R. (1985). A simple model relating yield loss to weed density. *Annals of Applied Biology* 107(2), 239-252.

Cousens, R., Brain, P., O'Donovan, J. T. and O'Sullivan, P. A. (1987). The use of biologically realistic equations to describe the effects of weed density and relative time of emergence on crop yield. *Weed Science* 35(5), 720-725.

Dieleman, J. A. and Mortensen, D. A. (1999). Characterizing the spatial pattern of Abutilon theophrasti seedling patches. *Weed Research* 39(6), 455–467.

Dyrmann, M., Karstoft, H. and Midtiby, H. S. (2016). Plant species classification using deep convolutional neural network. *Biosystems Engineering* 151, 72–80.

Eberlein, C. V., King, B. A. and Guttieri, M. J. (2000). Evaluating an automated irrigation control system for site-specific herbigation 1. *Weed Technology* 14(1), 182–187.

Eizenberg, H., Aly, R. and Cohen, Y. (2012). Technologies for smart chemical control of broomrape (*Orobanche* and *Phelipanche* spp.). *Weed Science* 60(2), 316–323.

Eizenberg, H. and Goldwasser, Y. (2018). Control of Egyptian broomrape in processing tomato: A summary of 20 years of research and successful implementation. *Plant Disease* 102(8), 1477–1488.

Fernández-Quintanilla, C., Peña, J. M., Andújar, D., Dorado, J., Ribeiro, A. and López-Granados, F. (2018). Is the current state of the art of weed monitoring suitable for site-specific weed management in arable crops? *Weed Research* 58(4), 259–272.

Franco, C., Pedersen, S. M., Papaharalampos, H. and Ørum, J. E. (2017). The value of precision for image-based decision support in weed management. *Precision Agriculture* 18(3), 366–382.

Gerhards, R. and Kühbauch, W. (1993). Dynamisches Entscheidungsmodell zur Lenkung von Unkrautkontrollmaßnahmen in Wintergetreide mit Hilfe digitaler Bildverarbeitung. *Journal of Agronomy and Crop Science* 171(5), 329–335.

Gerhards, R. and Oebel, H. (2006). Practical experiences with a system for site-specific weed control in arable crops using real-time image analysis and GPS-controlled patch spraying. *Weed Research* 46(3), 185–193.

Gerhards, R., Sökefeld, M., Schulze-Lohne, K., Mortensen, D. A. and Kühbauch, W. (1997). Site specific weed control in winter wheat. *Journal of Agronomy and Crop Science* 178(4), 219–225.

Gerowitt, B. and Heitefuss, R. (1990). Weed economic thresholds in cereals in the Federal Republic of Germany. *Crop Protection* 9(5), 323–331.

Ghormade, V., Deshpande, M. V. and Paknikar, K. M. (2011). Perspectives for nano-biotechnology enabled protection and nutrition of plants. *Biotechnology Advances* 29(6), 792–803.

Giraldo, J. P., Landry, M. P., Kwak, S. Y., Jain, R. M., Wong, M. H., Iverson, N. M., Ben-Naim, M. and Strano, M. S. (2015). A ratiometric sensor using single chirality near-infrared fluorescent carbon nanotubes: Application to in vivo monitoring. *Small* 11(32), 3973–3984.

Giraldo, J. P., Wu, H., Newkirk, G. M. and Kruss, S. (2019). Nanobiotechnology approaches for engineering smart plant sensors. *Nature Nanotechnology* 14(6), 541–553.

Gobor, Z., Schulze Lammers, P. and Martinov, M. (2013). Development of a mechatronic intra-row weeding system with rotational hoeing tools: Theoretical approach and simulation. *Computers and Electronics in Agriculture* 98, 166–174.

Gutjahr, C. and Gerhards, R. (2010). Decision Rules for site-specific weed management. In: Oerke, E. C., Gerhards, R., Menz, G. and Sikora, R. A. (Eds), *Precision Crop Protection - the Challenge and Use of Heterogeneity* (1st edn.), 345pp., Hardcover. Springer. ISBN: 978-90-481-9276-2, London, UK.

Gutjahr, C., Sokefeld, M. and Gerhards, R. (2012). Evaluation of two patch spraying systems in winter wheat and maize. *Weed Research* 52(6), 510–519.

Heijtimg, S., Werf, W. van der, Stein, A. and Kropff, M. J. (2007). Are weed patches stable in location? Application of an explicitly two-dimensional methodology. *Weed Research* 47(5), 381–395.

Hofmann, T., Lowry, G. V., Ghoshal, S., Tufenkji, N., Brambilla, D., Dutcher, J. R., Gilbertson, L. M., Giraldo, J. P., Kinsella, J. M., Landry, M. P., Lovell, W., Naccache, R., Paret, M., Pedersen, J. A., Unrine, J. M., White, J. C. and Wilkinson, K. J. (2020). Technology readiness and overcoming barriers to sustainably implement nanotechnology-enabled plant agriculture. *Nature Food* 1(7), 416-425.

Izquierdo, J., Blanco-Moreno, J. M., Chamorro, L., Recasens, J. and Sans, F. X. (2009). Spatial distribution and temporal stability of prostrate knotweed (Polygonum aviculare) and corn poppy (*Papaver rhoeas*) seed bank in a cereal field. *Weed Science* 57(5), 505-511.

Kaur, R., Das, T. K., Singh, R., Jaidka, M. and Shekhawat, K. (2020). Managing weeds using sequential herbicides in maize for improving crop growth and productivity under irrigated conditions in North-Western India. *Maydica* 65(2).

Kennedy, H., Fennimore, S. A., Slaughter, D. C., Nguyen, T. T., Vuong, V. L., Raja, R. and Smith, R. F. (2019). Crop signal markers facilitate crop detection and weed removal from lettuce and tomato by an intelligent cultivator. *Weed Technology* 34(3), 342-350.

Kounalakis, T., Triantafyllidis, G. A. and Nalpantidis, L. (2019). Deep learning-based visual recognition of rumex for robotic precision farming. *Computers and Electronics in Agriculture* 165, 104973.

Krebs, M., Held, K., Binder, A., Hashimoto, K., Den Herder, G., Parniske, M., Kudla, J. and Schumacher, K. (2012). FRET-based genetically encoded sensors allow high-resolution live cell imaging of Ca2+ dynamics. *The Plant Journal: For Cell and Molecular Biology* 69(1), 181-192.

Kropff, M. J. and Spitters, C. J. T. (1991). A simple model of crop loss by weed competition from early observations on relative leaf area of the weeds. *Weed Research* 31(2), 97-105.

Kunz, C., Weber, J. F. and Gerhards, R. (2015). Benefits of precision farming technologies for mechanical weed control in soybean and sugar beet - Comparison of precision hoeing with conventional mechanical weed control. *Agronomy* 5(2), 130 -142.

Kunz, C., Weber, J. F., Peteinatos, G. G., Sökefeld, M. and Gerhards, R. (2018). Camera steered mechanical weed control in sugar beet, maize and soybean. *Precision Agriculture* 19(4), 708-720.

Lambert, J. P. T., Hicks, H. L., Childs, D. Z. and Freckleton, R. P. (2018). Evaluating the potential of Unmanned Aerial Systems for mapping weeds at field scales: A case study with Alopecurus myosuroides. *Weed Research* 58(1), 35-45.

Larrieu, A., Champion, A., Legrand, J., Lavenus, J., Mast, D., Brunoud, G., Oh, J., Guyomarc'h, S., Pizot, M., Farmer, E. E., Turnbull, C., Vernoux, T., Bennett, M. J. and Laplaze, L. (2015). A fluorescent hormone biosensor reveals the dynamics of jasmonate signalling in plants. *Nature Communications* 6, 6043.

Lati, R. N., Filin, S., Aly, R., Lande, T., Levin, I. and Eizenberg, H. (2014). Using genetically modified tomato crop plants with purple leaves for absolute weed/crop classification. *Pest Management Science* 70(7), 1059-1065.

Lati, R. N., Filin, S. and Eizenberg, H. (2011). Robust methods for measurement of leaf-cover area and biomass from image data. *Weed Science* 59(2), 276-284.

Lati, R. N., Filin, S., Elnashef, B. and Eizenberg, H. (2019). 3-D image-driven morphological crop analysis: A novel method for detection of sunflower broomrape initial subsoil parasitism. *Sensors* 19(7), 1-32.

Li, Z., Paul, R., Ba Tis, T., Saville, A. C., Hansel, J. C., Yu, T., Ristaino, J. B. and Wei, Q. (2019). Non-invasive plant disease diagnostics enabled by smartphone-based fingerprinting of leaf volatiles. *Nature Plants* 5(8), 856-866.

Longchamps, L., Panneton, B., Simard, M.-J. and Leroux, G. D. (2014). An imagery-based weed cover threshold established using expert knowledge. *Weed Science* 62(1), 177-185.

López-Granados, F. (2011). Weed detection for site-specific weed management: Mapping and real-time approaches. *Weed Research* 51(1), 1-11.

Lotz, L. A. P., Christensen, S., Cloutier, D., Quintanilla, C. F., Legere, A., Lemieux, C., Lutman, P. J. W., Iglesias, A. P., Salonen, J., Sattin, M., Stigliani, L. and Tei, F. (1996). Prediction of the competitive effects of weeds on crop yields based on the relative leaf area of weeds. *Weed Research* 36(1), 93-101.

Louargant, M., Villette, S., Jones, G., Vigneau, N., Paoli, J. N. and Gée, C. (2017). Weed detection by UAV: Simulation of the impact of spectral mixing in multispectral images. *Precision Agriculture* 18(6), 932-951.

Martín, C. S., Andújar, D., Fernández-Quintanilla, C. and Dorado, J. (2015). Spatial distribution patterns of weed communities in corn fields of central Spain. *Weed Science* 63(4), 936-945.

Mathews, H., Clendennen, S. K., Caldwell, C. G., Liu, X. L., Connors, K., Matheis, N., Schuster, D. K., Menasco, D. J., Wagoner, W., Lightner, J. and Wagner, D. R. (2003). Activation tagging in tomato identifies a transcriptional regulator of anthocyanin biosynthesis, modification, and transport. *The Plant Cell* 15(8), 1689-1703.

McKenzie-Gopsill, A. G., Amirsadeghi, S., Earl, H. J., Jones, A. M. P., Lukens, L., Lee, E. and Swanton, C. J. (2019). Early physiological and biochemical responses of soyabean to neighbouring weeds under resource-independent competition. *Weed Research* 59(4), 288-299.

Mink, R., Dutta, A., Peteinatos, G. G., Sökefeld, M., Engels, J. J., Hahn, M. and Gerhards, R. (2018). Multi-temporal site-specific weed control of Cirsium arvense (L.) scop. and Rumex crispus L. in maize and sugar beet using unmanned aerial vehicle based mapping. *Agriculture (Switzerland)* 8(5), 65.

Moss, S., Ulber, L. and Hoed, I. den (2019). A herbicide resistance risk matrix. *Crop Protection* 115, 13-19.

Myers, R. G. (1985). Center-pivots and chemigation. Proceedings of 5th National Symposium of Chemigahon, Rural Development Center, Tifton, GA.

Nørremark, M., Griepentrog, H. W., Nielsen, J. and Søgaard, H. T. (2012). Evaluation of an autonomous GPS-based system for intra-row weed control by assessing the tilled area. *Precision Agriculture* 13(2), 149-162.

Pantazi, X. E., Moshou, D. and Bravo, C. (2016). Active learning system for weed species recognition based on hyperspectral sensing. *Biosystems Engineering* 146, 193-202.

Peteinatos, G. G., Reichel, P., Karouta, J., Andújar, D. and Gerhards, R. (2020). Weed identification in maize, sunflower, and potatoes with the aid of convolutional neural networks. *Remote Sensing* 12(24), 4185-4207.

Peteinatos, G. G., Weis, M., Andújar, D., Rueda Ayala, V. and Gerhards, R. (2014). Potential use of ground-based sensor technologies for weed detection. *Pest Management Science* 70(2), 190-199.

Piron, A., Heijden, F. van der and Destain, M. F. (2011). Weed detection in 3D images. *Precision Agriculture* 12(5), 607-622.

Pray, L. (2008). Recombinant DNA technology and transgenic animals. *Nature Publishing Education* 1, 51.

Raja, R., Nguyen, T. T., Slaughter, D. C. and Fennimore, S. A. (2020a). Real-time robotic weed knife control system for tomato and lettuce based on geometric appearance of plant labels. *Biosystems Engineering* 194, 152–164.

Raja, R., Nguyen, T. T., Slaughter, D. C. and Fennimore, S. A. (2020b). Real-time weed-crop classification and localisation technique for robotic weed control in lettuce. *Biosystems Engineering* 192, 257–274.

Raja, R., Nguyen, T. T., Vuong, V. L., Slaughter, D. C. and Fennimore, S. A. (2020c). RTD-SEPs: Real-time detection of stem emerging points and classification of crop-weed for robotic weed control in producing tomato. *Biosystems Engineering* 195, 152–171.

Raja, R., Slaughter, D. C., Fennimore, S. A., Nguyen, T. T., Vuong, V. L., Sinha, N., Tourte, L., Smith, R. F. and Siemens, M. C. (2019). Crop signalling: A novel crop recognition technique for robotic weed control. *Biosystems Engineering* 187, 278–291.

Rasmussen, J., Griepentrog, H. W., Nielsen, J. and Henriksen, C. (2012). Automated intelligent rotor tine cultivation and punch planting to improve the selectivity of mechanical intra-row weed control. *Weed Research* 52(4), 327–337.

Rasmussen, J., Nielsen, J., Streibig, J. C., Jensen, J. E., Pedersen, K. S. and Olsen, S. I. (2019). Pre-harvest weed mapping of Cirsium arvense in wheat and barley with off-the-shelf UAVs. *Precision Agriculture* 20(5), 983–999.

Rew, L. J. and Cousens, R. D. (2001). Spatial distribution of weeds in arable crops: Are current sampling and analytical methods appropriate? *Weed Research* 41(1), 1–18.

Rigoulot, S. B., Schimel, T. M., Lee, J. H., et al. (2019). Fluorescence-based whole plant imaging and phenomics. bioRxiv 865428.

Rozenberg, G., Kent, R. and Blank, L. (2021). Consumer-grade UAV utilized for detecting and analyzing late-season weed spatial distribution patterns in commercial onion fields. *Precision Agriculture* 22(4), 1317–1332.

Ruckelshausen, A., Klose, R., Linz, A., Marquering, J., Thiel, M. and Tölke, S. (2006). Autonome Roboter zur Unkrautbekämpfung. *Zeitschrift für Pflanzenkrankheiten und Pflanzenschutz, Sonderheft* XX, 173–180.

Sa, I., Popović, M., Khanna, R., Chen, Z., Lottes, P., Liebisch, F., Nieto, J., Stachniss, C., Walter, A. and Siegwart, R. (2018). WeedMap: A large-scale semantic weed mapping framework using aerial multispectral imaging and deep neural network for precision farming. *Remote Sensing* 10(9), 1423–1448.

Sarangi, D. and Jhala, A. J. (2017). Biologically effective rates of a new premix (atrazine, bicyclopyrone, mesotrione, and S-metolachlor) for preemergence or postemergence control of common waterhemp [Amaranthus tuberculatus (Moq.) Sauer Var. rudis] in corn. *Canadian Journal of Plant Science* 97, 1075–1089.

Shang, Y., Kamrul Hasan, M. K., Ahammed, G. J., Li, M., Yin, H. and Zhou, J. (2019). Applications of nanotechnology in plant growth and crop protection: A review. *Molecules* 24(14), 2558–2581.

Sharpe, S. M., Schumann, A. W., Yu, J. and Boyd, N. S. (2020). Vegetation detection and discrimination within vegetable plasticulture row-middles using a convolutional neural network. *Precision Agriculture* 21(2), 264–277.

Slaughter, D. C., Giles, D. K. and Downey, D. (2008). Autonomous robotic weed control systems: A review. *Computers and Electronics in Agriculture* 61(1), 63–78.

Su, W.-H. (2020a). Advanced machine learning in point spectroscopy, RGB- and hyperspectral-imaging for automatic discriminations of crops and weeds: A review. *Smart Cities* 3(3), 767–792.

Su, W.-H. (2020b). Systemic crop signaling for automatic recognition of transplanted lettuce and tomato under different levels of sunlight for early season weed control. *Challenges* 11(2), 23.

Su, W.-H. (2020c). Crop plant signaling for real-time plant identification in smart farm: A systematic review and new concept in artificial intelligence for automated weed control. *Artificial Intelligence in Agriculture* 4, 262–271.

Su, W. H., Fennimore, S. A. and Slaughter, D. C. (2019). Fluorescence imaging for rapid monitoring of translocation behaviour of systemic markers in snap beans for automated crop/weed discrimination. *Biosystems Engineering* 186, 156–167.

Su, W. H., Fennimore, S. A. and Slaughter, D. C. (2020). Development of a systemic crop signalling system for automated real-time plant care in vegetable crops. *Biosystems Engineering* 193, 62–74.

Thomas, J. G., Pennington, D. A., Schmidt, A. M. and Pringle, L. (2003). *Chemigation.* MS: Mississippi State Extension Service Publication 1551, Starkville, MS.

Thorp, K. R. and Tian, L. F. (2004). A review on remote sensing of weeds in agriculture. *Precision Agriculture* 5(5), 477–508.

Tillett, N. D., Hague, T., Grundy, A. C. and Dedousis, A. P. (2008). Mechanical within-row weed control for transplanted crops using computer vision. *Biosystems Engineering* 99(2), 171–178.

Tillett, N. D., Hague, T. and Miles, S. J. (2002). Inter-row vision guidance for mechanical weed control in sugar beet. *Computers and Electronics in Agriculture* 33(3), 163–177.

van Thor, J. J., Gensch, T., Hellingwerf, K. J. and Johnson, L. N. (2002). Phototransformation of green fluorescent protein with UV and visible light leads to decarboxylation of glutamate 222. *Nature Structural Biology* 9(1), 37–41.

Vuong, V. L., Slaughter, D. C., Nguyen, T. T., Fennimore, S. A. and Giles, D. K. (2017). An automated system for crop signaling and robotic weed control in processing tomatoes. ASABE Annual International Meeting, ASABE Paper No. 1700871.

Wang, A., Zhang, W. and Wei, X. (2019). A review on weed detection using ground-based machine vision and image processing techniques. *Computers and Electronics in Agriculture* 158, 226–240.

Wang, X., Zhu, H., Reding, M. E., Locke, J. C., Leland, J. E., Derksen, R. C., Spongberg, A. L. and Krause, C. R. (2009). Delivery of chemical and microbial pesticides through drip irrigation systems. *Applied Engineering in Agriculture* 25(6), 883–893.

Webster, T. and Culpepper, A. (2005). Eggplant tolerance to halosulfuron applied through drip irrigation. *HortScience* 40(6), 1796–1800.

Weis, M., Gutjahr, C., Rueda-Ayala, V., Gerhards, R., Ritter, C. and Schölderle, F. (2008). Precision farming for weed management: Techniques. *Gesunde Pflanzen* 60(4), 171–181.

Wilson, B. J. and Brain, P. (1991). Long-term stability of distribution of Alopecurus myosuroides Huds. within cereal fields. *Weed Research* 31(6), 367–373.

Wilson, R. H., Nadeau, K. P., Jaworski, F. B., Tromberg, B. J. and Durkin, A. J. (2015). Review of short-wave infrared spectroscopy and imaging methods for biological tissue characterization. *Journal of Biomedical Optics* 20(3), 030901.

Chapter 10

Advances in mechanical weed control technologies

*Bo Melander and Margaret R. McCollough, Aarhus University, Denmark**

1 Introduction

Mechanical weed control (MWC) for agricultural and horticultural crops encompasses various belowground soil cultivation techniques and aboveground cutting, mowing and weeding tactics. Mowing disrupts aboveground vegetation, immediately eliminates weed competition and hinders the shedding of weed seeds. Removal of vegetation is common practice in many orchards and nurseries where the wide spacing between rows of woody plants allows for the operation of mowers (Hammermeister, 2016). Mowing also plays a significant role in the control of perennial weeds, for example, *Cirsium arvense* in pastures and whole-year green manure crops (Melander et al., 2016). Repeated aboveground cutting of thistle plants depletes the sugars stored in belowground root structures over time, reducing their potential to infest succeeding crops (Graglia et al., 2006). Finally, the development of intra-row weed control tactics using air-propelled abrasive grit shows promise in crops tolerant to the treatment (Carlson et al., 2018). However, this chapter will not address aspects related to mowing and abrasive grit techniques any further. The main focus is on soil cultivation strategies for the mechanical control of weeds growing in annual field crops sown in narrow rows (cereals, pulses and oilseed crops) or wide rows (sugar beets, maize and many

* Aarhus University, Department of Agroecology, Research Centre Flakkebjerg, DK-4200 Slagelse, Denmark. E-mail bo.melander@agro.au.dk

http://dx.doi.org/10.19103/AS.2021.0098.11

vegetables). In this context, MWC is used when the upper 0–5 cm soil layer is cultivated to control weeds. The majority of technologies discussed in this chapter have little effect on the shoots of perennial weeds. Mechanical control of severe perennial infestations requires deeper and more intense cultivations between crop plantings (Melander et al., 2012).

Long before the invention of herbicides, MWC constituted the backbone of weed management. Mouldboard ploughing, seedbed cultivation prior to crop establishment, and inter-row hoeing and weed harrowing within established crops were the primary strategies for reducing weed infestations. However, other preventive and cultural measures were needed to supplement MWC to provide satisfactory control, among which the diversification of crop rotation was arguably most important. In modern times, an increase in conversion to organic farming and the imposition of herbicide restrictions in many European countries and elsewhere around the world have resulted in a revitalization of interest and investment in MWC. Older methods, such as weed harrowing and hoeing, have been the subject of new research to better understand their weeding mechanisms and strategic use in various crops (e.g. Melander et al., 2003; Kurstjens and Kropff, 2001; Rasmussen, 1991). This development began to take off in the 1990s and accelerated in the following years due to further restrictions on herbicide use, increasing problems with herbicide resistance, and poor prospects concerning the development of herbicides with new modes of action (Kudsk and Mathiassen, 2020).

In recent years, the exchange of knowledge and ideas among practitioners, consultants and researchers has increased immensely in countries restricting herbicide use and possessing vibrant organic sectors, such as Germany, Denmark, Austria and Switzerland. This change has led to improvements and many new crop-specific weed management strategies (e.g. Rasmussen et al., 2010; Melander et al., 2018; van der Weide et al., 2008). However, the continuous integration of electronics into mechanical devices for weed control has meant a significant step forward over the last 20 years. Mechanical solutions are now feasible in weed management programmes outside the organic sector. Particularly, the invention of GNSS (global navigation satellite system) and vision guidance technologies has helped automate and ease the task of steering mechanical tools, such as hoes and finger weeders (Machleb et al., 2020). In recent years, implements designed for automatic intra-row weed control in row crops have appeared on the market, and more are likely to come in the future. Intra-row weeds are defined as those growing in the crop line and few centimetres to either side. The prospect that row crops can be grown without herbicides and manual weeding could potentially solve urgent issues, such as herbicide resistance, the absence of effective herbicides, and lack of labour for hand weeding. Growers currently benefit from automatic intra-row weeders in transplanted crops through significant labour savings for manual weeding (Lati et al., 2016). In addition, the release of labour for other tasks makes the expansion of acreage with valuable row crops possible, thereby increasing farm income (Melander, 1998). Despite these obvious advantages with new technologies, limitations and drawbacks exist and must be addressed before the broader adoption of MWC can be achieved.

Today's market offers a wide range of weeding devices for the mechanical control of small-sized weeds that can be grouped into three categories: full-width cultivators, inter-row cultivators, and intra-row cultivators (Machleb et al., 2020). Several reviews on MWC methods have been published in recent years (e.g. Gallandt et al., 2018; Machleb et al., 2020; Melander et al., 2005; van der Weide et al., 2008). This chapter will highlight the most recent and relevant advances within each MWC category. The focus will be on novel inventions and developments of mechanical devices, designs, and the weed problems they are meant to solve. Moreover, automation technologies that assist weeding operations are becoming increasingly important and will be given special attention.

2 The mechanisms of mechanical weed control

Weeds that establish from seeds are vulnerable to mechanical control when small in size; they are most sensitive from the white thread stage until the first true leaf begins to unfold. Weeding efficacy declines as weeds develop; however, efficacy decreases at differing rates among weeding devices. The lethal effects of mechanical cultivators arise from the soil disturbance they cause; mechanical cultivation uproots weed plants and covers them with soil, both mechanisms working simultaneously during operation (Melander et al., 2017). Some cultivators also cut weeds, dissecting the roots from shoots or causing damage to the roots, stem, or leaves, contributing to an increased desiccation rate. Uprooting occurs when roots are displaced from their original position, causing them to tear apart. Uprooting reduces root function and increases desiccation rate if soil conditions are dry. Soil burial excludes light and prevents photosynthesis in green plant tissue, becoming lethal if weeds cannot grow through the soil layer due to insufficient energy reserves. Rasmussen (1991) described crop and weeds effects following light tine cultivation, in the form of weed harrowing, by quantifying the amount of soil thrown onto the crop plants. The percentage of crop soil cover provided a reasonable relationship with crop response and weeding effectiveness. However, Rasmussen's studies did not clarify the exact mechanisms responsible for weed mortality when operating a weed harrow. Kurstjens and Kropff (2001) got closer to understanding the mechanisms of tine cultivation using a laboratory weed harrowing setup; this enabled careful assessments of weed size and position and the degree of uprooting and burial damage. Results showed that uprooting is the primary lethal mechanism of tine cultivation when weed seedlings are weakly anchored in soil, typical from the white thread stage until the first true leaves start to unfold. Therefore, soil covering becomes an increasingly important mechanism of weed mortality as rooting, and thus anchoring, improves with growth. Even relatively large weeds can be killed through soil burial; however, partial burial increases the likelihood of survival (Merfield et al., 2020). Melander (1997) observed this when covering *Sinapis arvensis* at the zero to two true leaf stage and the two to four leaf stage with 5 cm of soil and achieved approximately 80% and 40% control, respectively. Merfield et al. (2020) suggest that a burial depth of 6 cm will kill most plants regardless of species or growth stage. Weed plants that have surpassed the seedling stage would therefore require cultivation to a greater total soil depth to achieve 6 cm of soil cover. The effects of soil covering described above hold true for tines and weeding devices that provide a ridging action. Notably, hoe shares and other blades possessing a cutting action can uproot or sever weed plants at more advanced growth stages with several true leaves (Melander et al., 2005)

3 Full-width cultivation

Harrowing effectively controls weeds when they are small, before the first true leaves become visible. Post-emergence weed harrowing treats both the crop and weeds uniformly. Therefore, successful harrowing occurs when the increased crop yield attributed to reduced competition from effective weed control is greater than the yield losses resulting from the crop damage and burial inflicted. Selective harrowing typically requires a size difference between the crop and weeds, where crop plants are large enough to withstand uprooting and soil covering, while weed plants are smaller and more vulnerable to mechanical impact (Fig. 1). Several studies have focused on improving the selectivity of full-width weed harrowing in small grain cereals, pulses, maize and vegetables (Melander et al., 2017). The strategic use of weed harrowing and guidelines for appropriate settings during operation have been improved thanks to research and the exchange of knowledge among practitioners. Attempts have been made to adjust the aggressiveness of weed harrowing in real-time according to online weed detection using ultrasonic sensors mounted at the front of the tractor (Rueda-Ayala et al., 2015). Gerhards et al. (2021) determined the intensity of weed harrowing in real-time by computing crop soil cover using digital cameras mounted before and after the harrow. Harrowing intensity was continuously adjusted to achieve 10% crop soil cover; being the pre-set threshold for the decision algorithm, it was expected to maximize weed control efficacy while limiting crop injury. These examples of improving harrowing performance by employing advanced technologies have not yet resulted in the commercialization of equipment, but the potential for improved operation is evident.

Figure 1 A well-anchored barley crop with few and relatively small weed plants - successful weed harrowing possible. Courtesy of Bo Melander, Aarhus University, Denmark.

As implements, tine harrows have not improved to a noteworthy degree, with the exception of the newly introduced Treffler harrow (www.treffler.net/en/products/agricultural-machinery/precision-tine-harrow, accessed 27 December 2020). The Treffler harrow has not resolved the fundamental problem of low selectivity, that is, treating both crops and weeds. Instead, Treffler has markedly improved the mechanisms for adjusting tine aggression and suspension. Each tine is able to move independently on the frame and is individually preloaded with a spring. Tines can therefore adjust to within-field contours while maintaining constant down-pressure regardless of their position. The Treffler harrow has also demonstrated its advantages for weed harrowing along ridges, such as potato ridges, with the ability to cultivate the plateau-like profile with relative uniformity (Fig. 2). However, following several passes with the harrow, the ridge will have to be re-established.

Ridging potatoes generally offers an excellent opportunity for intense cultivation until the potato shoots start emerging. Potato ridgers, rolling

Figure 2 Weed harrowing on potato ridges with a Treffler harrow. Courtesy of Bo Melander, Aarhus University, Denmark.

cultivators and weed harrows are all proven effective for weed management in potatoes, but drawbacks have also been encountered. Crop injuries, insufficient working capacities, and forming ridges off centre from crop rows are emphasized among others (Melander et al., 2011). A new invention was introduced recently to resolve some of the problems mentioned called the OptiWeeder (https://msrplanttechnology.dk, accessed 27 December 2020). OptiWeeder does not use modern vision or GNSS technologies to assist the steering task. Instead, units following each row are flexible at their toolbar attachment point, allowing each unit to align independently while following along the ridges. Weed control is achieved by running angled knives on either side of the ridge and on the top that function to undercut weeds at a depth of 2 cm; knives are followed by a set of plates that re-build the ridge. (Fig. 3). Driving speeds of 15 km/h are possible; however, the width of the machine requires further expansion to achieve working rates desired by conventional potato growers. The first tests with OptiWeeder showed high weeding effectiveness and no noteworthy crop injuries, though documentation of its weeding potential is still limited (Fig. 4).

4 Inter-row cultivation

Weed harrowing used to be the principal physical weed control method applied in organic cereals. However, the adoption of weed harrowing in

Figure 3 A unit of the OptiWeeder for treating one potato ridge. Courtesy of Bo Melander, Aarhus University, Denmark.

practice has been difficult in many cases. There seems to be a steady move away from the sole use of this technology and towards other methods and strategies. Optimal timing, settings and execution are the main challenges of weed harrowing mentioned by practitioners, leading to poor weed control and occasionally substantial crop yield loss. Erect dicotyledonous weed species with taproots and tall-growing annual grasses are particularly difficult to control; in addition, perennial weed species are not affected much by harrowing (Rasmussen, 1998). Species such as *S. arvensis*, *Brassica rapa* and *Raphanus raphanistrum* are troublesome because they establish quickly, have fast initial growth rates and can emerge in series of cohorts (Rasmussen et al., 2010).

Because of the disadvantages of full-width weed harrowing in cereals and other crops grown at narrow row spacing, growers have turned to inter-row cultivation with steerable hoes. Hoeing between crop rows is widely applied in traditional row crops where the operation is straightforward (Melander et al.,

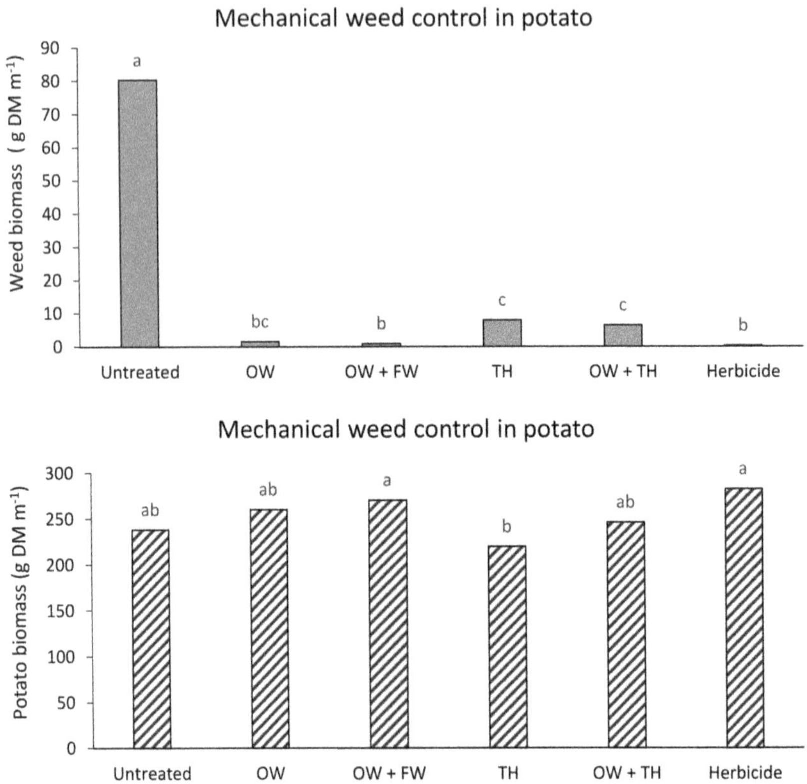

Figure 4 Mechanical weed control in potato with OptiWeeder (O.W.), Treffler harrow (T.H.), and finger weeding (F.W.) – three passes were implemented for each mechanical treatment. Effects are shown for weed and crop biomasses. Columns with similar letters are not statistically different ($P < 0.05$). (Melander, B., unpublished data).

2005; Machleb et al., 2020). The inter-row weeding device typically employed is the goosefoot share, providing a cutting action that nearly removes all inter-row weeds unless soil conditions are wet or weeds have become too large for control (Melander et al., 2005). Inter-row hoeing also has application in cereals grown at an increased inter-row spacing to make room for the operation of a goosefoot share between crop rows (Jabran et al., 2017). Hoeing is most effective against annual weeds but may also have some effect on perennials (Graglia et al., 2006). Belowground propagules are not directly affected by hoeing; however, shoot removal will stimulate re-sprouting, depleting belowground food reserves over time. Shoot removal interrupts the translocation of photosynthetic assimilates to roots and rhizomes; overall, these effects can impede perennial weeds' regenerative capacity.

Renewed interest in inter-row hoeing for cereals and pulses may also be attributed to recent and substantial innovations that ease the task of steering, namely, automated systems based on camera and GNSS technologies (Kunz et al., 2018). These technologies remove the need for manual steering and enable inter-row hoeing with greater operational capacity since implement width and driving speed can both be increased (Kunz et al., 2015). Vision-based steering systems typically consist of one or more cameras mounted on the hoeing implement to detect crop lines (Fig. 5). The imaging information is computed to signal actuators that align the hoe with crop rows while driving. Some hoes have a hydraulic side-shift between the hoe and the tractor, enabling the hoe to move right or left; for example, see Garford Robocrop System (https://garford.com/products/robocrop-guided-hoes/, accessed 27 December 2020), which is explained in detail by Connolly (2003). Danish organic growers report that inter-row hoeing in cereals works well with driving speeds of 5-10 km/h and 25 cm inter-row spacing, a doubling of the traditional 12.5 cm inter-row spacing. Manufacturers of vision guidance technologies and hoes claim that inter-row hoeing down to 15 cm inter-row spacing is possible at reasonable forward speeds, but this option is not purchasable yet (Agrointelli, personal communication). Vision guidance technologies are currently dominating the market for automatic steering systems sold alongside well-known hoe brands across Europe (Fernández-Quintanilla et al., 2018). RTK-GPS (real-time kinematic global positioning system) steering systems can also be used for precise inter-row hoeing if the crop rows' positions are recorded during seeding. RTK-GPS does not require crop-specific knowledge but relies on the expected location of the crop rather than real-time information delivered by cameras. However,

Figure 5 Camera-steered inter-row hoeing in spring barley. Courtesy of Bo Melander, Aarhus University, Denmark.

camera-steered side shifting units can change lateral position instantly and directly in response to the actual conditions in the field, which is a clear advantage over GNSS solutions.

Autonomous tool carriage systems have recently become available on the European market, offering an alternative to automatic tractor-mounted cultivators. Compared to tractor-based MWC, autonomous weeding robots reduce labour requirements and soil compaction; however, they rely on similar methods for tracking and following crop rows. Naïo Technologies (https://www.naio-technologies.com/, accessed 27 December 2020) combines camera-vision and RTK-GPS or sensor-based guidance in their models designed for operation in vineyard and vegetable cropping systems. Agrointelli (https://www.agrointelli.com/robotti/, accessed 27 December 2020) utilizes RTK-GPS and possesses a standard three-point hitch with power take off (PTO). While the designs of Naïo Technologies' and Agrointelli's autonomous weeding robots undoubtedly represent a significant step forward, the tools responsible for weed control remain simple, including selective inter-row tools (shares and knives) and non-selective intra-row tools (finger, torsion, and brush weeders, as well as tine harrows).

Compared to weed harrowing, inter-row hoeing in cereals is more effective against problematic weed species, such as grasses and tap-rooted broadleaved species with an erect growth (Melander et al., 2003, 2018). Moreover, efficacy increases with the proportion of the surface area being cultivated (Fig. 6). Timing of treatment is less crucial with inter-row hoeing than weed harrowing because the shares' cutting action also controls weeds with more than two or three true leaves (Fig. 7). Intra-row weeds are not directly affected by hoe shares

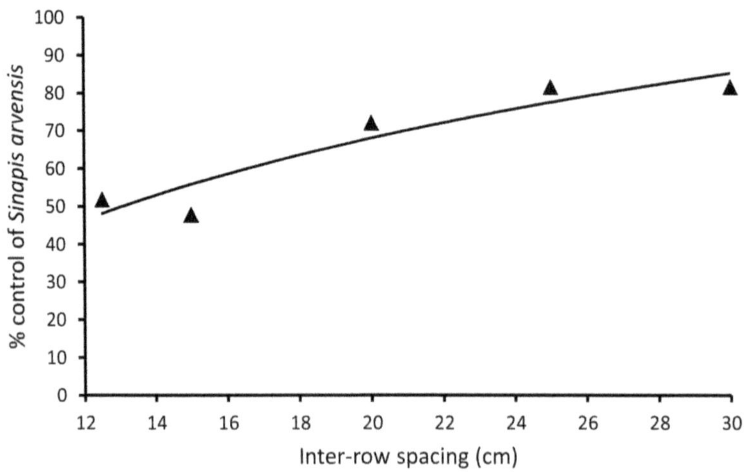

Figure 6 Relationship between % control of *Sinapis arvensis* and inter-row hoeing at increasing inter-row spacing in organic spring barley (Melander, B., unpublished data).

Figure 7 Effective inter-row hoeing is still possible despite large-sized weeds. Courtesy of Bo Melander, Aarhus University, Denmark.

and are not controlled unless sideways soil movement causes some burial. This ridging action is determined by driving speed and share configuration, and ridging may cause some adverse crop effects if exaggerated (Melander et al., 2018; Wiltshire et al., 2003). Fast driving speed is desirable for the achievement of high work rates but is risky at small crop growth stages when crop leaves are easily buried (Melander et al., 2003). Risk can be alleviated by reducing the share blade angle, making the tool's configuration flatter (Znova et al., 2018). Machleb et al. (2018) observed less sideward soil movement with a flatshare versus the traditional goosefoot share when hoeing in cereals at narrow inter-row spacings of 12.5 cm and 15 cm. Crop yields also tended to be higher with the flatshare, while efficacy was slightly lower than the goosefoot share, which caused more intra-row soil coverage. Flatshares need to work closer to the crop row to achieve similar efficacies as shares with a greater blade angle (Fig. 8). Steering accuracy then becomes particularly crucial with a flatshare to avoid crop injuries. Maintaining a constant and stable position of the shares in relation to the crop rows is another critical factor in ensuring uniform hoeing treatments. Share edges should be kept at the desired distance from the crop row to avoid crop injuries. Apart from accurate steering, the stiffness of shanks onto which the shares are mounted is important to obtain uniformity and reliability. An example of a new shank and share, designed for stiffness and flatness, is shown in Fig. 9.

Intra-row weeds remain a problem when inter-row hoeing, especially tall-growing cruciferous species that can reduce crop yields markedly, as shown in Table 1 (Melander and McCollough, 2020). Mixed intra-row weed populations with a greater proportion of weed species short in stature may not be as competitive as seen in a Danish study with inter-row hoeing, performed in 11 weedy fields

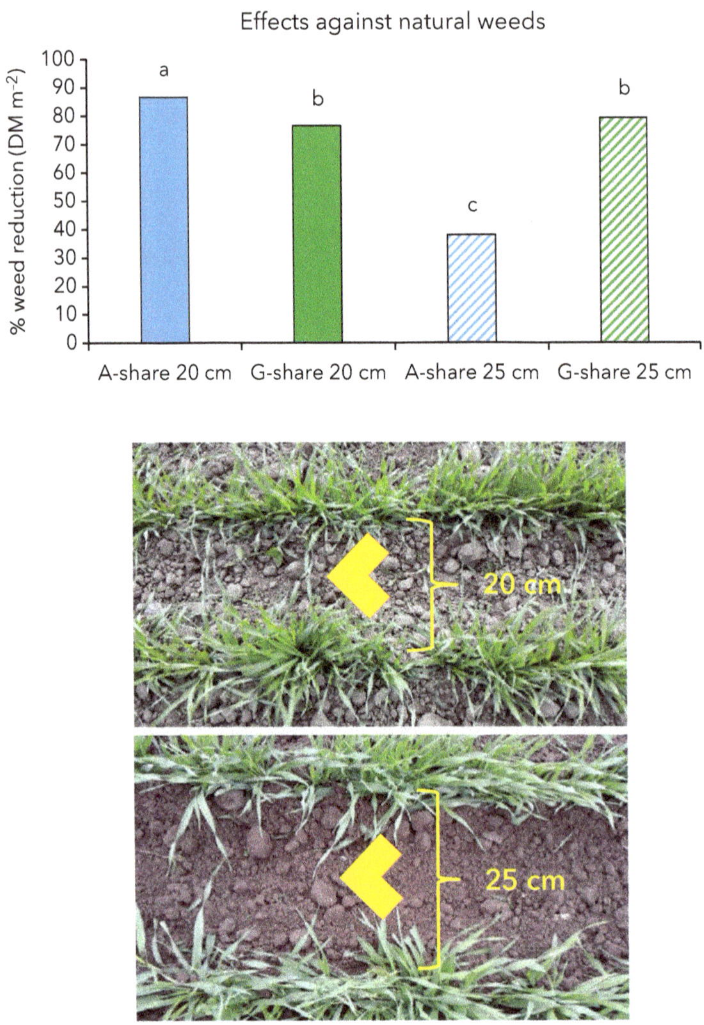

Figure 8 Weeding efficacy of inter-row hoeing in spring barley using a 13 cm wide goosefoot share (G-share) and flat share (A-share, see Figure 9) at 20 cm and 25 cm inter-row spacings. Columns with similar letters are not statistically different ($P < 0.05$). (Melander, B., unpublished data).

with organic spring cereals. Yields were on average only 7% lower with inter-row hoeing versus inter-row hoeing plus hand-weeding of surviving intra-row weeds (Theilgaard and Bertelsen, 2017). Nevertheless, competitive intra-row weeds need to be managed by other means, such as increased weed suppression through band sowing (McCollough et al., 2020a,b) and/or an increase of within-row crop density (Jabran et al., 2017). Supplementary herbicide application or

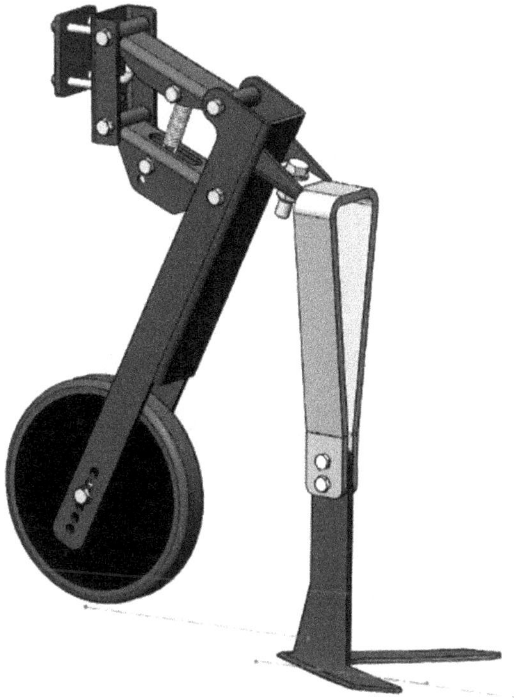

Figure 9 New share and shank design from AgroIntelli (www.AgroIntelli.com, accessed 27 December 2020).

weed harrowing applied pre- and post-crop emergence can reduce intra-row weed numbers and eliminate or mitigate potential yield losses.

Another drawback seen with inter-row hoeing is a yield penalty of 11–12 % in conventional cereals arising from the widening of inter-row spacing from the standard 12.5 cm to 25 cm (Melander et al., 2003). Interestingly, the same yield penalty was not observed in organic spring cereals where wide inter-row spacings (up to 30 cm) yielded the same as narrow spacings (down to 12.5 cm). Lower yields in organically grown crops and the use of manures, from which nutrients are released more slowly and are less abundant, are probable reasons for this discrepancy between the conventional and organic scenarios (Melander et al., 2018).

5 Intra-row cultivation

Crop stands are typically very dense in the intra-row zone of cereals, pulses, oilseed rape and some horticultural crops such as carrot and direct-sown onion and leek. High-density planting makes the selective operation of mechanical

Table 1 Ranges of yield losses resulting from two years of experiments on intra-row weed competition in organic spring barley and spring wheat, grown at 15 and 25 cm inter-row spacings. White mustard (*Sinapis alba*) was used to simulate cruciferous intra-row weed growth typical for *Raphanus raphanistrum*, *Sinapis arvensis*, and *Brassica rapa*. Intra-row surrogate weeds *Sinapis alba* (plants m^{-2}) are defined as those plants growing in the uncultivated area 2.5 cm to either side of the crop row's center (Melander and McCollough, 2020).

Crop	Intra-row density of *Sinapis alba* Plants m^{-2}	% yield loss
Spring barley	20	12-25
	100	28-70
	500	38-99
Spring wheat	20	13-49
	100	38-86
	500	60-99

tools very difficult, especially if individual crop plants are to be left untouched. Cereal rows can be ridged slightly to control intra-row weeds that are much smaller than crop plants. Any other operation of a mechanical device in the intra-row zones will negatively affect the crop plants, which may result in yield loss. Thus, intra-row weeds cannot be mechanically controlled to a satisfactory degree in densely planted crops.

The operation of mechanical intra-row cultivators such as finger-weeders, torsion weeders, brush weeders become more relevant when within-row crop spacing increases. Finger-weeders steered by automatic guidance systems can be used in many row crops, notably transplanted vegetables (cabbages, onion, leek, celery, etc.). Intra-row cultivators can also be employed in direct-sown row crops when conditions favour effective weed control without crop injuries. This typically happens when there is a marked size-difference between weeds and crop plants, and soil conditions are relatively dry, loose and workable.

5.1 Stacking tools for intra-row cultivation

Intra-row weed control efficacy increases with additional passes and heightening intensity at which each pass is conducted (Melander et al., 2005). Finger weeders and tine-based cultivators work the soil differently; combining or 'stacking' different tools into one pass may improve overall efficacy when compared to single passes with the same tool. Brown and Gallandt (2018) equipped an implement with three intra-row tools in sequence: torsion weeder, finger weeder and tine rake. This three-tool combination resulted in a synergistic effect on surrogate weed mustard (*Sinapis alba*), comparing to treatments using single tools. A range of tool combinations was studied, and not all had a synergistic effect; rather, several were additive. Stacking tools also means that the intensity of cultivation increases, and severe crop injuries become more likely since the crop is also treated. The most obvious advantage of stacking tools is that weed problems requiring several intense passes with a single tool might be controlled in one pass when employing the stacking concept. Stacking becomes particularly relevant in well-anchored and robust crop stands that can withstand intense cultivation. Tool stacking may help control weeds in situations where precipitation has delayed field operations, resulting in weeds too large to be effectively controlled with individual tools; however, a favourable outcome is not achieved if the crop is badly injured.

5.2 Automatic intra-row weeding

Intra-row weeds in row crops pose a unique challenge because of their close proximity to the crop. In sugar beet, greater yield reductions result from weeds growing 2 cm from crop plants than from weeds 8 cm away (Heisel et al., 2002). Yield loss caused by intra-row weeds is strongly dependent on the crop species. While intra-row weeds growing within 2 cm of transplanted white cabbage did not reduce marketable yield, intra-row weeds growing the same distance from transplanted onion reduced yield by 60 % (Fig. 10) (Melander et al., 2015). For most row crops, automatic intra-row weeding machines must operate as close to the crop plants as possible to minimize yield loss and the need for manual removal of surviving weeds (Lati et al., 2016; Fennimore et al., 2014). As weeds are most vulnerable when small in size, the same is true for the establishing

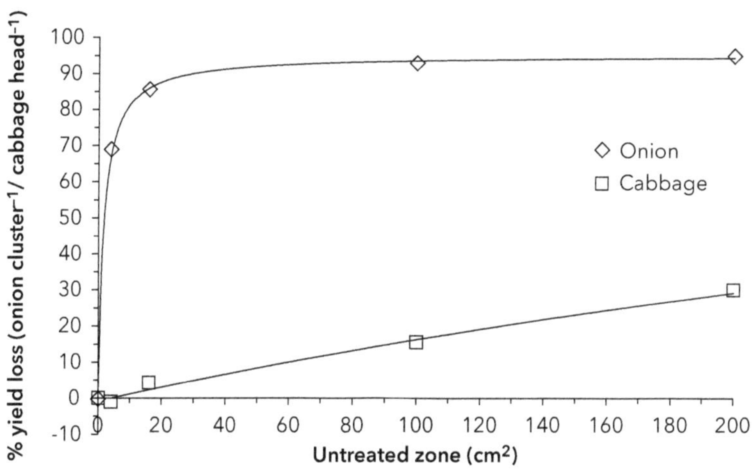

Figure 10 Percent weight loss per onion cluster and per cabbage head as affected by the size of a non-weeded zone around the crop plant (Melander et al., 2015, with permission from Crop Protection).

crop. Balancing the efficacy of weeding near crop plants while minimizing crop injury is another important consideration; selectivity must be considered while implementing automated post-emergence treatments.

In transplanted crops, automated intra-row weeders outfitted with vision-guidance systems are capable of cultivating between crop plants within the row without reducing crop stands or yields (Lati et al., 2016). Currently, five automatic intra-row weeders are available for practical use in the European market: Robovator (www.visionweeding.com, accessed 27 December 2020), Robocrop InRow (www.garford.com, accessed 27 December 2020), Steketee IC (www. steketee.com, accessed 27 December 2020), Ferrari Removeed (www.ferraricostruzioni.com, accessed 27 December 2020) and Farmdroid (www.farmdroid.dk/en, accessed 27 December 2020). The Ferrari Removeed uses infrared light sensors to detect crop plants, while Robovator, Robocrop, and Steketee IC-weeder use cameras to detect crop plants, distinguishing them from weeds. The website mentioned for each weeder contains excellent images and video clips that visualize the working principles of these intelligent cultivators.

The Robovator consists of a pair of rigid tines, each equipped with a flat knife-like blade that operates horizontally to the soil's surface at a depth of 1-2 cm, removing weeds by cutting (Fig. 11). Additional hoe shares treat the inter-row zone on either side of the crop row. Automated blades function in the intra-row zone until they approach a crop plant. At that point, the computer settings determine when to move the blades apart to avoid crop injury. When the crop plant has passed, the blades close and continue cultivating the intra-row. The movement in and out of the crop row is performed by a hydraulic actuator that responds to information produced by a camera mounted directly in front of it (Fig. 11). For each crop row, there is a camera that detects every crop plant based on the size differential between crop and weeds. Images are processed by a computer that calculates when the actuator must be activated according to driving speed and proximity to crop plants. The Steketee IC-weeder also has cameras that detect crop plants

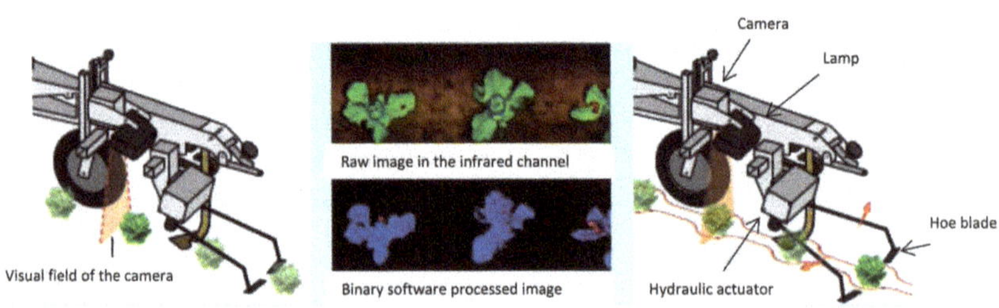

Figure 11 The working principles of the Robovator, intelligent mechanical intra-row weeder (Melander et al., 2015, with permission from Crop Protection and Enginøren).

Published by Burleigh Dodds Science Publishing Limited, 2022.

to provide visual information for computation. The subsequent guidance of a mechanical weeding device selectively controls for intra-row weeds. The device consists of a pair of sickle-shaped knives that move in and out of the crop row by pneumatic pressure created from a compressor. In contrast, the Robocrop InRow weeder employs a crescent-shaped disc that rotates about an axis. The tool is set to cultivate at a shallow depth of 1 cm to 2 cm within the crop row. The crescent-shaped disc is designed to arc around crop plants, cutting between the plants as it rotates. Rotation of the disc is synchronized with forward movement and informed by crop plant positional information delivered from the imaging camera. The disc is coupled directly to a hydraulic motor, driven by a proportional hydraulic valve controlled by the Robocrop computer.

The Farmdroid is an entirely different concept based on GNSS technology for marking a single crop plant's position. The machine is designed to perform both crop sowing and mechanical intra-row weeding. The placement of every crop seed is recorded during sowing; this geographical information is used to guide knife-like blades, weeding around the area where the crop plants are expected to establish. The blades move in and out of the intra-row zone, similar to Steketee and Robovator. In contrast to machines based on canopy monitoring, intra-row weeding can begin before crop emergence. The futuristic

Figure 12 Farmdroid working in newly established winter oilseed rape. The oilseed rape was sown by Farmdroid and is now being inter-row cultivated - another possible application with the machine. Courtesy of Sven Hermansen, SEGES, Denmark.

design of Farmdroid becomes apparent by its unmanned autonomous operation, powered by solar panels charging four batteries (Fig. 12). Currently, Farmdroid is the only machine that offers a selective autonomous intra-row weeding solution for direct-sown crops.

5.3 Experiences with automatic intra-row weed control

All the vision-guided machines mentioned above are best suited for use in crop stands where a clear crop-weed distinction is present. Crop recognition, and thus weeding accuracy, becomes more precise and reliable when crop plants are distinctly larger than the weeds and when there is abundant spacing between crop plants within the row (Frank Poulsen Engineering, personal communication).

There are relatively few scientific evaluations of the weeding performance of new automatic weeders. One study evaluating the performance of Robocrop in transplanted cabbage showed that under normal commercial growing conditions, crop damage levels are low, with weed reductions in the range of 62–87%, measured within a 24 cm radius zone around treated crop plants (Tillett et al., 2008). Fennimore et al. (2014) compared the performance of Robocrop with a standard inter-row cultivator in transplanted vegetables. As expected, intelligent weeding was more effective than the standard cultivator at reducing intra-row weed density and subsequent hand weeding times; this was mainly because the standard inter-row cultivator could not remove intra-row weeds. Lati et al. (2016) also compared automatic intra-row weeding using Robovator to a standard inter-row cultivator without the ability to control intra-row weeds in transplanted lettuce and direct-seeded broccoli. Despite the standard cultivator only leaving a 10.2 cm wide non-cultivated band centred over the crop line, automatic weeding was superior when weed pressure was moderate to high. The Robovator removed between 18% and 41% more intra-row weeds, resulting in up to 45% saving of hand-weeding labour compared to the standard cultivator. However, Robovator was not superior to non-intelligent intra-row weeding tools, such as the finger-weeder, weed harrow, and torsion weeder when operating in transplanted onion and white cabbage (Melander et al., 2015). Robovator removed between 54% and 86% of intra-row weeds, and only minor differences in efficacy were found among intelligent and non-intelligent cultivation treatments. Robovator works around a 'safety zone' encompassing the base of each crop plant, within which the decision algorithm prevents any hoeing from taking place to avoid crop injuries. In Melander et al. (2015), uncultivated safety zones of 4 cm and 6 cm were tested; however, zone size was found to have negligible effects. Tools without intelligence cultivate the entire area around crop stems, therefore, damaging crop plants. Theoretically, intelligent weeding should result in lower intra-row weed control than non-intelligent tools, but there are no indications of that. The weeding mechanism of Robovator is more about cutting (and partly uprooting) the weeds rather than covering them with soil, typical of the tine-based weed harrow and the finger-weeder. The effect of cutting weeds rather than burying them is more aggressive and less sensitive to weed growth stage at the time of treatment (Jones et al., 1996). Robovator may also be used later than most non-intelligent tools, allowing more weeds to germinate before cultivation and resulting in more weeds being controlled than with earlier treatments. Although the Robovator cultivates a smaller percentage of the intra-row area than the non-intelligent tools, Robovator's improved weeding efficacy may offset assumed adverse effects. As emphasized in Melander et al. (2015) and Lati et al. (2016), intelligent weeding has many other benefits over non-intelligent tools, including increased hours of operation (which is possible at night), ease of implementation, reduced risk of crop injury, need of only one operator, greater flexibility in treatment timing in relation to weed growth stage, and being the only alternative to manual intra-row hand weeding in lettuce.

The performance of Farmdroid has not yet been documented; however, some experiences have been garnered from operating units in commercial sugar beet fields over the last 2 years (Hermansen, 2020; personal communications with project manager Otto Nielsen at Nordic Beet Research (https://www.nordicbeet.nu/en/, accessed 27 December 2020) and farm manager Tom Ellerød Hansen at Oremandsgaard, Denmark). Farmdroid runs at a forward speed of only 0.8 km/h, weeding six rows simultaneously, resulting in low work rates. However, the machine can operate 24 h a day due to continuous battery charging during the daytime hours via attached solar panels. The crop seed-mapping feature makes intra-row weeding possible shortly after crop sowing and onwards, thanks to its autonomous operation. Large areas may require the simultaneous operation of several units, increasing investment costs markedly. Similar to camera-based intra-row weeders, the proximity at which knife-like blades can operate relative to crop plants without injury has a significant influence on the success of weed control. Fields with low weed pressure will have fewer weeds establish in the uncultivated safety zone around crop plants; whereas, fields with high weed pressure will inevitably have more survivors, requiring subsequent treatment measures, such as hand-weeding, to achieve satisfactory control. Practitioners have reported that the slow forward speed employed during crop sowing results in reliable positioning of the emerged crop plants. This enables intra-weeding as close as 1 cm from each plant's centre, especially if crop rows are treated from both directions; the knife-like blades are adjusted to weed closer to the crop plant upon passing. Therefore, the weeding action is performed in a movement away from, rather than towards, the crop plant.

One pass from each direction is needed to treat one row from both sides. The period in which effective weeding can take place is quite broad since the cutting action of tools can control weeds beyond the cotyledon stage. More importantly, Farmdroid can operate continuously, preventing weeds from becoming particularly large. Intra-row weeding machines reliant upon GNSS references do have the disadvantage of not cultivating areas where a seed has been planted, but a crop plant failed to establish, whereas camera-guided implements avoid all established crop plants and treating everything else.

The Farmdroid and the camera-guided solutions all undergo continuous improvement, receiving both hardware and software upgrades as these technologies continue to evolve. Changes to construction and design are also made; for example, the first version of Farmdroid was very light, which limited its function on heavy soils. Such experiences from the field have necessitated a heavier version with more robust components, including the frame, toolbar, shanks, weeding devices and wheels. Thus, the performance of an automatic intra-row weeder observed in one growing season may not hold true in the next due to continuous upgrades.

5.4 Perspectives for automatic intra-row weeding in direct-sown row crops

Industry representatives, advisory bodies and the research community all agree that the adaption of intelligent intra-row weeding technologies for operation in direct-sown row crops would constitute a major step forward (Utstumo et al., 2018; Melander et al., 2015). With seeding and weeding capabilities integrated into the same machine, Farmdroid is the only on-market implement specifically designed for operating in direct-sown crops. Sole reliance on GNSS technology for crop plant detection may be upgraded in the future and supplemented by vision guidance, helping to solve the problem of missing crop plants within the row and enhancing crop detection in general.

By using artificial intelligence and machine learning, significant progress is being made in developing vision-based technologies for selective intra-row weeding in direct-sown row crops. Machine learning is an iterative process; when the model does not detect crop plants accurately, previous images are re-assessed, and the model is revised to handle new data with greater accuracy (Fig. 13). Detection models are continuously rebuilt to handle crop plants' varying in appearance among different sites and growth stages. Eventually, comprehensive training across many scenarios will lead to a reliable crop detection system. Robovator is currently capable of adequate intra-row weeding in direct-seeded sugar beet fields with weeds overlapping the crop plants

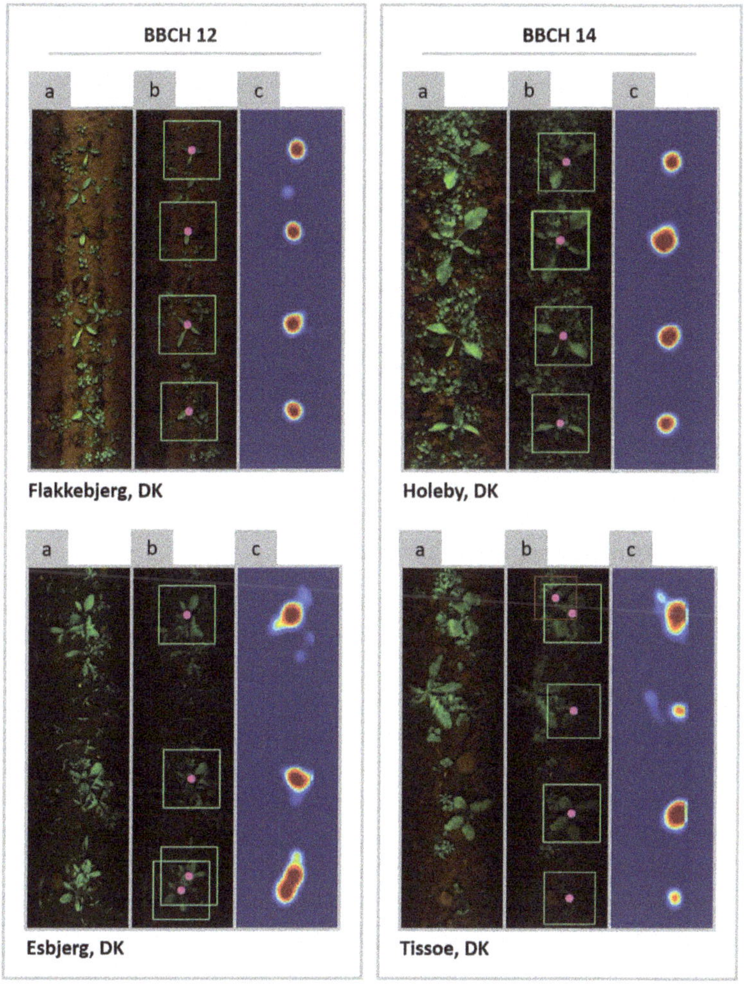

Figure 13 The ability of artificial intelligence (A.I.) to identify young direct-sown sugar beet plants in multiple varying scenarios. Examples include instances where weed pressure can be characterized as moderate to heavy. Successful crop detection is depicted across four sites in Denmark and at two early crop growth stages; the two true leaf stage (BBCH 12, left) and the four true leaf stage (BBCH 14, right). Images show (a) the raw image captured by the camera, (b) an A.I. output pinpointing the centre of each detected sugar beet plant, and (c) a second A.I. output depicting a heat map, showing the probability of sugar beet plant presence. Courtesy of Frank Poulsen Engineering.

(Fig. 14). Steketee IC has also taken on the challenge of achieving precise and reliable crop recognition in direct-seeded sugar beet, however, their current minimum requirement of 21 cm within-row spacing makes it difficult to achieve desired crop densities per hectare.

Figure 14 Robovator operating in weedy sugar beets. Courtesy of Frank Poulsen Engineering.

6 Future trends and conclusion

Full-width cultivation suffers from the fact that crops and weeds are treated simultaneously. New implements have emerged in recent years, and knowledge about the operation of full-width cultivators is continuously improving. Equipment design and the ease of making adjustments are also progressing; it is impressive to watch skilled growers operating these tools and the effects they can achieve with them. Nevertheless, the fundamental problem of non-selective implements remains a barrier for broader application and popularity; this issue is unsolvable as long as tools do not discriminate crop plants from weeds.

The increasing interest in inter-row cultivators does not stem from an ambition to solve the intra-row weed problem. Instead, the aim is to simplify and improve the control of inter-row weeds directly affected by the weeding device. Automatic steering systems constitute a major step forward in this regard, but the refinement of tools is still pertinent. The concept of stacking tools is an option with most commercial inter-row cultivators, although the solutions are often a compromise between cost and necessity. Inexpensive solutions comprised of inter-row tines mounted behind shares are often seen; however, the addition of tines may only contribute limited effects to work already done by aggressive shares. Given soil conditions prone to aggregate formation, hoeing efficacy may be diminished due to the survival of weeds attached to soil clods following cultivation. Weeds that remain upright and whose roots are protected from desiccation are likely to survive in a clod of soil if soil moisture remains adequate (Fig. 15). Mounting a device with a rotating and crushing action behind hoe shares is an appealing idea for breaking apart clods, resulting in weed roots' exposure. The split-hoe demonstrates such a

Figure 15 Weed seedlings attached to a clod. Courtesy of Bo Melander, Aarhus University, Denmark.

feature; however, the current iteration of the machine is designed for high-value specialty crops only (Pannacci et al., 2017).

Intra-row weeds remaining in the hoed cereal system pose a problem for the preservation of crop yields. The within-row crop stand is too dense for the operation of intelligent in-row weeding devices without inflicting crop injury. Preventive and cultural strategies, as well as the inclusion of tine harrowing, can provide some additional control of intra-row weeds; however, some weed species may escape these measures and reduce crop yields. Organic growers usually accept surviving weeds after mechanical interventions. Conventional growers expect cleaner fields; weedy crop lines may hinder the broader acceptance of the hoed cereal system. Other considerations, such as work rate and investment costs, may impede adoption among conventional growers. Band-spraying may be a viable solution to the intra-row weed problem. Preliminary results from the United Kingdom suggest that compared to full-width spraying, a 60% reduction in herbicide use is achievable when band spraying in cereals grown at a 16 cm row spacing (Cussans, J., personal communication). Results are undoubtedly in line with EU policies on integrated pest management, but feasibility relies on the practicalities of integrating band-spraying with inter-row cultivation.

Significant progress has been made in recent years regarding intelligent intra-row weeding in row crops that leave enough space for the selective operation of a weeding tool. Both vision and GNSS technologies are continuously being improved for plant detection, and automated weeding technologies are expected to become more affordable over time. Geo-referencing technology may soon lead to the establishment of crops in a grid-like arrangement, with even spacing between individual plants (Machleb et al., 2020). The GeoSeed by Kverneland (2020, https://be.kverneland.com/Actualites/Product-news/Archive-2015/Electric-drive-GEOSEED-offers-new-opportunities, accessed 23 July 2021) aims to sow crops in a pattern that allows for crosswise inter-row hoeing in opposing directions. If successful, this might lead to selective and crosswise weed harrowing in cereals established within a grid. However, seeding technology requires further improvement before precision planting becomes possible. A challenge shared by the developers of vision- and GNSS-based crop and weed detection systems is improving accuracy, so automated selective intra-row cultivation can be implemented in closer proximity to crop plants. By minimizing the uncultivated 'safety zone' surrounding individual crop plants, remaining intra-row weeds may be reduced to densities of insignificant concern; indeed, this scenario is already a reality in some transplanted row crops (Melander et al., 2015). To apply automated precision weeding in direct-sown crops, several issues must be addressed in the future. For example, the trade-off that exists when reducing operation distance between weeding tool and crop, between the crop injuries resulting from physical disturbance, and the yield benefits associated with weeding a greater area of the soil's surface. As automatic intra-row weeders are developed to function in direct-sown crops, it is essential to parameterize the crop-related effects of mechanical and thermal weeding devices across early growth stages, at multiple intensities, and multiple working distances from crop; such research is currently underway in Denmark. The benefits of MWC in close proximity to crop plants are obvious for the organic sector, as well as conventional specialty crops lacking effective herbicides (Fennimore et al., 2014). For conventional row crops where effective herbicides are still available, spot-spraying of close-to-crop weeds in combination with intelligent intra-row weeding could minimize herbicide consumption immensely and live up to the intentions of IPM.

7 Where to look for further information

The following chapters in textbooks provide useful introductions to the subject:

Cloutier, D. C., van der Weide, R. Y., Peruzzi, A. and Leblanc, M. L. (2007) *Mechanical Weed Management*. In: Non-Chemical Weed Management: Principles, Concepts and Technology, (Editors: M. K. Upadhyaya & R. E. Blackshaw). CAB International (www.cabi.org), Wallingford (U.K.), 111-134.

Melander, B., Liebman, M., Davis, A. S., Gallandt, E. R., Bàrberi, P., Moonen, A. C., Rasmussen J., von der Weide, R. and Vidotto, F. (2017). *9 Non-Chemical Weed Management*. In: Weed Research. Expanding Horizons, (Editors: P.

E. Hatcher & R. Froud-Williams). John Wiley & Sons Ltd, West Sussex (U.K.), 245-270.

Gallandt, E. R., Brainard, D. and Brown, B. (2018) *Developments in physical weed control*. In: Integrated weed management for sustainable agriculture, (Editor: R. L. Zimdahl). Burleigh Dodds Science Publishing, Cambridge (U.K.), 261-279.

Important research on mechanical weed control is currently conducted in the ongoing EU Horizon2020 project with the acronym IWMPRAISE grant agreement No 727321 (https://iwmpraise.eu/).

8 References

Brown, B. and Gallandt, E. R. (2018). Evidence of synergy with 'stacked' intrarow cultivation tools. *Weed Research* 58(4), 284-291.

Carlson, M., Forcella, F., Wortman, S. E. and Clay, S. A. (2018). Using abrasive grit for weed management in field crops. In: *Physical Methods for Stimulation of Plant and Mushroom Development*, El-Esawi, M. A. (Ed) Books on Demand GmbH, Norderstedt, German, 41-64.

Connolly, C. (2003). Vision guidance system facilitates high-speed inter-row weeding. *Industrial Robot: An International Journal* 30(5), 410-413.

Fennimore, S. A., Smith, R. F., Tourte, L., LeStrange, M. and Rachuy, J. S. (2014). Evaluation and economics of a rotating cultivator in bok Choy, celery, lettuce, and radicchio. *Weed Technology* 28(1), 176-188.

Fernández-Quintanilla, C., Peña, J. M., Andújar, D., Dorado, J., Ribeiro, A. and López-Granados, F. (2018). Is the current state of the art of weed monitoring suitable for site-specific weed management in arable crops? Weed Research 58, 259-272.

Gallandt, E. R., Brainard, D. and Brown, B. (2018). Developments in physical weed control. In: *Integrated Weed Management for Sustainable Agriculture* Zimdahl, R. L. (Ed). Burleigh Dodds Science Publishing, Cambridge, UK, 261-279.

Gerhards, R., Späth, M., Sökefeld, M., Peteinatos, G. G., Nabout, A. and Ayala, V. R. (2021). Automatic adjustment of harrowing intensity in cereals using digital image analysis. *Weed Research* 61(1), 68-77.

Graglia, E., Melander, B. and Jensen, R. K. (2006). Mechanical and cultural strategies to control *Cirsium arvense* in organic arable cropping systems. *Weed Research* 46(4), 304-312.

Hammermeister, A. M. (2016). Organic weed management in perennial fruits. *Scientia Horticulturae* 208, 28-42.

Heisel, T., Andreasen, C. and Christensen, S. (2002). Sugarbeet yield response to competition from *Sinapsis arvensis* or *Lolium perenne* growing at three different distances from the beet and removed at various times during early growth. *Weed Research* 42(5), 406-413.

Hermansen, S. (2020). Markrobotterne bliver bedre og bedre. *Økologi, Inspiration Til Jordbruget* 10, 3.

Jabran, K., Mahmood, K., Melander, B., Bajwa, A. A. and Kudsk, P. (2017). Weed dynamics and management in wheat. *Advances in Agronomy* 145, 97-166.

Jones, P. A., Blair, A. M. and Orson, J. H. (1996). Mechanical damage to kill weeds. In: *Proceedings of the Second International Weed Control Congress*, Copenhagen, Denmark (vol. 3), 949-954.

Kudsk, P. and Mathiassen, S. K. (2020). Pesticide regulation in the European Union and the glyphosate controversy. *Weed Science* 68(3), 214-222.

Kunz, C., Weber, J. F. and Gerhards, R. (2015). Benefits of precision farming technologies for mechanical weed control in soybean and sugar beet–comparison of precision hoeing with conventional mechanical weed control. *Agronomy* 5(2), 130-142.

Kunz, C., Weber, J. F., Peteinatos, G. G., Markus Sökefeld, M. and Gerhards, R. (2018). Camera steered mechanical weed control in sugar beet, maize and soybean. *Precision Agriculture* 19(4), 708-720.

Kurstjens, D. A. G. and Kropff, M. J. (2001). The impact of uprooting and soil-covering on the effectiveness of weed harrowing. *Weed Research* 41(3), 211-228.

Lati, R. N., Siemens, M. C., Rachuy, J. S. and Fennimore, S. A. (2016). Intrarow weed removal in broccoli and transplanted lettuce with an intelligent cultivator. *Weed Technology* 30(3), 655-663.

Machleb, J., Kollenda, B. L., Peteinatos, G. G. and Gerhards, R. (2018). Adjustment of weed hoeing to narrowly spaced cereals. *Agriculture* 8(4), 54; doi: 10.3390/agriculture8040054.

Machleb, J., Peteinatos, G. G., Kollenda, B. L., Andújar, D. and Gerhards, R. (2020). Sensor-based mechanical weed control: present state and prospects. *Computers and Electronics in Agriculture* 176, doi: 10.1016/j.compag.2020.105638.

McCollough, M. R., Gallandt, E. R., Darby, H. M. and Molloy, T. (2020a). Band sowing with hoeing in organic grains: I. Comparisons with alternative weed management practices in spring barley. *Weed Science* 68(3), 285-293. doi: 10.1017/wsc.2019.64.

McCollough, M. R., Gallandt, E. R. and Molloy, T. (2020b). Band sowing with hoeing in organic grains: II. Evidence of improved weed management in spring wheat, oats, field peas, and flax. *Weed Science* 68(3), 294-300. doi: 10.1017/wsc.2020.18.

Melander, B. (1997). Optimization of the adjustment of a vertical axis rotary brush weeder for intra-row weed control in row crops. *Journal of Agricultural Engineering Research* 68(1), 39-50.

Melander, B. (1998). Economic aspects of physical intra-row weed control in seeded onions. *Proceedings of the 12th International IFOAM Scientific Conference*, Mar del Plata, 180-185.

Melander, B., Cirujeda, A. and Jørgensen, M. H. (2003). Effects of inter-row hoeing and fertiliser placement on weed growth and yield of winter wheat. *Weed Research* 43(6), 428-438.

Melander, B., Rasmussen, I. A. and Barberi, P. (2005). Integrating physical and cultural methods of weed control–examples from European research. *Weed Science* 53(3), 369-381.

Melander, B., Ørum, J. E. and Thomsen, H. C. (2011). Mekanisk ukrudtsbekæmpelse i kartofler. *Plantekongres 2011(11-13 Januar), Herning Kongrescenter*, 239-240.

Melander, B., Holst, N., Rasmussen, I. A. and Hansen, P. K. (2012). Direct control of perennial weeds between crops–implications for organic farming. *Crop Protection* 40, 36-42.

Melander, B., Lattanzi, B. and Pannacci, E. (2015). Intelligent versus non-intelligent mechanical intra-row weed control in transplanted onion and cabbage. *Crop Protection* 72, 1-8.

Melander, B., Rasmussen, I. A. and Olesen, J. E. (2016). Incompatibility between fertility building measures and the management of perennial weeds in organic cropping systems. *Agriculture, Ecosystems and Environment* 220, 184-192.

Melander, B., Liebman, M., Davis, A. S., Gallandt, E. R., Bàrberi, P., Moonen, A. C., Rasmussen, J., von der Weide, R. and Vidotto, F. (2017). 9 non-chemical weed management. In: *Weed Research. Expanding Horizons* Hatcher, P. E. and Froud-Williams, R. (Eds). John Wiley & Sons Ltd, West Sussex, UK, 245-270.

Melander, B., Jabran, K., De Notaris, C., Znova, L., Green, O. and Olesen, J. E. (2018). Inter-row hoeing for weed control in organic spring cereals–influence of inter-row spacing and nitrogen rate. *European Journal of Agronomy* 101, 49-56.

Melander, B. and McCollough, M. R. (2020). Influence of intra-row cruciferous surrogate weed growth on crop yield in organic spring cereals. *Weed Research* 60(6), 464-474; doi: 10.1111/wre.12452.

Merfield, C. N., Bennett, J. R., Berry, N. A., Bluon, A., O'Connell, D. M. and Hodge, S. (2020). The potential of mini-ridging for controlling intrarow weeds: estimating minimum lethal burial depth. *Weed Research* 60(5), 353-362.

Pannacci, E., Lattanzi, B. and Tei, F. (2017). Non-chemical weed management strategies in minor crops: a review. *Crop Protection* 96, 44-58.

Rasmussen, J. (1991). A model for prediction of yield response in weed harrowing. *Weed Research* 31(6), 401-408.

Rasmussen, J. (1998). Ukrudtsharvning i vinterhvede. (Weed harrowing in winter wheat. With English summary). *Proceedings of the 15th Danish Plant Protection Conference–Weeds*. DJF-report no. 2. Danish Institute of Agricultural Sciences, Nyborg, Denmark, 179-189.

Rasmussen, J., Mathiasen, H. and Bibby, B. M. (2010). Timing of post-emergence weed harrowing. *Weed Research* 50(5), 436-446.

Rueda-Ayala, V., Peteinatos, G., Gerhards, R. and Dionisio Andújar, D. (2015). A non-chemical system for online weed control. *Sensors* 15(4), 7691-7707.

Utstumo, T., Urdal, F., Brevik, A., Dørum, J., Netland, J., Overskeid, Ø., Berge, T. W. and Gravdahl, J. T. (2018). Robotic in-row weed control in vegetables. *Computers and Electronics in Agriculture* 154, 36-45.

Theilgaard, M. and Bertelsen, I. (2017). Radrensning giver merudbytte i vårbyg. *Magasinet Mark Marts* 2017, 42-43.

Tillett, N. D., Hague, T., Grundy, A. C. and Dedousis, A. P. (2008). Mechanical within-row weed control for transplanted crops using computer vision. *Biosystems Engineering* 99(2), 171-178.

van der Weide, R. Y., Bleeker, P. O., Achten, V. T. J. M., Lotz, L. A. P., Fogelberg, F. and Melander, B. (2008). Innovation in mechanical weed control in crop rows. *Weed Research* 48(3), 215-224.

Wiltshire, J. J. J., Tillett, N. D. and Hague, T. (2003). Agronomic evaluation of precise mechanical hoeing and chemical weed control in sugar beet. *Weed Research* 43(4), 236-244.

Znova, L., Melander, B., Lisowski, A., Klonowski, J., Chlebowski, J., Edwards, G. T. C., Nielsen, S. K. and Green, O. (2018). A new hoe share design for weed control: measurements of soil movement and draught forces during operation. *Acta Agriculturae Scandinavica, Section B–Soil and Plant Science* 68(2), 139-148.

Part 3

Case studies

Chapter 11

On-farm implementation of integrated weed management

M. M. Riemens and M. Elings, Wageningen University and Research, The Netherlands

1 Introduction

The focus of twentieth-century agriculture was on the increase of productivity. Pesticides were a major driver for agricultural development and increase of crop yields. Today, the major challenge is to (a) increase the sustainability of our production systems to halt biodiversity losses and (b) reduce the environmental impact, while (c) attaining food security and food safety. Governmental policies and the agricultural value chain respond to this challenge.

The European Union (EU) installed pesticide legislation that is considered the most comprehensive and stringent in the world. Directive 2009/118, the so-called Sustainable Use Directive (SUD), provides a framework to achieve sustainable pesticide use and promotes low-pesticide farming in the EU.

http://dx.doi.org/10.19103/AS.2021.0098.12

The SUD introduced the term integrated pest management (IPM) into the EU legislation and member states were required to develop national action plans to reduce the risks of pesticides to human health and the environment. The following eight principles of IPM were introduced: (1) Prevention and suppression, (2) Monitoring, (3) Decision-making based on monitoring and economic thresholds, (4) Non-chemical methods, (5) Pesticide selection, (6) Reduced pesticide use, (7) Anti-resistance strategies, and (8) Evaluation (Barzman et al., 2015).

In the farm to fork strategy, the EU Commission presents its ambition to reduce the overall use and risk of chemical pesticides by 50% and the use of more hazardous pesticides by 50% in 2030 (https://ec.europa.eu/food/system/files/2020-05/f2f_action-plan_2020_strategy-info_en.pdf). The commission will revise the above-mentioned SUD and regards IPM as one of the main tools in reducing the use and dependency on pesticides. Agricultural value chains comprise input suppliers, farms, processing and retailing firms. Demands from processing or retailers can either improve implementation of IPM or hamper further uptake of IPM in cropping systems (https://doi.org/10.1080/09670874.2018.1435924). They can impose requirements on the production process that involve IPM standards and make a positive contribution. However, some processors demand counterproductive cropping practices that increase farmers' dependency on pesticides, such as the requirement of specific varieties or zero tolerance of microorganisms that improve the processing and shelf life of the product.

Oerke (2006) showed that the potential yield loss of weeds was 34% without effective weed control. Since the discovery of active ingredients such as 2,4-D and MCPA in the 1940s, herbicides have been the preferred weed control option of farmers in conventional agriculture (Kudsk and Streibig, 2003). This has led to production systems that rely on herbicides for weed control. More than 10 years after the instalment of the SUD, and despite the demands of retail and processing firms, the reliance of farmers on herbicides remains high and exceeds the use of other pesticides in many EU countries (https://appsso.eurostat.ec.europa.eu/nui/submitViewTableAction.do).

Since integrated weed management (IWM) is a part of IPM, the adoption of IWM is an important driver for IPM implementation and an important prerequisite for the increase of the sustainability of production systems. IWM is an integral, holistic approach in which cultural, physical, genetic, mechanical, biological as well as chemical tactics are combined in a diversified weed management strategy (Moss, 2018). Although individual tactics can be successful in managing weeds in the short term, they may select for species that tolerate the tactic or have low susceptibility. To obtain sustainable weed control in the long term, a combination of several tactics is necessary.

Within the project IWMPRAISE (www.iwmpraise.eu), we developed a novel framework for integrating and implementing existing and novel tactics into IWM strategies (https://iwmpraise.eu/wp-content/uploads/2018/06/WP1_IS_1_IWMPRAISE_eng.pdf). The IWM framework distinguishes five different pillars of management tactics for IWM. To be able to make an informed decision on what tactics to combine into a weed management strategy that can manage weed populations on a time scale exceeding the current growth season, successful IWM strategies must combine tactics from all or most of these five pillars. The five pillars are (a) diverse cropping system (e.g. rotation, cover crops), (b) cultivar choice and establishment (e.g. suppressive and tolerant varieties, seed rate), (c) field/soil management (e.g. nutrient placement, tillage systems), (d) direct control (e.g. mechanical control, flame weeding), (e) monitoring and evaluation (e.g. farm management systems and decision support systems). A successful IWM strategy combines multiple tactics by selecting tools from all or most of the five pillars for IWM (Riemens et al., 2021). In this way pillar-based IWM framework translates the eight IPM principles into pragmatic IWM approaches.

Although governmental policies and agricultural value chains provide the context for weed management, farmers and their advisors are responsible for on-farm implementation. They decide, on a day-to-day basis, how they manage their crops and control pests, diseases and weeds impacting crop yield. For IWM to contribute to the increased sustainability of crop production systems, it is highly important that it is implemented by farmers. Understanding the drivers of decision-making by farmers about their choice of weed management tools to implement on their farms is pivotal for a successful on-farm IWM. Dessart et al. (2019) presented a framework of behavioural factors affecting the adoption of sustainable farming practices by farmers. They distinguished 1) cognitive factors relating to learning and reasoning and comprising aspects such as the perceived level of control, perceived risks, perceived costs and benefits and the knowledge of sustainable farming practices, 2) social factors relating to interactions with other individuals and including the need for social approval, social comparison and the need for social status and 3) dispositional factors such as personality, environmental concern, risk tolerance and aversion, farming objectives and resistance to change.

Several of these factors have been specifically mentioned to affect the low adoption of IWM and continued reliance on herbicides of farmers by others (Hillocks and Cooper, 2012, Lefebvre et al, 2015, Liebman et al., 2016, Moss, 2018), and these can be categorized as cognitive, social and dispositional factors:

Cognitive factors are:

1 lack of available knowledge on IWM;

2 limited evidence of efficiency, reliability and cost-effectiveness of IWM;
3 trade-offs with other attributes of the cropping system;
4 increased complexity involved in IWM; and
5 insufficient infrastructure to support relevant learning and decision-making by farmers and land managers.

Social factor includes:

6 differences in individual values and beliefs between farmers that cause differences in their attitudes to IWM.

Dispositional factor includes:

7 resistance to change and farm objectives.

In this chapter, we review these factors affecting IWM adoption and decision-making by farmers.

2 Lack of available knowledge on integrated weed management

Until recently, research on weed management has primarily focused on the increased efficiency of herbicides and on the substitution of some herbicide treatments with non-chemical control methods, and relatively little attention was paid to IWM. This is reflected by the number of papers published on weed control (mostly single tactic, herbicide-based weed control), weed management (combining herbicide treatments with one or a few other non-chemical weed control tactics) and IWM (holistic approach) (Harker and O'Donavon, 2013, Table 1).

These numbers indicate that research on IWM was relatively limited until recently. The knowledge available to farmers and advisors in an easily excisable and applicable form can be estimated as a smaller proportion of the knowledge available in these research papers. Further redesign of crop production systems

Table 1 Number of weed research articles published with weed control (WC), weed management (WM), or integrated weed management (IWM) listed in the title, abstract or key words from 1995 to 2011 (Harker and O'Donovan, 2013)

Term	Number of articles
Weed control	9964
Weed management	2708
Integrated weed management	697

with reduced dependency on pesticides, and specifically herbicides, requires further development of knowledge on IWM.

3 Limited evidence of efficiency, reliability and cost-effectiveness of integrated weed management

The concept of integrating multiple tactics to manage pests with a reduced reliance on pesticides is not new. It was already described in the 1950s (Stern and Van den Bosch, 1959) for insect pests, and in 1991, a similar concept for IWM was presented (Swanton and Weise, 1991). Since then, limited evidence for the efficiency, reliability and cost-effectiveness at the cropping system level has been presented. Good examples have been published in maize-based systems in the United States (Davis et al., 2012) and Canada (Swanton et al., 2002) as well as Europe (Vasileiadis et al., 2015). The studies showed that increasingly diverse cropping systems maintained or exceeded yield levels of less diverse systems, gave similar economic returns and a significant reduction in herbicide use compared to conventional less-diverse systems. Similar results were found in a survey among Dutch organic arable farmers. On farms where farmers implemented an increasingly diversified weed management strategy, weed seed bank density was significantly lower than on-farms with a less-diverse weed management system (Riemens et al., 2010).

A meta-analysis on the effects of cultivation techniques, sowing date, crop density and cultivar choice on blackgrass infestations in cereal crops showed inconsistent effects and highly variable outcomes. For instance, a cultural tactic such as mouldboard ploughing reduced, on average, blackgrass populations by 69%, but with a variation from −82% to 95%. Variability found for this and other non-chemical tools is probably caused by the many parameters (e.g. weather conditions) that will affect a weed population in the field (Lutman et al., 2013).

IWM is typically based on the combination of several tactics that individually do not meet the required levels of efficiency. Together, combined they can however manage weed populations effectively. Liebman and Gallandt (1997) called this the 'many little hammers approach'. The approach will take time, and effects will not be visible within one growing season. The beneficial impact of a diverse IWM strategy on the weed population will become visible in the course of a rotation, after several years. The economic costs of such a strategy are visible at once. Contrarily, the effect of herbicide treatments on the weed population is visible immediately, and in the short term, they are economically beneficial. Ghersa et al. (2020) called this the prisoners' dilemma for weed management strategies: weed management strategies with herbicides as the principal component have higher economic returns in the short term. Continued use of these herbicide-based strategies will lead to reduced yields and less stability, and in the long run, IWM becomes the better economic alternative.

At present, we do need better predictions of the outcome of individual cultural measures (Moss, 2018) and the effect of a combination of these and other non-chemical tactics at the individual field level to design IWM strategies that are cost-effective in the long run.

4 Trade-offs with other attributes of the cropping system

Increased research efforts on IWM strategies are essential to show that it is possible to change the management system and reduce the impact on both the environment and human health (Lefebvre et al., 2015, Lechenet et al., 2014, 2017, Vasileiadis et al., 2015). This will also involve engagement with agronomists, plant pathologists and agricultural entomologists, who are developing recommendations for integrated crop management (ICM) or integrated management of diseases and/or pests that may complement or conflict with some of the IWM approaches discussed above. Implementation of IWM strategies will in general encompass an increased use of non-chemical weed control tactics such as mechanical weed control and cover crops. A trade-off of mechanical weed control may be soil erosion or damage to soil structure when performed under suboptimal conditions (Van der Weide et al., 2008), damage to soil biota such as earthworms (Andersen et al., 2013, Faber et al., 2017, Schreck et al., 2012), ground-dwelling insects (Dierauer and Pfiffner, 1994, Holland and Luff, 2000, Kromp, 1999, Lorenz et al., 1994, Navntoft et al., 2016), and arbuscular mycorrhizae and soil microbiota in general (Douds et al., 2018, Rego et al., 2004, Zaller et al., 2018). Reduced tillage is generally perceived to be beneficial for soil health but may lead to an increased weed pressure or dependency on herbicides compared to traditional tillage systems (Melander et al., 2013, Moonen and Barberi, 2004).

5 Implementation of integrated weed management is complex

One of the barriers to the uptake of IWM is that it requires the transition from a simple system based on herbicides to a more complex, knowledge-intensive system. In the twentieth century, new technology and insights into weed control were mainly based on increased efficiency of herbicide use or other single tactic solutions. The focus was on short-term strategies and optimization of the profit of the present crop. IWM strategies, on the other hand, are more related to the long-term results. Decisions made in one season may affect the weed population, crop growth and income in the following season(s). A farmer may have many concerns and needs to acquire new skills and knowledge before an effective IWM strategy can be adopted on the farm. Most farmers prefer to take smaller steps and make gradual changes in their crop management systems (Chantre and Cardona, 2014). A recent study amongst arable farmers in the United Kingdom and Ireland found that all farmers implemented some IPM tactics. However, only 6% of the farmers adopted more than 85% of the possible IPM tactics (Creissen et al., 2019). Farmers in a French study were found to use a step-by-step approach during the adoption of more integrated farming practices on their farms. They learned, step by step, by trial and error (Chantre and Cardona, 2014).

Studies from the United Kingdom and Australia showed that farmers are more likely to build on the development of their IWM strategy through the adoption of a diverse set of weed management tactics if they already implemented tactics in the past (Sharma et al., 2011) or when they had problems with resistant weeds or did not expect new herbicide modes of action entering the market soon (Llewellyn, 2007). Hence, it may be more effective to communicate on each of the IWM or IPM tactics separately and present step-by-step changes to farmers rather than communicating about the complete redesign of farming systems. A good starting point would be to make an inventory of the type of tactics a farmer has adopted in the past, main weed-related management issues on the farm, such as perennial weed infestations or high levels of weed resistance, economic consequence of the proposed changes and farmers' beliefs on future herbicide availability. Based on this information, targeted messages can be brought across either through specific information, filling the knowledge gap, or through collaborative learning or co-innovation trajectories (see Section 6 for more information on co-innovation).

6 Infrastructure to support relevant learning by farmers

In general, two approaches for education and extension can be distinguished (Liebman et al., 2016): (1) developing information packages, delivered to farmers by experts to fill the gap in farmers knowledge and (2) participatory learning (Meir and Williamson, 2005) to co-produce innovation in dealing with local, specific agro-ecosystems. Several concepts and associated infrastructures have been developed that use either one of these approaches or both, namely farmer field school (FFS), pilot farms and on-farm demonstrations and lighthouse farm concept.

FFS is a widely used concept that is used to educate farmers to adapt their agricultural decisions (FAO, 2016). Within a FFS, farmers are brought together for hands-on field-based learning over a production cycle. FFS has been widely used in Asia, and FFS that focused on IPM decreased the pesticide use of participants by 17% on average (Waddington et al., 2014, Van den Berg et al., 2020). In the past, as an example FFS that focused on the transition of new technologies from research to farmers, but today, an important aspect of learning within FFS is the participation of farmers in the innovation process and facilitation of experimentation amongst the communities themselves. Its focus has shifted from the development of standardized information packages to participatory learning activities.

As described earlier in this chapter, the next step towards increased sustainability in EU agriculture is to redesign farming systems. Vereijken (1997) described a methodological way that has been used to prototype new farming systems and test these by farmers. It combines the development of farming system prototypes with trials on commercial pilot farms. The approach consists of five steps: (1) establishing a hierarchy of objectives considering the current farming systems in the region, (2) transforming the objectives in a set of multi-objective parameters, to quantify them and establishing a set of multi-objective farming methods to achieve them, (3) designing a theoretical prototype by linking parameters to farming methods and designing the methods in this context until they are ready for initial testing, (4) laying out the prototype on at least 10 pilot farms in appropriate variants and testing and improving the prototype (variants) until the objectives, as quantified in the set of parameters, have been achieved (after repeated layout), (5) disseminating the prototype (variants) to other farms with a gradual shift in supervision from researchers to extensionists. The critique of this approach is that it does not take the existing diversity in farmers into account and is strongly dominated by researchers (Leeuwis, 1999). Dogliotti et al. (2014) took the methodology to the next level and included farmers and technical advisors in the first steps, creating a collective learning process of all stakeholders involved: co-innovation (Rossing et al., 2010).

The widespread use of new technology depends on the adoption of these tactics by a small group of innovative farmers (Rogers et al., 2008). Traditionally, experimental farms are a source of information to these innovative commercial farmers. During field days and excursions at the experimental farms, farmers get demonstrations of new tactics and strategies and are informed about the latest insights. Other farmers will learn about the techniques and will follow the example set by their peers. The Lighthouse Farm Network (https://www.wur.nl/en/Research-Results/Chair-groups/Plant-Sciences/Farming-Systems-Ecology-Group/Lighthouse-project.htm) builds upon this idea. A lighthouse farm is an existing, commercially viable farm in the real world that acts as a positive deviant and is 'already in 2050' in terms of providing sustainably produced food and ecosystem services. The farms demonstrate what can be achieved within the local circumstances, both agronomically as well as socially and economically.

Relevant learning by farmers must be supported by demonstrations of new techniques during field days at experimental farms, networks of farmers to support peer-to-peer learning and co-innovation to enable the redesign of IWM strategies.

7 Individual values and beliefs of farmers

IWM typically involves complex risk management decisions. It comprises preventive, curative and control measures that require decisions on crop choice and sequence, intercropping, fertilisation, cultivation type and frequency. IWM can therefore not be seen as a set of weed control tactics alone, it is a complex system approach in which many different risks and benefits need to be considered. Current decision support systems and extension work for IWM often focus on pure economics. They assume that farmers will solely use economic rationality when they evaluate trade-offs between weed management strategies. Although paid agri-environmental schemes are effective, as long as they are running, they currently fail to create a long-lasting change in farmers' attitude and subsequent behaviour (Van Dijk et al., 2016). One example is the Dutch cross-compliance subsidy (Van Zeeland et al., 2009). Maize growers received a financial benefit when they replaced herbicides with mechanical weed control tactics such as hoeing or herbicides with a reduced impact on the environment. Once the subsidy scheme was terminated, herbicide use increased by 20% and farmers reverted to the use of less environmentally benign herbicides again. The financial focus of these subsidies may even lead to a reduction in the farmers' intrinsic motivation to protect the environment and increase their extrinsic motivation to earn money (Van Dijk et al., 2016).

Previous studies showed that farmers' decisions on weed management are not solely based on cognitive factors with economic impact but are also influenced by social factors such as their personal values and beliefs (Zwickle, 2011). Similarly, factors such as social pressure, prestige and self-identity play an important role in farmers' decision-making in general (Lokhorst et al., 2011). Self-identity is the theory that 'one's self consists of several identities dependent on the social roles that one occupies. In different situations, other identities may be more or less salient to affect the intention to perform a certain behaviour (Van Dijk et al., 2016). This means that a farmer who is for instance concerned about the environmental impact of herbicides may implement non-chemical weed control tactics, even though these may cause financial costs counteracting the identity of a farmer as an entrepreneur. This phenomenon can be used by benchmarking (De Snoo et al., 2010). Benchmarking can set a descriptive norm, where farmers may tend to conform to. Creating benchmarks on IWM (for instance, the number of applied weed management tactics), in combination with tailored information on IWM, may increase the number of farmers willing to implement IWM practices, although these may not always be more cost-effective or may even be more costly. Another way to use the concept of self-identity is to make use of labels. Van Dijk et al. (2016) describe that labelling is emphasizing a certain identity by using a positive trait label, while referring to a certain behaviour that a person performs. For IWM, this could be done by labelling farmers who implement IWM as responsible farmers, for example, 'You perform IWM on your farm, this shows you are a farmer that cares about the environment.'

8 Resistance to change and farm objectives

Resistance to change is thought to be one of the most important factors hampering the adoption of sustainable practices by farmers. A study amongst farmers in Germany indicated that farming communities are not likely to change (Burton et al., 2008) and, therefore, rejection of change may be one of the major factors for the lack of adoption of sustainable farming practices (Dessart et al., 2019) and more specifically IWM. Farmers, in general, are rather risk averse and, therefore, not likely to quickly adopt new practices (Pennings and Garcia, 2001, https://www.jstor.org/stable/1244709). Zwickle et al. (2016) stressed the importance to focus on the long-term benefits of this sustainable weed management approach instead of focusing on the potential risks.

Only on rare occasions, farmers may redesign their complete farming system overnight. Farmers do so usually convert from conventional to organic farming. Interestingly, two types of rationale lie behind the radical changes: either an economic incentive or a sudden awareness of the potential impact of pesticides on the environment (Chantre and Cardona, 2014, Darnhofer et al., 2005). Both factors are known to be embraced by farmers and can cause conflicting objectives: objectives related to entrepreneurship and economics and objectives related to conservation and lifestyle (Dessart et al., 2019). Most farmers set goals on both objectives, although to a varying degree. Weed management practices that are both economically and environmentally favourable are most likely to be adopted. Consequently, paid agri-environmental schemes may be a successful tool to promote IWM implementation.

9 Case study: Decision process for on-farm integrated weed management amongst conventional European farmers

In the previous sections, we stated that IWM as a part of IPM is an important driver for the increased sustainability of plant production systems. Farmers' decisions on weed management are key to the successful implementation of IWM. To increase our understanding of farmers' decision-making with respect to weed management, within the IWMPRAISE project: (1) a decision process for European on-farm IWM was described that includes farmers' knowledge on IWM and perceived risks and benefits of IWM tactics and weeds and (2) knowledge of farmers and experts on IWM was compared for different crop production systems.

The decision process was based on an adapted version of the decision process described in the study by Jabbour et al. (2014) and Zwickle (2011) for weed management by organic farmers in the United States. According to this scheme, the farmer's perception of weeds and IWM is influenced by the individual characteristics of the farmers and their farms (farm and farmer attributes), by benefits and risks of weeds themselves and by benefits and risks of weed management tools. These factors are underlying any decision that farmers make on IWM. The IWM behaviour itself consists of the applied tools and tactics from the IWM framework described in the introduction section. The experience of the farmer with the selected IWM strategy will influence his/her knowledge and experience and, by nature, therefore changes the farmers' characteristics. The factors influencing the decision process can therefore be categorized as either impacting the decision before it is made, after it is made (possible behaviours) or after the IWM action is implemented (experience trial/ error) (Fig. 1).

Over 35 experts from the Netherlands, Denmark, United Kingdom, France, Slovenia, Italy and Spain were interviewed to get insight on expert knowledge on IWM. Experts were selected from different backgrounds (academia, farmers' organizations, companies and extensionists). The interviews with the experts from the different partner countries were fully transcribed in Word and afterwards coded by the researchers with a common coding framework.

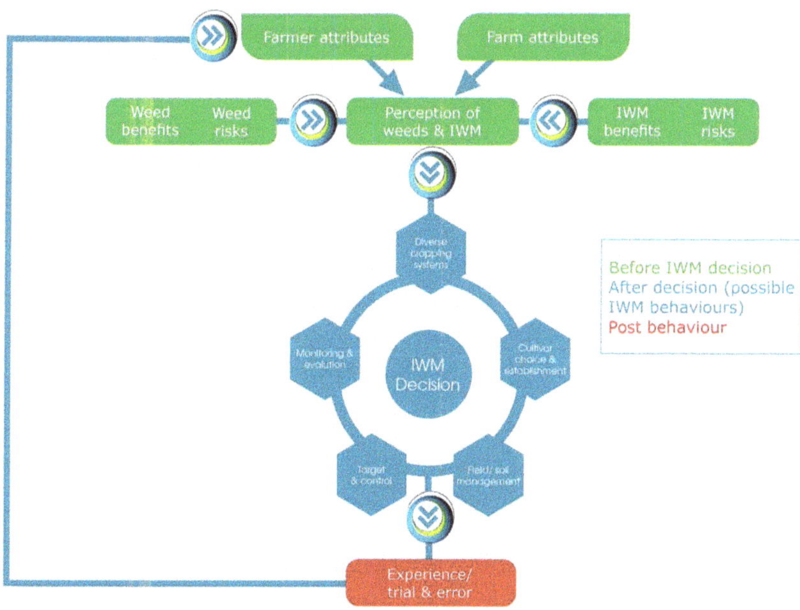

Figure 1 Decision process for IWM behaviour by European farmers.

By coding the data, a clear picture of the experts' knowledge on tactics important to on-farm integrated weed management was obtained. Experts found scouting, cover crops, rotation, cultivar choice, sowing date, tillage type and mechanical weeding as important aspects for on-farm IWM. When asked about the importance of specific tactics and tools for successful IWM strategies to specific production systems, experts stressed the importance of cover crops and herbicides more frequently for weed management in perennial woody crops. The importance of tactics such as rotation, mechanical weeding, cultivar choice and tillage type was mentioned more frequently for arable systems.

In each country, farmers from different weed management systems (annual narrow row crops, annual row crops and perrenial woody crops systems) were interviewed. In total, 131 arable farmers and 20 farmers for the perennial woody crop scenario were interviewed. The analysis was done by coding the transcribed interviews with a common coding framework.

The farmers' responses showed clear differences among farmers' thinking on the relevance of the five pillars of IWM between the different cropping systems. Furthermore, differences were found between experts and farmers.

Tactics related to the IWM pillar field and soil management (e.g. nutrient placement, dead mulching) were seen as important by perennial woody crop growers but hardly mentioned by experts. While growers frequently mentioned chemical control tactics (e.g. herbicides) in the direct control pillar, experts mentioned non-chemical tactics (e.g. mowing and grazing by livestock) more frequently. Growers and expert responses for the IWM pillars monitoring and evaluation and diverse cropping systems were similar.

Both experts and farmers highlighted the importance of the use of tactics from all pillars of the IWM framework in arable systems. However, they differed in the perceived importance of individual tactics belonging to these pillars. For instance, within the field and soil management pillar, almost all farmers found the choice of tillage type an important tactic, while only half of the experts mentioned this tactic. Within the pillar cultivar choice and establishment, experts mentioned the benefits of machine hygiene and the prevention of seed spread, while farmers found tactics such as sowing depth and transplanting of greater value.

Arable farmers, as was the case for woody perennial crop growers, mentioned direct chemical control tactics frequently (e.g. different functions of herbicides in a weed management strategy), while experts talked about herbicides in general and mentioned non-chemical tactics such as flame weeding and mechanical control more often than farmers.

In terms of risks posed by weed management tactics, farmers perceive risks of tactics related to diverse cropping systems such as intercropping, field margins and the risk of direct control (e.g. failed mechanical control and the risk of hand weeding) larger than experts. Experts emphasized the benefits of these tactics more often than the risks of these tactics, compared to the farmers.

These results indicate a mismatch in the perceived relevance and risks of weed management tactics between farmers and experts. During linear knowledge transfer activities (e.g. farm demonstrations, articles in farmer journals, fact sheets, informative videos), it is important that experts such as researchers and advisors are aware of these differences and adjust the message of their findings accordingly. During co-innovation processes, awareness about this mismatch is important to understand the preferences of participants for specific weed management strategies.

10 Conclusion

IWM is a part of IPM, and on-farm implementation of IWM is an important driver for IPM and the transition to sustainable farming in Europe. Within the project IWMPRAISE, we developed a novel framework for integrating and implementing existing and novel approaches to IWM. The framework can be used to redesign on-farm weed management to reduce agriculture's dependency on herbicides. Understanding the drivers of decision-making by farmers is essential for the successful implementation of on-farm IWM. In this chapter, we reviewed cognitive, social and dispositional factors often associated with the lack of IWM adoption by farmers: lack of available knowledge on IWM; limited evidence of efficiency, reliability and cost-effectiveness of IWM; trade-offs with other parts of the cropping system; increased complexity involved in IWM; infrastructure to support relevant learning and decision-making by farmers; individual values and beliefs of farmers affecting their attitudes to integrated weed management; resistance to change and farm objectives.

From the review, it becomes apparent that there is a lack of research on IWM and limited information on the reliability and cost-effectiveness of IWM in the long term. The knowledge available to farmers and advisors is potentially even lower. Relevant infrastructure to support learning by farmers should be focused on co-innovation of IWM strategies, supported by farmers' networks for peer-to-peer learning and demonstrations of new techniques on experimental farms. Since most farmers are adverse to change, prefer to take small steps and in general tend to focus on short-term economic benefits, it may be more effective to focus on communicating on the benefits of individual tactics and present step-by-step changes to farmers, rather than communicate about system redesign. Farmers who have implemented IWM tactics will be more likely to implement the next set of IWM tactics. Therefore, although paid agri-environmental schemes will not alter behaviour in the long term, they may be used to jump-start the implementation of on-farm IWM. Combined with labelling farmers to stimulate certain behaviour and the use of benchmarks for IWM, the first small steps towards on-farm IWM may be taken. Finally, the presented case study indicates a mismatch between the perceived relevance and risks of weed management tactics between farmers and experts such as advisors and applied researchers. This mismatch must be considered when advisors and researchers organise field demonstration, instruction material or present research results on IWM.

11 Future trends in research

The many-faceted aspects of barriers for IWM implementation call for transdisciplinary research to address both the fundamental ecological mechanisms, the practical management strategies and the socio-economical aspects (Neve et al., 2018, Jordan et al., 2016). As mentioned before in this chapter, future research should shift focus from weed control and weed management to IWM. Collaborative research projects combining knowledge of social scientists, economists, agronomists, engineers and weed scientists are needed to develop novel plant production systems that are based on IWM instead of herbicide. Fundamental research should be paralleled with applied research and co-innovation processes to enable on-farm implementation.

12 Where to look for further information

12.1 Review articles

- Dessart, F. J., Barreiro-Hurlé, J. and Van Bavel, R. (2019). Behavioural factors affecting the adoption of sustainable farming practices: A policy-oriented review. *European Review of Agricultural Economics*, 46(3): 417-471. https://doi.org/10.1093/erae/jbz019.
- Liebman, M., Baraibar, B., Buckley, Y., Childs, D., Christensen, S., Cousens, R., Eizenber, H., Heijting, S., Loddo, D., Merotto, A., Renton, M. and Riemens, M. (2016). Ecologically sustainable weed management: How do we get from proof-of-concept to adoption? *Ecological Applications* 25(5): 1352–1369. https://doi.org/10.1002/15-0995.
- Wigboldus, S., Klerkx, L., Leeuwis, C., et al. Systemic perspectives on scaling agricultural innovations: A review. *Agron. Sustain. Dev.* 36, 46 (2016). https://doi.org/10.1007/s13593-016-0380-z.

12.2 Research projects

- IWMPRAISE: www.iwmpraise.eu.
- IPMWORKS: https://www.ipmdecisions.net/about-the-project/ipmworks/.

- IPMDECISIONS: https://www.ipmdecisions.net/.
- REMIX: https://www.remix-intercrops.eu/The-Project.
- DIVERIMPACTS: https://www.diverimpacts.net/.

12.3 Key societies

- www.ewrs.org.
- www.wssa.net.
- www.ecaf.org.
- www.croplifeeurope.eu.
- www.agroecology-europe.org/.

12.4 Research centres

- Aarhus University, https://international.au.dk/.
- Sant' Anna Scuola Universitaria Superiore Pisa, https://www.santannapisa.it/it.
- Rothamsted Research, https://www.rothamsted.ac.uk/.
- INRAE, https://www.inrae.fr/en.
- Wageningen University and Research, www.wur.nl.
- Spanish National Research Council, https://www.csic.es/en/csic.

13 Acknowledgements

Thanks are due to Veronica Pedraza Jimenez, Camilla Moonen, Sabine Gennai Schott, Mette Sonderskov, Richard Hull and Robert Leskovsek for data collection. The data presented in this chapter were collected during research funded by EY grant agreement No 727321.

14 References

Andersen, L., Kuhn, B. F., Bertelsen, M., Bruus, M., Larsen, S. E. and Strandberg, M. (2013). Alternatives to herbicides in an apple orchard, effects on yield, earthworms and plant diversity. *Agriculture, Ecosystems and Environment* 172: 1-5. https://doi.org/10.1016/j.agee.2013.04.004.

Barzman, M., Barberi, P., Birch, A. N. E., Boonekamp, P., Dachbrodt-Saaydeh, S., Graf, B., Hommel, B., Jensen, J. E., Kiss, J., Kudsk, P., Lamichhane, J. R., Messéan, A., Moonen, A., Ratnadass, A., Ricci, P., Sarah, J. and Sattin, M. (2015). Eight principles of integrated pest management. *Agronomy for Sustainable Development* 35(4): 1199-1215. https://doi.org/10.1007/s13593-015-0327-9.

Burton, R., Kuczera, C., Schwarz, G. (2008). Exploring farmers' cultural resistance to voluntary agri-environmental schemes. *Sociologia Ruralis* 2008(48). https://doi.org/10.1111/j.1467-9523.2008.00452.x.

Chantre, E. and Cardona, A. (2014). Trajectories of French field crop farmers moving toward sustainable farming practices: change, learning, and links with the advisory services. *Agroecology and Sustainable Food Systems* 38(5): 573–602. https://doi.org/10.1080/21683565.2013.876483.

Creissen, H. E., Jones, P. J., Tranter, R. B., Girling, R. D., Jess, S., Burnett, F. J., Gaffney, M., Thorne, F. S. and Kildea, S. (2019). Measuring the unmeasurable? A method to quantify adoption of integrated pest management practices in temperate arable farming systems. *Pest Management Science* 75(12): 3144–3152. https://doi.org/10.1002/ps.5428.

Darnhofer, I., Schneeberger, W. and Freyer, B. (2005). Converting or not converting to organic farming in Austria: farmer types and their rationale. *Agriculture and Human Values* 22(1): 39–52. https://doi.org/10.1007/s10460-004-7229-9.

Davis, A. S., Hill, J. D., Chase, C. A., Johanns, A. M. and Liebman, M. (2012). Increasing cropping system diversity balances productivity, profitability and environmental health. *PLoS ONE* 7(10): e47149. https://doi.org/10.1371/journal.pone.0047149.

De Snoo, G. R., Lokhorst, A. M., Van Dijk, J., Staats, H. and Musters, C. J. M. (2010). Benchmarking biodiversity performances of farmers. *Aspects of Applied Biology* 100: 311–318. ISSN: 0265-1491.

Dessart, F. J., Barreiro-Hurlé, J. and Van Bavel, R. (2019). Behavioural factors affecting the adoption of sustainable farming practices: a policy-oriented review. *European Review of Agricultural Economics* 46(3): 417–471. https://doi.org/10.1093/erae/jbz019.

Dierauer, H. U. and Pfiffner, L. (1994). Effects of flame weeding in organic vegetable growing on ground beetles (Coleoptera: Carabidae). Paper presented at the Maitrise des adventices par voie non chimique. Communications de la Quatrieme Conference Internationale I.F.O.A.M., Dijon, France, 5–9 July 1993 (2nd edn.), pp. 137–140.

Dogliotti, S., García, M. C., Peluffo, S., Dieste, J. P., Pedemonte, A. J., Bacigalupe, G. F., Scarlato, M., Alliaume, F., Alvarez, J., Chiappe, M. and Rossing, W. A. H. (2014). Co-innovation of family farm systems: a systems approach to sustainable agriculture. *Agricultural Systems* 126: 76–86, https://doi.org/10.1016/j.agsy.2013.02.009.

Douds, D. D., Lee, J., McKeever, L., Duffield, R., Ziegler-Ulsh, C. and Moyer, J. (2018). Stale seedbed technique for weed control negatively impacts the indigenous AM fungal population. *Biological Agriculture and Horticulture* 34(3): 199–210. https://doi.org/10.1080/01448765.2018.1426493.

Faber, F., Wachter, E. and Zaller, J. G. (2017). Earthworms are little affected by reduced soil tillage methods in vineyards. *Plant, Soil and Environment* 63(6): 257–263. https://doi.org/10.17221/160/2017-pse.

FAO (2016). Farmer field schools guidance document, planning for quality programmes. Food and Agriculture Organization of the United Nations, Rome. Available at: http://www.fao.org/3/i5296e/i5296e.pdf.

Ghersa, F., Pessah, S., Vera, A. C. D. and Ferraro, D. O. (2020). Theory and practice for environmental risk assessment: understanding the trade-off Between the benefits and risks Behind herbicide use as tool for designing sustainable weed management systems. In: *Decision Support Systems for Weed Management* Chantre, G. and González-Andújar, J. (Eds), Springer, Cham. https://doi.org/10.1007/978-3-030-44402-0_8.

Harker, K. N. and O'Donovan, J. T. (2013). Recent weed control, weed management, and integrated weed management. *Weed Technology* 27(1): 1–11. Available at: http://www.jstor.org/stable/23358301; https://doi.org/10.1614/WT-D-12-00109.1.

Hillocks, R. J. and Cooper, J. E. (2012). Integrated pest management - can it contribute to sustainable food production in Europe with less reliance on conventional pesticides? *Outlook on Agriculture* 41(4): 237–242. https://doi.org/10.5367/oa.2012.0107.

Holland, J. M. and Luff, M. L. (2000). The effects of agricultural practices on Carabidae in temperate agroecosystems. *Integrated Pest Management Reviews* 5(2): 109–129.

Jabbour, R., Zwickle, S., Gallandt, E. R., McPhee, K. E., Wilson, R. S. and Doohan, D. (2014). Mental models of organic weed management: comparison of New England US farmer and expert models. *Renewable Agriculture and Food Systems* 29(4): 319–333. https://doi.org/10.1017/s1742170513000185.

Jordan, N., Schut, M., Graham, S., Barney, J. N., Childs, D. Z., Christensen, S. B., Cousens, R. D., Davis, A. S., Eizenberg, H., Ervin, D. E., Fernandez-Quintanilla, C., Harrison, L. J., Harsch, M. A., Heijting, S., Liebman, M., Loddo, D., Mirsky, S. B., Riemens, M., Neve, P., Peltzer, D. A., Renton, M., Williams, M., Recasens, J. and Sønderskov, M. (2016). Transdisciplinary weed research: new leverage on challenging weed problems? *Weed Research* 56(5): 345–358. https://doi.org/10.1111/wre.12219.

Kromp, B. (1999). Carabid beetles in sustainable agriculture: a review on pest control efficacy, cultivation impacts and enhancement. *Agriculture, Ecosystems and Environment* 74(1–3): 187–228. https://doi.org/10.1016/S0167-8809(99)00037-7.

Kudsk, P. and Streibig, J. C. (2003). Herbicides - a two-edged sword. *Weed Research* 43(2): 90–102. https://doi.org/10.1046/j.1365-3180.2003.00328.x.

Lechenet, M., Bretagnolle, V., Bockstaller, C., Boissinot, F., Petit, M. S., Petit, S. and Munier-Jolain, N. M. (2014). Reconciling pesticide reduction with economic and environmental sustainability in arable farming. *PLoS ONE* 9(6): e97922. https://doi.org/10.1371/journal.pone.0097922.

Lechenet, M., Dessaint, F., Py, G., Makowski, D. and Munier-Jolain, N. (2017). Reducing pesticide use while preserving crop productivity and profitability on arable farms. *Nature Plants* 3: 17008. https://doi.org/10.1038/nplants.2017.8.

Leeuwis, C. (1999). Integral design: innovation in agriculture and resource management. *Mansholt Studies Series* (nr. 15). Available at: https://edepot.wur.nl/319159, Mansholt Institute, Backhuys Publishers, p. 277. ISBN 90-6754-585-6.

Lefebvre, M., Langrell, S. R. H. and Gomez-y-Paloma, S. (2015) Incentives and policies for integrated pest management in Europe: a review. *Agronomy for Sustainable Development* 35(1): 27–45 https://doi.org/10.1007/s13593-014-0237-2.

Liebman, M. and Gallandt, E. R. (1997). Chapter 9. Many little hammers: ecological approaches for management of crop-weed interactions. In: *Ecology in Agriculture* Jackson, L. E. (Ed.), Academic Press, pp. 291–343. ISBN 0-12-378260-0.

Liebman, M., Baraibar, B., Buckley, Y., Childs, D., Christensen, S., Cousens, R., Eizenber, H., Heijting, S., Loddo, D., Merotto, A., Renton, M. and Riemens, M. (2016). Ecologically sustainable weed management: how do we get from proof-of-concept to adoption? *Ecological Applications* 26(5): 1352–1369. https://doi.org/10.1002/15-0995.

Llewellyn, R. S. (2007). Information quality and effectiveness for more rapid adoption decisions by farmers. *Field Crops Research* 104(1–3): 148–156, https://doi.org/10.1016/j.fcr.2007.03.022.

Lokhorst, A. M., Staats, H., Van Dijk, J., Van Dijk, E. and De Snoo, G. (2011). What's in it for me? Motivational differences between farmers' subsidised and non-subsidised conservation practices. *Applied Psychology* 60(3): 337–353. https://doi.org/10.1111/j.1464-0597.2011.00438.x.

Lorenz, E., Ulber, B. and Poehling, H. (1994). Influence of different mechanical weed control techniques on the abundance of carabid beetles (Coleoptera, Carabidae) in sugar beets. *Sonderheft* 14: 635–644.

Lutman, P. J. W., Moss, S. R., Cook, S. and Welham, S. J. (2013). A review of the effects of crop agronomy on the management of *Alopecurus myosuroides*. *Weed Research* 53(5): 299–313. https://doi.org/10.1111/wre.12024.

Meir, C. and Williamson, S. (2005). Farmer decision-making for ecological pest management. In: *Book Chapter: The Pesticide Detox: Towards a More Sustainable Agriculture* Pretty, J. (Ed.), pp. 83–96. ISBN: 1844071421.

Melander, B., Munier-Jolain, N., Charles, R., Wirth, J., Schwarz, J., Van der Weide, R., Bonin, L., Jensen, P. K. and Kudsk, P. (2013). European perspectives on the adoption of nonchemical weed management in reduced-tillage systems for arable crops. *Weed Technology* 27(1): 231–240. https://doi.org/10.1614/WT-D-12-00066.1.

Moonen, A. C. and Barberi, P. (2004). Size and composition of the weed seedbank after 7 years of different cover-crop-maize management systems. *Weed Research* 44(3): 163–177. https://doi.org/10.1111/j.1365-3180.2004.00388.x.

Moss, S. (2018). Integrated weed management (IWM): why are farmers reluctant to adopt non-chemical alternatives to herbicides? *Pest Management Science* 75(5): 1205–1211. https://doi.org/10.1002/ps.5267.

Navntoft, S., Kristensen, K., Johnsen, I., Jensen, A. M. M., Sigsgaard, L. and Esbjerg, P. (2016). Effects of weed harrowing frequency on beneficial arthropods, plants and crop yield. *Agricultural and Forest Entomology* 18(1): 59–67. https://doi.org/10.1111/afe.12134.

Neve, P., Barney, J. N., Buckley, Y., Cousens, R. D., Graham, S., Jordan, N. R., Lawton-Rauh, A., Liebman, M., Mesgaran, M. B., Schut, M., Shaw, J., Storkey, J., Baraibar, B., Baucom, R. S., Chalak, M., Childs, D. Z., Christensen, S., Eizenberg, H., Fernández-Quintanilla, C., French, K., Harsch, M., Heijting, S., Harrison, L., Loddo, D., Macel, M., Maczey, N., Merotto, A., Mortensen, D., Necajeva, J., Peltzer, D. A., Recasens, J., Renton, M., Riemens, M., Sønderskov, M. and Williams, M. (2018). Reviewing research priorities in weed ecology, evolution and management: a horizon scan. *Weed Research* 58(4): 250–258. https://doi.org/10.1111/wre.12304.

Oerke, E.-C. (2006). Crop losses to pests. *Journal of Agricultural Science* 144(1): 31–43, http://doi.org/10.1017/S0021859605005708.

Pennings, J.M.E. and Garcia, P. (2001): Measuring producers' risk preferences: a global risk-attitude construct. *American Journal of Agricultural Economics* 83(4). https://doi.org/10.1111/0002-9092.00225.

Rego, I. A. C., Trindade, A. V. and Carvalho, J. E. B. d. (2004). Community and efficiency of arbuscular mycorrhizal fungi under different weed control managements in a citrus orchard. [Comunidade e eficiencia de fungos micorrizicos arbusculares sob diferentes manejos da vegetacao espontanea em pomar de citros.]. *Magistra* 16(2): 96–104.

Riemens, M. M., Groeneveld, R. M. W., Kropff, M. J. J., Lotz, L. A. P., Renes, R. J., Sukkel, W. and Van der Weide, R. Y. (2010). Linking farmer weed management behavior with weed pressure: more than just technology. *Weed Science* 58(4): 490–496. https://doi.org/10.1614/WS-09-048.1.

Riemens, M. M., Sønderskov, M., Moonen, A.-C., Storkey, J., Kudsk, P. (2021). An integrated weed management framework: a pan-European perspective. *European Journal of Agronomy* 133. https://doi.org/10.1016/j.eja.2021.126443.

Rogers, E. M., Singhal, A. and Quinlan, M. M. (2008). Diffusion of innovations. In: *An Integrated Approach to Communication Theory and Research* Stacks, D. W. and Salwen, M. B. (Eds). https://doi.org/10.4324/9780203887011.

Rossing, W. A. H., Dogliotti, S., Bacigalupe, G. F., Cittadini, E., Mundet, C., Mariscal, A. V., Douthwaite, B. and Alvarez, S. (2010). Project design and management based on a co-innovation framework. In: *Building Sustainable Rural Futures: The Added Value of Systems Approaches in Times of Change and Uncertainty*. Institute for Study Abroad Foundation, Viena, Austria, pp. 402–412.

Schreck, E., Gontier, L., Dumat, C. and Geret, F. (2012). Ecological and physiological effects of soil management practices on earthworm communities in French vineyards. *European Journal of Soil Biology* 52: 8–15. https://doi.org/10.1016/j.ejsobi.2012.05.002.

Sharma, A., Bailey, A. and Fraser, I. (2011). Technology adoption and pest control strategies among UK cereal farmers: evidence from parametric and nonparametric count data models. *Journal of Agricultural Economics* 62(1): 73–92. https://doi.org/10.1111/j.1477-9552.2010.00272.x.

Stern, V. M. and Van den Bosch, R. (1959). The integration of chemical and biological control of the spotted alfalfa aphid: field experiments on the effects of insecticides. *Hilgardia* 29(2): 103–130. https://doi.org/10.3733/hilg.v29n02p103.

Swanton, C. J., Shrestha, A., Clements, D. R., Booth, B. D. and Chandler, K. (2002). Evaluation of alternative weed management systems in a modified no-tillage corn-soybean-winter wheat rotation: weed densities, crop yield, and economics. *Weed Science* 50(4): 504–511. https://doi.org/10.1614/0043-1745(2002)050[0504:EOAWMS]2.0.CO;2. Available at: https://www.jstor.org/stable/4046534.

Swanton, C. J. and Weise, S. F. (1991). Integrated weed management: the rationale and approach. *Weed Technology* 5(3): 657–663. https://doi.org/10.1017/s0890037x00027512.

Van den Berg, H., Phillips, S., Dicke, M. and Fredrix, M. (2020). Impacts of farmer field schools in the human, social, natural and financial domain: a qualitative review. *Food Security* 12(6): 1443–1459. https://doi.org/10.1007/s12571-020-01046-7.

Van der Weide, R. Y., Bleeker, P. O., Achten, V. T. J. M., Lotz, L. A. P., Fogelberg, F. and Melander, B. (2008). Innovation in mechanical weed control in crop rows. *Weed Research* 48(3): 215–224. https://doi.org/10.1111/j.1365-3180.2008.00629.x.

Van Dijk, W. F. A., Lokhorst, A. M., Berendse, F. and de Snoo, G. R. (2016). Factors underlying farmers' intentions to perform unsubsidised agri-environmental measures. *Land Use Policy* 59: 207–216. Available at: https://www.sciencedirect.com/science/article/pii/S0264837716309346.

Van Zeeland, M., Spits, H. G., Kroonen-Backbier, B. and van de Weide, R. (2009). Invloed van het afschaffen van de cross-compliance regeling op de onkruidbestrijdingsstrategieën in maïs. *Research Report*. Praktijkonderoek Plant & Omgeving B.V., Wageningen UR, Wageningen, Netherlands. Available at: http://edepot.wur.nl/152333250031200.

Vasileiadis, V. P., Otto, S., Van Dijk, W., Urek, G., Leskovšek, R., Verschwele, A., Furlan, L. and Sattin, M. (2015). On-farm evaluation of integrated weed management tools for maize production in three different agro-environments in Europe: agronomic efficacy, herbicide use reduction, and economic sustainability. *European Journal of Agronomy* 63: 71–78. https://doi.org/10.1016/j.eja.2014.12.001.

Vereijken, P. (1997). A methodical way of prototyping integrated and ecological arable farming systems (I/EAFS) in interaction with pilot farms. *European Journal of Agronomy* 7(1–3): 235–250. https://doi.org/10.1016/S1161-0301(97)00039-7.

Waddington, H., Snilstveit, B., Hombrados, J. G., Vojtkova, M., Anderson, J. and White, H. (2014). Farmer field schools for improving farming practices and farmer outcomes in low-and middle-income countries: a systematic review. *Campbell Systematic Reviews* 10(6). Available at: https://www.3ieimpact.org/sites/default/files/2019-05/srs1_ffs_revise_060814_final_web_2.pdf.

Zaller, J. G., Cantelmo, C., Dos Santos, G. D., Muther, S., Gruber, E., Pallua, P., Mandl, K., Friedrich, B., Hofstetter, I., Schmuckenschlager, B. and Faber, F. (2018). Herbicides in vineyards reduce grapevine root mycorrhization and alter soil microorganisms and the nutrient composition in grapevine roots, leaves, xylem sap and grape juice. *Environmental Science and Pollution Research International* 25(23): 23215–23226. https://doi.org/10.1007/s11356-018-2422-3.

Zwickle, S. (2011). Weeds and organic weed management: investigating farmer decisions with a mental models approach. Available at: https://etd.ohiolink.edu/apexprod/rws_olink/r/1501/10?clear=10&p10_accession_num=osu1316453516.

Zwickle, S., Wilson, R., Bessette, D., Herms, C. and Doohan, D. (2016). Facilitating ecological weed management decisions by assessing risk-benefit tradeoffs. *Agroecology and Sustainable Food Systems* 40(7): 635–659. https://doi.org/10.1080/21683565.2016.1166170.

Chapter 12

Optimising integrated weed management in narrow-row crops

L. Bonin, ARVALIS-Institut du Végétal, France; R. Leskovšek, Agricultural Institute of Slovenia, Slovenia; C. Moonen, Institute of Life Science, Italy; W. Smith, NIAB, UK; and M. Sønderskov, Aarhus University, Denmark

1 Introduction

Irrespective of crop or crop type, the sustainable management of weeds represents a major pest challenge as weeds are a primary yield-reducing factor (Oerke, 2006). Weeds are currently viewed by farmers and advisors as the principal obstacle in the transition to low-input cropping systems. Moreover, integrated weed management (IWM) practices should be viewed on a long-term basis and should consider all agricultural practices such as changes in crop rotation, tillage type, sowing date and/or the use of cover crops. Herbicides offer an effective curative technique, as long as resistance is avoided, and their use over the last 50 years has resulted in a significant decrease in the labour required for weed control, at the same time increasing the crop yields. However, reducing the reliance on herbicides is necessary to avoid human health issues, protect the environment and preserve the efficacy of the available modes of action, as herbicide resistance is rapidly developing for many modes of action (Peterson et al., 2018). To date, alternative methods of managing crops sown in narrow rows have mainly been based on short-term changes, such as changes in sowing time or in the soil tillage system (Weiner et al., 2001; Lutman et al., 2013). Changes in crop rotation, with more diverse crop selection and longer rotation, have slowly gained support in

http://dx.doi.org/10.19103/AS.2021.0098.13

many countries and are often driven by the development of herbicide resistance (Peterson et al., 2018). The implementation of cover crops or intercropping has mainly been encouraged for purposes other than weed control, such as reducing nitrogen leaching and control of diseases or insect pests (Adetunji et al., 2020). Biological control has shown little success for weed management. When considering alternative weed management strategies, efficacy and economic aspects are essential factors for the farmer, as IWM strategies will only be adopted if they are effective and can secure yield and gross margin.

This chapter focuses on narrow-row crops, mostly small-seeded crops, which create a dense crop stand. The spatial layout of all narrow-row crops is a key determinant of potential weed control options and therefore, these crops can be considered collectively. For example, mechanical weeding is limited to harrowing at the very early crop growth phase, as machinery cannot operate when the crop canopy has closed without seriously damaging the crop plants. The most important narrow-row crops belong to the group of small-grain cereals, pulses, forage and oilseed crops. Besides the common row width and physical similarity in the early growth season, these crops vary considerably in below and above ground crop architecture during the rest of the growing season, which in turn affects their competitive ability against weeds. Species such as grain legumes are often extremely sensitive to weed interference, whereas cereals are more competitive. Independent of competitive ability, the farmers' objective is to control weeds as effectively as possible, in order to avoid a build up in the weed seedbank. The above-mentioned crops are discussed as one crop group in this chapter.

The IWM framework developed in IWMPRAISE (Kudsk et al., 2020) is comprised of five pillars, which detail the tactics to prevent weed establishment, reduce the impact of weeds on the crop and reduce seed return. The five pillars are (1) diverse crop rotation, (2) cultivar choice and establishment, (3) field/soil management, (4) direct control and (5) monitoring and evaluation. Each of the pillars constitutes an important part of a successful IWM strategy, as every alternative to a purely herbicide-driven strategy needs to consider a suite of tactics to achieve a sufficient level of weed control. The pillars are described in more detail in the following sections. Monitoring and evaluation are, in practice, integrated into the other four pillars, as this is the starting point for any IWM strategy and needs to be carried out following the application of any weed control tactics. This means that throughout the growing season, the effect of the applied tactics must be monitored and the strategy must be continuously re-evaluated. Furthermore, a long-term perspective is needed to achieve adequate weed control, as the effects of IWM strategies are more complex and often require a transition phase when changing from a conventional weed management system. In theory, the systems equilibrate but under changing climatic conditions, as we are experiencing currently, continuous monitoring and adaptation is necessary.

2 Cropping system diversification

The first pillar of the IWMPRAISE framework includes diversification of the cropping system and is generally considered the backbone of all IWM strategies (Barzman et al., 2015). The most consolidated tactics are diversification of crop rotation and the use of intercrops and cover crops, although field margin and crop edge management diversification may provide additional support to IWM. Alterations to crop rotation, using crop species with different life cycles (e.g. annual vs. perennial crops and summer vs. winter annual crops), leads to crops with differing competitive ability and creates diversification in soil disturbance patterns, in micro-climatic conditions and in the use of active ingredients and mechanical weeding tactics. An example is winter rye, which is often successfully utilised in IWM programmes due to its weed suppressive and allelopathic potential, regardless of use as a cash or cover crop (Mwendwa et al., 2020; Werle et al., 2017). Diverse crop rotation, and the concurrent changes in management practices, strongly affects the composition of the weed population emerging from the seed- and bud bank. In addition, variability will limit the build-up of species-poor weed communities, dominated by species that are well adapted to the specific conditions created by monoculture or short crop rotations with similar crops (Ryan et al., 2010). The same tactics are also effective in preventing the build-up of resistance, as they often result in a higher diversity in the use of herbicide mode of actions. Despite its efficacy for weed management, economic or market-related drivers may hinder the adoption of crop diversification by farmers, who may persist with short crop rotations based on cereals, alternating winter cereals with a few summer annual crops (Lefebvre et al., 2014; Meynard et al., 2018).

Agricultural landscape in France in spring. Photo credits: ARVALIS.

The possibilities and effects on weeds of changing crop rotation largely depend on the regional climatic conditions and on market opportunities. In northern Europe, the inclusion of spring crops, including spring cereals, leads to the extremely efficient control of winter annual grass weeds, for example, *Lolium multiflorum*, *Alopecurus myosuroides* and *Vulpia muyros*. Depending on the severity of the grass weed problem, farmers can change the length of crop rotation and the proportion of spring crops needed. Decision Support Systems (DSSs) may help farmers to plan their crop rotation and maximise weed control, and a few European DSSs have been introduced considering the effects of crop rotation changes (Parsons et al., 2009; Sønderskov et al., 2020). In the Mediterranean region, where the climate is not suitable for spring cereals, perennial forage crops can be used to reduce weed populations. For example, if *Medicago sativa* (lucerne) is sown after winter wheat and regularly cut for 3–4 years, it will provide farmers with forage, while controlling emerging weeds and exhausting the seedbank. This practice is only advantageous in regions where the market for forage is good, since transportation of hay bales or silage is very expensive and cumbersome. Another issue relates to the length of crop rotation, which increases due to the introduction of forage crops. Farmers cannot produce a cash crop in the years forage crops are grown. An interesting approach to limit the time of presence of forage crop is the application of relay intercropping. In southern Europe, this technique traditionally consists of sowing lucerne in the cereal crop field at the end of winter, just before stem elongation. Lucerne will establish in the cereal crop field but will not compete with the cash crop. After the harvest of the cereal in early summer, minimal rainfall or remaining soil humidity will help lucerne to grow and cover the soil at the end of summer and in the following autumn and winter months. In this way, the farmer can perform a first cut in the autumn of the same year as harvesting the cereal crop and at the same time control weeds and improve soil fertility. Other legumes have recently been tested for their suitability to serve as relay intercrops with cereals. Some of these legumes are annuals or annual self-seeding legumes. Using annual species would increase the feasibility of this technique being adopted by the farmers, who prefer short crop rotation dominated by cereals, because they cannot afford the loss in income of growing a 4-year lucerne crop (Vrignon-Brenas et al., 2016; Leoni et al., 2019).

The introduction of cover crops and various intercropping methods has been found to support IWM (Dorn et al., 2015; Kumar et al., 2020), mainly through direct weed suppression in terms of competition for light, space and nutrients, but also through the creation of a physical barrier in the case of dead mulches. The main advantage of these tactics is that they provide multiple ecosystem services to the cropping system, besides suppressing weeds. Legume crops increase soil fertility, winter cover crops decrease soil erosion and nutrient leaching, and species of the Brassicaceae family may provide plant material for biofumigation, which suppresses soil-borne diseases and nematodes, as well as improving the soil structure (Schipanski et al., 2014; Blanco-Canqui et al., 2015).

Diversification of field margins and crop edge management have received much less attention, although the benefits and opportunities have been described and demonstrated over a long period. For example, protecting the field margins from soil disturbance or fertiliser input substantially reduces the development of typical weeds found in and near cereal fields, such as *Bromus sterilis* and *Gallium aparine*. In this way, the field margin will not harbour crop weeds, but will develop perennial vegetation that has minimal adaptation to a field environment dominated by annual crops (e.g. Cirujeda et al., 2019). Another approach is to focus on effective weed management in the field margins by preventing the spread of weed species from the field margin to the main area of the field. It has been demonstrated that different types of field margins provide different ecosystem services (Van Vooren et al., 2018; Mkenda et al., 2019); hence, it is important to diversify the field edge typologies in farmed landscapes. Hedgerows, ditches, grass strips and sown wild-flower strips should coexist to optimise their benefits and mitigate potential negative effects on the weed population at a landscape level, which will benefit not only narrow-row crops but all crops.

3 Cultivar choice and establishment

The second pillar covers a range of cultural tactics to increase crop competitiveness and reduce the impact of weeds on the crop. These cultural weed practices include high planting densities, narrow row spacing, modification of sowing pattern and spatial arrangement, delayed sowing and the selection of cultivars and mixtures with strong weed suppressive ability (Olsen et al., 2005a,b; De Vita et al., 2017).

Narrow-row crops are usually sown in a relatively uniform row pattern, where crop arrangement is defined by row distance and seeding rate. Increasing crop seeding rate and narrowing row spacing can contribute to weed suppression via faster canopy closure and enhanced crop-weed competition (Lemerle et al., 2006). Crops can, however, also be sown in a different spatial distribution than uniform rows. Sowing pattern and spatial arrangement have an impact on weed suppression (Olsen et al., 2005a; Weiner et al., 2001), but due to complexity of using sowing machinery and mechanical weeding tools for altered sowing patterns, these strategies are rarely implemented in practice.

Delayed sowing is a simple strategy, applicable under various environmental conditions and cropping systems, where part of the weed population germinates before crop establishment, enabling weed control before crop sowing. As late sowing can involve a risk of yield penalty, increased sowing rates are required to attain the same number of heads per unit area at harvest. Late autumn sowing also increases the risk that weather conditions are unfavourable for sowing and thus increases the risk of being unable to establish a crop in the autumn. Like other IWM strategies, the adoption of delayed sowing requires highly adaptable farmers who are willing to accept the complexity of this weed management system (Moss, 2019). The risk of missing the optimum sowing window is more pronounced if it is applied across the whole farm and can be mitigated with more field-specific weed management plans, for example, by postponing the sowing of fields with large grass weed infestation to the end. In this way, the risk of yield losses can be limited to a smaller part of the farm. The benefit of this strategy when managing *A. myosuroides* is most consistent when the sowing of winter cereals is delayed until the end of October or even the beginning of November under northern European conditions (Lutman et al., 2013).

Comparison of two sowing dates (cv. Orvantis). Sowing on 13/11 on the left and 13/10 on the right. Despite less biomass, the left plot is also less infested (A. myosuroides) and will prove to be more productive. Photo credits: ARVALIS.

Cultivar selection or cultivar mixtures can also contribute to weed management strategies. There are several plant traits, such as early vigour, plant height, canopy architecture, tillering capacity or higher growth rate at various development stages, which provide improved competitiveness against weeds (Aharon et al., 2021; Fontaine et al., 2009; Wicks et al., 2004; Hansen et al., 2008; Korres and Froud-Williams, 2002). Andrew et al. (2015) found that a combination of traits was more likely to be responsible for a cultivar's ability to tolerate or compete against weeds than a single trait. Competitive cultivars or targeted cultivar mixtures do not require any changes in current farming practice and are therefore very cost effective. However, the benefits of competitive cultivars are likely to decrease at higher crop densities and delayed sowing dates (Andrew and Storkey, 2017). Mixing different cultivars is predominately aimed at reducing disease pressure, while maintaining yields at acceptable levels. In less competitive crops, however, this strategy can contribute to enhanced weed suppression ability and, in turn, reduced herbicide rates (Snyder et al., 2020; Oveisi et al., 2021).

Sowing date, sowing density, cultivar choice and other IWM cultural methods and strategies are generally effective when used in a complementary way and are as efficient as herbicide-based systems, even in managing troublesome grass species such as *A. myosuroides* (Zeller et al., 2021; Chikowo et al., 2009).

4 Field and soil management

The third pillar relates to managing the field in terms of soil management. Since the beginning of agriculture, soil tillage has been an important component of crop cultivation (e.g. soil loosening, manure and residue incorporation), and weed control has been a key objective of soil tillage, although its role was overshadowed by the introduction of herbicides (Mazoyer and Roudart, 1997). Soil tillage largely influences weed biology and ecology, and in recent years a shift in soil tillage systems, towards less soil disturbance, has increased interest and research in the effect of soil tillage. The choice of method, depth, timing and frequency of soil cultivation may influence the composition, density and long-term persistence of the weed population (Mohler and Galford, 1997). Ploughing is the most commonly used practice to control perennial weeds - bury weed seeds and prepare the soil for a new crop (Padel et al., 1999; Melander et al., 2012) - and provides an effective way of managing weeds (Håkansson, 2003; Lutman et al., 2013). The annual loss of seeds from a natural soil weed seedbank with no addition of fresh seeds and no cultivation is roughly 20%, but it depends on the composition of the weed species (Roberts and Dawkins, 1967; Barralis et al., 1988). If the soil is cultivated twice a year, the depletion of the seedbank can reach 30%, and when cultivated four times it reaches nearly 40%. The vertical distribution of weed seeds in the soil is significantly affected by the type of tillage operation. In a ploughed system, less than 30% of seeds were present in the top 5 cm (Barralis et al., 1988). With a chisel plough and no-till system, 66% and 90% of seeds, respectively, were present in the upper 10 cm soil layer, with nearly all seeds in the top layer (0–5 cm) for the no-till system (Swanton et al., 2000). In the absence of tillage, weed seeds are mostly present close to the soil surface, enabling them to emerge in the succeeding crop and lead to very high levels of infestation

(Bàrberi and Lo Cascio, 2001; Vullioud et al., 2006). Leaving weed seeds on the soil surface will also decrease the viability of a long list of weed species, including grass weeds (Jensen, 2009, 2010a,b). Annual broadleaved weeds are less influenced by tillage than annual grass weeds (Lutman et al., 2013). Annual meadow grass (*Poa annua*), wild oat (*Avena fatua*), bromes (*Bromus* sp.) and blackgrass are all favoured by non-ploughing techniques. These effects were also observed in an organic farming system, with an increase of grass weed populations in no-till systems (Dittmann, 2012). However, it is possible to reduce the frequency of ploughing without increasing grass weeds or herbicide use if modification of the system is well planned. More generally, when establishing a crop that provides the most favourable conditions to the dominant weed in the rotation the best rule is ploughing; in addition to ploughing, the weed infestation exhibits a substantial increase over the years (Colbach et al., 2013).

Ploughing of a cover crop during November in France. Photo credits: ARVALIS.

The timing of seedbed preparation considerably affects weed populations and provides an opportunity to reduce weed emergence in the crop. A false seedbed, where emerging weeds are controlled with mechanical tools prior to crop establishment, is a common technique (Johnson and Mullinix, 1995; Benvenuti and Macchia, 2006; Bonin and Labreuche, 2007). Stale seedbeds are similar, but emerged weeds are killed without disturbing the soil, for example, by herbicides or flaming. These techniques are based on the principle of depleting the seedbank in the upper soil layer and reducing the subsequent emergence of weed seedlings and can be an effective method of decreasing the density of annual weeds (Leblanc and Cloutier, 1996; Bonin and Labreuche, 2007); however, it is highly dependent on the prevailing conditions. The most important factor determining the timing of a flush of weed emergence is adequate soil moisture (Colbach and Mézière, 2013). Consequently, in dry years the false seedbed method is less likely to provide substantial weed control. If there is no rain during the false seedbed period, the preceding tillage operations can increase soil drying due to the bare soil (Johnson and Mullinix, 1995). In addition, many weed seeds require light activation to become sensitive to the environmental stimuli (e.g. rain and tillage) that trigger germination, and this photo-activation is only possible if the seeds are imbibed. To meet these conditions, it is important to delay the preparation of a false seedbed until after the first rains to allow seed imbibition and photo-activation. Otherwise, the false seedbed may induce dormancy and contribute to a build up of the seedbank. Besides soil moisture, timing is essential for success. Removal of emerged weeds has to be delayed until after the main flush of emergence or will need to be repeated. Farmers may be reluctant to delay crop establishment if soil conditions are good and there is a risk of heavy rain preventing future operations. It is important not to cultivate below the top 2–5 cm of soil when killing emerged weeds otherwise a subsequent flush may emerge (Bonin and Labreuche, 2007). In either case, no-till or reduced tillage, it is essential to optimise all cultivation techniques in terms of timing and combinations of tactics to maximise weed control.

5 Direct control

The fourth pillar is most employed in conventional cropping systems, as herbicides belong to direct control tactics. Herbicides are the most commonly used tool to manage weeds in narrow-row crops. Before crop establishment, non-selective herbicides, mainly glyphosate, can be used to control weeds emerging between crops as part of stubble management practices or to terminate cover crops. In no-till systems, in particular, there is a need for weed control between crops, as this is usually provided by tillage operations in other systems. Pre-emergence herbicides are applied after sowing but prior to crop emergence. Post-emergence herbicides are the most important group of herbicides and have different application timings during the season related to season and crop developmental stage. Herbicide resistance affects the efficacy of some herbicides, and human and environmental risks have resulted in the bans and the removal of several active ingredients. This means that herbicides must be used with care and consideration to prevent further development of resistance and to protect human health and the environment. Several tactics can assist targeted herbicide application and reduce herbicide use without hampering efficacy. It is possible to reduce the herbicide dose compared with the labelled rate in some situations, if the actual weed population composition in a field is taken into consideration (Kudsk, 2008). DSSs can help determine the most effective herbicides and the minimum required dose to control the observed weed flora under the prevailing climatic conditions at field level (Been et al., 2009). A further step towards site-specific management is patch or spot application, where only a fraction of a field is sprayed. This limits the application to spots with high densities of detrimental weeds (Somerville et al., 2020). A different type of patch spraying is the diversification of field margin management. This may be interesting if there are significant differences in the composition or abundance of weed species between the field margin and the central part of the field. In this case, weed management can be adapted and a different herbicide can be selected, or the dose can be reduced in the field margin. Farmers may even choose to establish a different crop in the field margin, for example, cereals surrounding sugar beets or field peas. This often enables the use of a wider selection of herbicides for a specific weed species. On the other hand, when weeds do not pose a severe threat to the crop, farmers may decide to leave the field margin unsprayed, saving herbicides and protecting the field margins and water courses, while having a relatively low yield as the crop margin is normally less productive than the main field (Marshall, 1989). Currently, technological advances in camera-guided weed recognition and mapping are a fast-developing area, which will enable identification and spray application in the same pass across the field. In addition, the development of automated carriers for a range of tools will mitigate the higher labour demand, which can be a barrier for differentiated spraying programmes.

In addition to a more targeted use of herbicide application, mechanical tools can provide direct weed control (Melander et al., 2005; Peruzzi et al., 2017; McCollough et al., 2020). Both weed hoeing and harrowing can provide weed control in narrow-row crops, but the efficacy of hoeing is higher than harrowing. Weed harrowing mainly controls small weed plants with shallow root systems, whereas weed hoeing can control larger weeds with a more developed root system. Both types of mechanical weeding are sensitive to timing and soil conditions. Previously, weed hoeing was only possible with row distances greater than 15 cm, but camera-guided systems have enabled this practice to be used on narrow row distances down to 12.5 cm (Gerhards et al., 2020). The seeding density can be increased to optimise crop competitiveness within the row (Melander et al., 2018).

In heavily infested fields, weed seed collection and destruction is being developed to control grass weeds in particular. This is a potential method for weed species retaining the seeds until harvest. The best-studied example is *Lolium rigidum* in Australian wheat fields (Guzzomi et al., 2017; Soni et al., 2020).

Hoeing (self-guided with camera) in wheat, during spring. Photo credit: ARVALIS.

6 Case studies

IWM requires farmers to integrate tactics from all pillars of the IWM framework as described above. The individual tactics may vary in effectiveness but, in combination, can provide weed control efficacies in line with standard herbicide strategies. In the four case studies involving cereals, weed management strategies were compared to standard herbicide strategies in the United Kingdom, France, Slovenia and Denmark.

6.1 Individual tactics for integrated weed management strategies

The individual tactics in an IWM strategy depend on local conditions and the crop rotations in which they are applied. Two examples of individual tactics for grass weed management (delayed sowing date and soil tillage system) were studied in spring barley with blackgrass in the UK and in triticale with ryegrass in France.

6.1.1 Sowing date of spring cereals in the UK

Spring barley is a popular crop choice for the diversification of crop rotations in northern Europe, as it is highly competitive against any weeds. However, in an attempt to improve yields of spring barley, farmers are moving to earlier spring sowing dates. In the UK, a field trial was established to understand the relationship between blackgrass density and sowing date in the spring. The trial included herbicide input as a factor to demonstrate the potential of reducing herbicide usage following delayed sowing (Table 1).

Table 1 List of treatments in a demonstration trial on weed control in response to delayed sowing of spring barley in the UK (Hardwick, 2020).

Treatment	Sowing date	Herbicide application	Application date
1. Early Sow_No Herb	5/03	Untreated (no herbicide)	7/03 (+ 20/03 for treatment 5)
2. Early Sow_Low Herb		Hurricane 0.25 l/ha (low herbicide)	
3. Early Sow_Med Herb		Liberator 0.3 l/ha (medium herbicide)	
4. Early Sow_High Herb		Liberator 0.3 l/ha + Crystal 2 l/ha (high herbicide)	
5. Early Sow_High Herb split		Liberator 0.3 l/ha fb Crystal 2 l/ha (high herbicide split)	
6. Med Sow_No Herb	23/03	Untreated (no herbicide)	26/03 (+ 10/04 for treatment 10)
7. Med Sow_Low Herb		Hurricane 0.25 l/ha (low herbicide)	
8. Med Sow_Med Herb*		Liberator 0.3 l/ha (medium herbicide)	
9. Med Sow_High Herb		Liberator 0.3 l/ha + Crystal 2 l/ha (high herbicide)	
10. Med Sow_High Herb split		Liberator 0.3 l/ha fb Crystal 2 l/ha (high herbicide split)	
11. Late Sow_No Herb	8/04	Untreated (no herbicide)	10/04 (+ 22/04 for treatment 15)
12. Late Sow_Low Herb		Hurricane 0.25 l/ha (low herbicide)	
13. Late Sow_Med Herb		Liberator 0.3 l/ha (medium herbicide)	
14. Late Sow_High Herb		Liberator 0.3 l/ha fb Crystal 2 l/ha (high herbicide split)	
15. Late Sow_High Herb split		Liberator 0.3 l/ha + Crystal 2 l/ha (high herbicide)	

Notes: Hurricane: 500 g/l diflufenican, Liberator: 400 g/l flufenacet + 100 g/l diflufenican, Crystal: 60 g/l flufenacet + 300 g/l pendimethalin. fb: followed by, *: standard used for economic calculations related to costs and gross margin.

The blackgrass head density in untreated plots decreased with delayed drilling time from 40 heads/m² at the early sowing date to 2 heads/m² at the late sowing date (Fig. 1). For comparison, in a 2018 trial in the UK, in the standard autumn sown winter barley crop, blackgrass head density reached 502 heads/m². Just changing the establishment timing (from autumn to spring) led to a 12-fold reduction in the blackgrass population.

The later sowing dates were associated with slightly higher crop yields, mainly as a result of better establishment conditions as time progressed (Fig. 2). This was mainly evident in the late March timing, although the use of 0.25 l/ha Hurricane in early March had a similar effect. The most effective herbicide application was achieved with the early sowing date and Liberator with Crystal at the same time or applied later. For later sowing dates (23/3 and 8/4), little effect was achieved with herbicide application. It is, therefore, questionable whether the use of herbicides for blackgrass is warranted in the spring, particularly in competitive crop species such as barley or oats. The highest gross margin (income from yield subtracting the cost of treatments) was reached in the 'no herbicide' application and late drilling (treatment 11) followed by intermediate or late sowing date and the lowest herbicide application rates (treatments 8, 9, 12 and 13).

Yield tended to be higher with delayed sowing dates. The herbicidal effect can be a double-edged sword: preserving the yield potential of the crop (by removing weed competition), but at a cost (reduction in gross margin). The results showed that the effect of sowing date exceeded the effect of herbicide application choices in the control of blackgrass, while cost-effectiveness

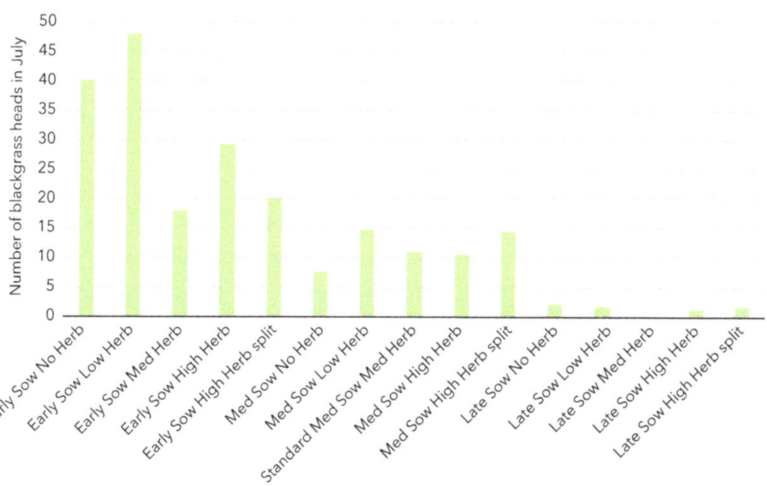

Figure 1 Impact of sowing date and herbicide treatments on blackgrass heads in July. (Hardwick, UK, 2020).

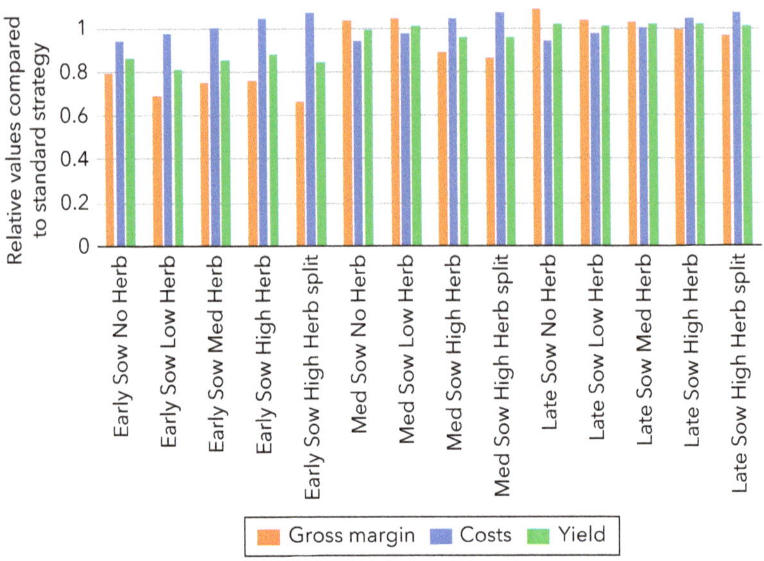

Figure 2 Relative index for costs, gross margin and yield compared with the standard strategy (MedSow_MedHerb, treatment 8) for the spring barley trial (Hardwick, UK, 2020). Yield: physical yield; Costs: all costs related to the crop inclusive of seeds, fertiliser, plant protection products and expenses related to machinery for treatments; Gross margin: income from yield subtracting the costs.

depends on the trade-off between the cost of herbicide use and the increase in yield obtained by weed control.

6.1.2 Soil tillage system in France

To illustrate the effect of soil tillage on Italian ryegrass (*L. multiflorum*) in combination with herbicide use, a trial was established with triticale in France. The trial field was managed using conservation tillage (no-till) until 2017. One half of the field was ploughed (P) at the beginning of the trial in 2017 and the other half was maintained as a conservation tillage (CT) system. The crop grown in 2017–18 was linseed. The field had a high infestation of Italian ryegrass and the weed control strategies targeted this grass weed problem. In 2018–19, the same weed control strategies were conducted in each part of the field (Table 2).

All plots were sown at a row spacing of 15 cm to allow the hoeing machine (Garford model self-guided by a camera) to pass between the rows. Hoeing was performed on 21 February (treatments 1, 2, 4, 5 and 7), 28 March (treatments 1, 4 and 7) and 12 April 2019 (treatment 7). Only the first two passes were followed by a harrow, as the triticale had reached the one node stage at the third pass on 12 April. The harrowing operation after hoeing would have broken up the soil

Table 2 List of treatments in a French trial with triticale (Boigneville, France, 2018). The trial was sown on 10 October 2018 with the variety RGT OMEAC

N°	Name of the strategy	Herbicide applications	Mechanical weed control
1	Pre-em + Post-em + 2H	Trooper 2 l/ha (*pre-emergence*) ± Defi 2.5 l/ha (*1-2 leaves)*	Hoeing (2 passes)
2	Pre-em+Post-em+1H	Trooper 2 l/ha (*pre-emergence*) ± Defi 2.5 l/ha(*1-2 leaves)*	Hoeing (1 pass)
3	Pre-em+Post-em *	Trooper 2 l/ha (*pre-emergence*) ± Defi 2.5 l/ha (*1-2 leaves)*	None
4	Post-em+2H	Defi 2.5 l/ha (*1-2 leaves)*	Hoeing (2 passes)
5	Post-em+1H	Defi 2.5 l/ha (*1-2 leaves)*	Hoeing (1 pass)
6	Post-em	Defi 2.5 l/ha (*1-2 leaves)*	None
7	Harrow+3H	None	Tine harrowing pre-emergence + 3 passes of hoeing
C	CONTROL	None	None

Notes: Trooper: 300 g/l pendimethalin + 60 g/l flufenacet, Defi: 800 g/l prosulfocarb. *: standard used for economic calculations related to costs and gross margin.

clods and prevented regrowth of the weed plants pulled up at hoeing. Climatic conditions before and after each pass of mechanical weeding were optimal.

Two years after ploughing half of the field, the difference between the CT and the P parts of the field was clearly visible in terms of number of ryegrass plants. It should be noted that the cultivation of linseed during the 2017-18 season did not promote ryegrass management, but rather exacerbated the differences between the two parts of the field. There were around 1800 Italian ryegrass individuals/m² in control plots in the CT part and only 76 Italian ryegrass individuals/m² in the P part.

This basic difference in the level of emerged ryegrass plants was evident in all treatments with herbicides as the efficacy levels were higher in the P part for all strategies. This was probably due to plant cover in the CT part (high density of Italian ryegrass), which prevented the spray from reaching all seedlings (Fig. 3). Without mechanical weed control, the intensive control programme (treatment 3) reached high levels of control in the P part. The less intensive programme achieved 75% control (treatment 6). This was not the case in the CT part, where both herbicide programmes without mechanical weeding provided insufficient control (67% and 43%). The addition of hoeing increased efficacy in both the CT and P parts. In the P part, the less-intensive programme with two hoeing operations almost reached the level of the intensive programme, whereas the CT part consistently provided lower efficacy. Adding a second hoeing to the intensive programme minimally increased efficacy in both parts of the trial.

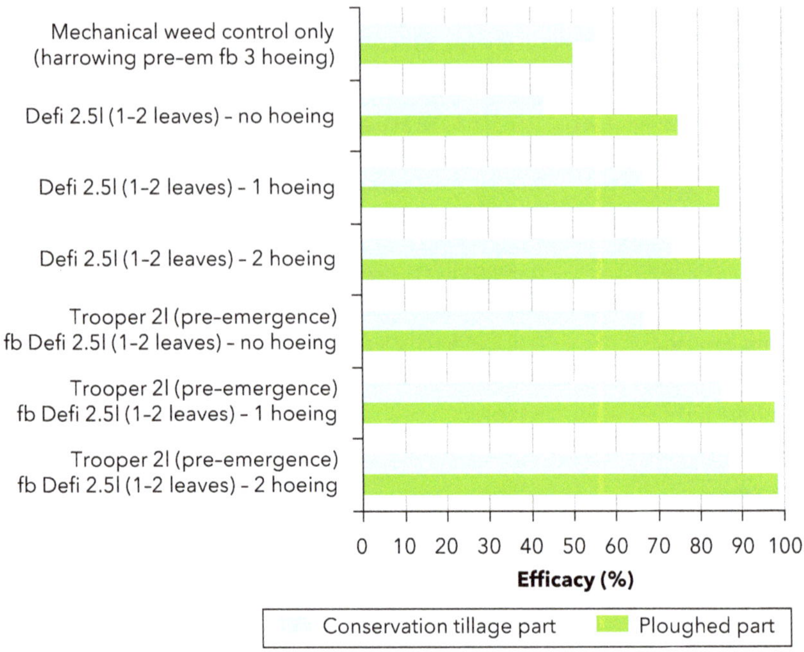

Figure 3 Efficacy (per cent reduction of biomass), on Italian ryegrass, of treatments in the ploughed and conservation tillage part of the trial. The reference for all treatments was the corresponding control plots (Arvalis, Boigneville, France, 2019). Treatments are described in Table 2. fb: followed by.

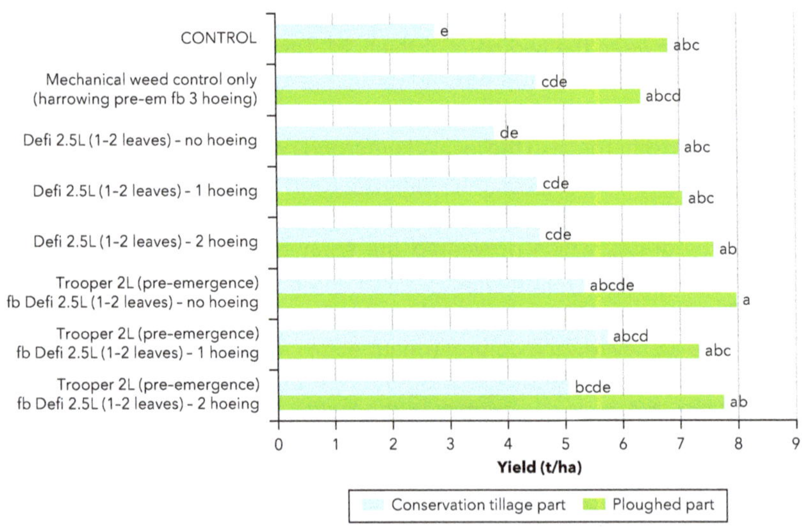

Figure 4 Yield (t/ha) of treatments in the ploughed (P) and conservation tillage (CT) part of the trial (Arvalis, Boigneville, France, 2019). Letters indicate levels of significance within the P and CT systems, respectively (Tukey Test, 5% Homogeneous Group (HG)).

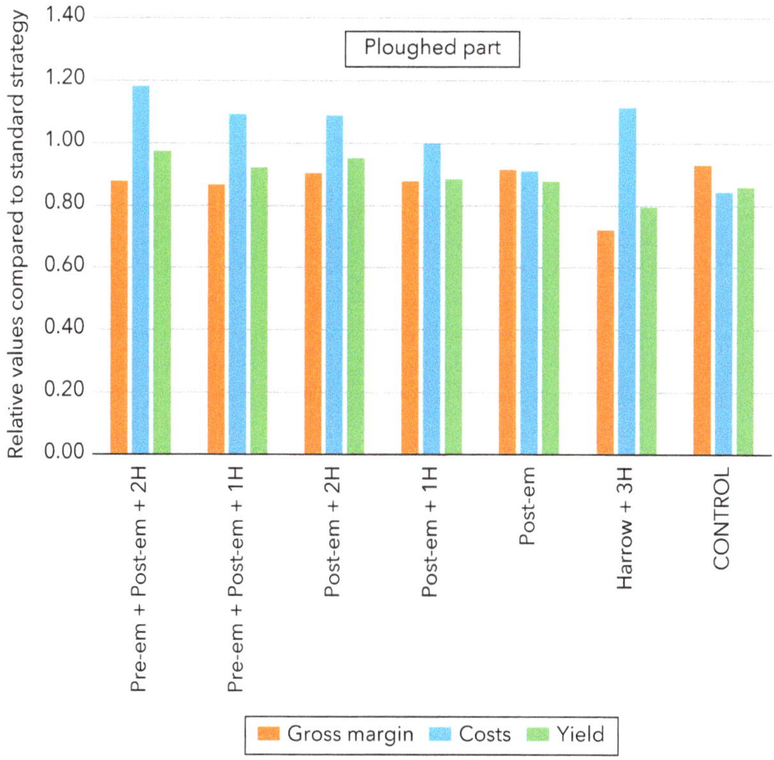

Figure 5 Relative index for costs, gross margin and yield compared to the standard strategy (pre-em + post-em, treatment 3) for the triticale trial, in the ploughed part (Boigneville, France, 2019). Yield: physical yield; Costs: all costs related to the crop inclusive of seeds, fertiliser, plant protection products and expenses related to machinery for treatments; Gross margin: income from yield subtracting the costs.

This trial highlighted the strong impact tillage has on Italian ryegrass management. Even 2 years after ploughing, the effects on the grass weed were clearly visible. The control plot in the P part had fewer Italian ryegrass plants than the intensive herbicide programme plot in the CT part (data not shown). The integration of occasional ploughing, as an IWM tool in no-till systems, is considered important in reducing the reliance on herbicides.

In the P part, the differences in yield between treatments with or without hoeing were not significant (Fig. 4). Only the mechanical weed control strategy tended to be less effective with a non-significant difference of 1.7 t/ha compared with the intensive programme without hoeing. The effect of hoeing, in addition to herbicide treatments, was limited but did not significantly affect yield.

Yields in the CT part were much lower than in the P part. The Italian ryegrass population had a strong negative impact on the crop; hence, the yield of the control in the CT part was 3.0 t/ha (compared with 6.8 t/ha in the P part). In the

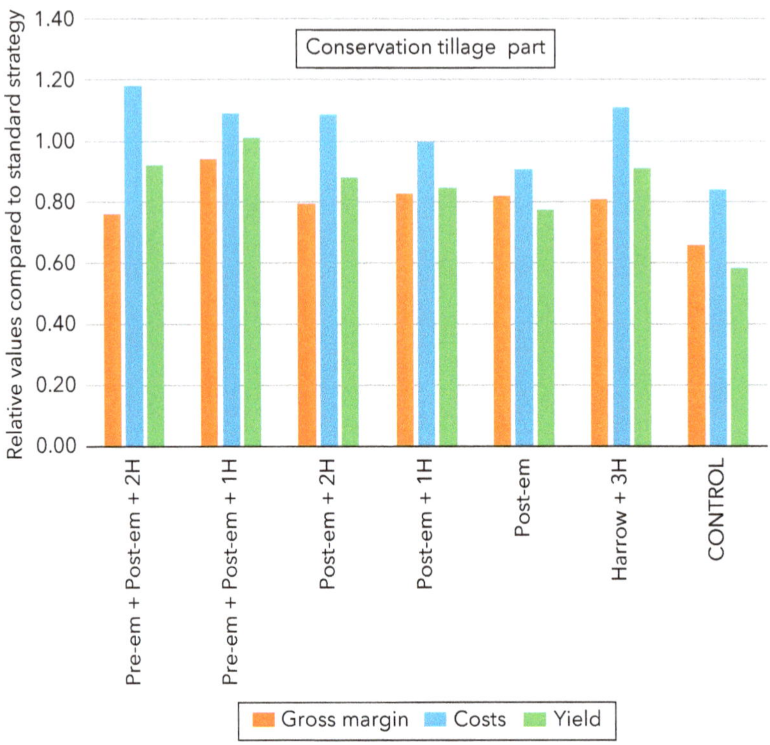

Figure 6 Relative index for costs, gross margin and yield compared to the standard strategy (pre-em + post-em, treatment 3) for the triticale trial, in the conservation tillage part (Boigneville, France, 2019). Yield: Physical yield; Costs: all costs related to the crop inclusive of seeds, fertiliser, plant protection products and expenses related to machinery for treatments; Gross margin: income from yield subtracting the costs.

CT part, only the intensive programme, combined with hoeing (treatment 5), significantly outperformed the control (increase of 3.0 t/ha). The less-intensive treatment, whether combined with hoeing or not, tended to achieve higher yields than the control (increase of 1 to 1.5 t/ha) but was not significantly different from the control.

In the P part, the gross margin of the control was equivalent or even higher than the other strategies studied due to the limited costs (Fig. 5); it is, however, lower than the standard strategy. The control only achieved a ratio of 0.66 of the standard strategy in the CT part (Fig. 6). Even after subtracting the cost of ploughing the field 2 years before the trial (€45–50/ha), the economic outcome was higher for the P part. Economically, the methods without hoeing were superior. Besides the interest in occasional ploughing to decrease grass weed pressure, the trial highlighted the potential for combining low herbicide levels with hoeing.

The conclusions are different for conservation tillage. The effect of rye grass is so important that the gross margin is highly correlated to efficacy (Fig. 6). Treatments with one hoeing are more effective with limited cost differences; however, this remains poor compared with the control in the ploughed situation.

6.2 Combination of tactics for integrated weed management strategies

6.2.1 Slovenia

In Slovenia, the conditions for winter cereal cropping differ between regions. The most favourable conditions are found in the northern part of Slovenia, represented by Rakičan, where a mild continental climate and loamy soil types enable intensive and high-quality winter cereal production. In the central region, the conditions are more challenging, and this is represented by Jablje. This region has a much heavier soil type and lower average temperatures. The average yields are lower, especially due to a shorter tillering period. Moreover, the grain quality is often decreased due to intensive rainfall periods in spring and summer. Therefore, the success of weed management strategies can differ between regions.

The aim was to establish a potential alternative to the standard strategy for winter cereals to minimise the use and reliance on herbicide applications. Spring or autumn herbicide application was considered the standard strategy. Alternative strategies aimed to reduce weed establishment in autumn with delayed sowing or reducing weed interference in the crop by mechanical weeding in the form of tine harrowing.

The strategies tested at the two locations were designed to consider local characteristics. They are not exactly the same but the overall principles were similar. For example, tine harrowing in autumn was not considered relevant at Jablje since the weather and soil conditions do not usually allow this operation (Tables 3 and 4). For the presentation of the results, the standard strategy was considered herbicide application only in the spring for both locations (strategy 1).

Weather conditions in 2019 were not favourable for winter cereals and relatively poor yields of winter barley were generally achieved in the central region of Slovenia. The results showed that in terms of winter barley yield, alternative strategy 4 (Late_SpHerb_Tine) and strategy 2 with autumn herbicide application (AutumnHerb) gave a slightly higher yield (just above 1.0) than the standard spring herbicide application (6.1 for both strategies compared with 5.6 t/ha of the standard strategy) (Fig. 7). Strategy 3 (SpHerb_springTine) resulted in a lower yield (5.0 t/ha). The weed control efficacy of strategy 4 was slightly lower than the standard strategy, whereas strategy 2 provided higher efficacy (Fig. 8). Even though the costs were slightly higher for strategy 2, the

Table 3 Strategies in winter barley (Jablje) in 2019. Crop seeding density was 420 plants/m^2 (variety Sandra)

2019-2020	Standard	AutumnHerb	SpHerb_springTine	Late_SpHerb_Tine
Jablje	Strategy 1	Strategy 2	Strategy 3	Strategy 4
	Reference/ standard	Ploughing Autumn herbicides	Ploughing Herbicides + spring tine harrow	Ploughing Late sowing + Herbicides + spring tine harrow
	Winter barley	Winter barley	Winter barley	Winter barley
Soil tillage	Ploughed 1/10 Seedbed prep. 5/10	Ploughed 1/10 Seedbed prep. 5/10	Ploughed 1/10 Seedbed prep. 5/10	Ploughed 1/10 Seedbed prep. 5/10
False seedbed	No	No	No	Tine harrowing 18/10
Sowing time	Normal sowing time 3/10	Normal sowing time 3/10	Normal sowing time 3/10	Late sowing time 18/10
Row width	12 cm	12 cm	12 cm	12 cm
Herbicides	Spring application 29/3: 10 g Iodosulfuron per ha	Autumn application 24/10: 2400 g prosulfocarb per ha	Spring application 29/3: 6 g Iodosulfuron per ha (60% reduction)	Spring application 29/3: 6 g Iodosulfuron per ha (60% reduction)
Mechanical weeding	No	No	Spring tine harrow: 27/2	Spring tine harrow: 27/2

higher yield compensated and resulted in a gross margin higher than the standard strategy (€17/ha compared with a negative gross margin at -€56/ha for the standard strategy) (Fig. 7). In terms of weed control and gross margin, autumn herbicide application gave the best outcome resulting in good residual weed control visible until harvest, with costs equal to the standard strategy and a higher yield. Autumn application can be considered as a standard for some farmers and even though the strategy performed well, it will not reduce the reliance on herbicides. In terms of alternative strategies, which are less dependent on herbicide use, strategy 4 was the best. Significantly, greater dry weed biomass was measured in strategy 3 established at the normal sowing time, followed by spring harrowing and reduced herbicide application. In strategy 3, yield was lower and the gross margin was far below the standard strategy (Fig. 7).

For winter wheat, none of the strategies gave positive gross margins and none of the alternative strategies came close to the standard strategy, which had a negative gross margin of -€30.5/ha (Fig. 7). Despite intensive investments in

Table 4 Strategies in winter wheat (Rakican) trials in 2019. Crop seeding density was 470 plants m² for the normal sowing date and 530 plants m² for the late sowing date (variety *Falado)*

2018-2019	Standard	AutumnHerb	SpHerb_Tine	Late_NoHerb_Tine
Rakican	Strategy 1	Strategy 2	Strategy 3	Strategy 4
	Reference/ standard	Ploughing Autumn herbicides	Ploughing Herbicides + spring tine harrow	Ploughing Late sowing + Herbicides + spring tine harrow
	Winter wheat	Winter wheat	Winter wheat	Winter wheat
Soil tillage	Ploughed 12/10 Seedbed prep. 15/10	Ploughed 12/10 Seedbed prep. 15/10	Ploughed 12/10 Seedbed prep. 15/10	Ploughed 12/10 Seedbed prep. 15/10
Sowing time	Normal sowing time 18/10	Normal sowing time 18/10	Normal sowing time 18/10	Late sowing time 29/10
Row width	12 cm	12 cm	12 cm	12 cm
Herbicides	Spring application 29/3: 250 g pyroxulam per ha	Autumn application 5/11: 600 g chlortuloron + 500 g diflufenican + 80 g pendimethalin per ha	Spring application 29/3: 175 g pyroxulam per ha (60% reduction)	No
Mechanical weeding	No	No	Tine harrow: autumn 16/11 + spring 16/3	Spring tine harrow: 16/3 + 22/3

terms of fertilisation, pest and disease protection, the yields obtained were below the average of the region, explaining the negative gross margins. For strategy 3, reduced herbicide dose and spring tine harrowing, the costs were higher and the yield was lower than for the standard strategy (€1361/ha and 7.8 t/ha, respectively, for the standard strategy), which resulted in a low gross margin. An even lower gross margin was achieved in strategy 4 with no herbicides and spring tine harrowing. Here, the main cause was a reduction in yield caused by high weed pressure. Strategies 1-3 were highly effective in terms of weed control.

These examples of alternative weed management show that with reductions in herbicide dose or the absence of herbicides, it was difficult to maintain yield. However, with the right combination, for example late sowing after a false seedbed and spring application of herbicides supported by spring tine harrowing (strategy 4 at Jablje), it was possible to achieve results similar to standard strategy. The replacement of spring herbicide application by autumn application was also successful and might be a tool in resistance prevention as

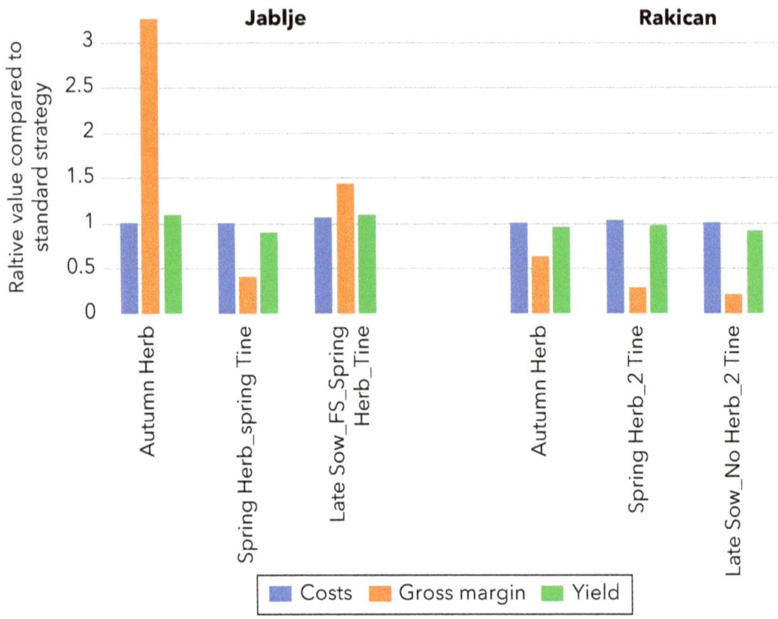

Figure 7 Relative index for costs, gross margin and yield compared to the standard strategy at both locations (winter barley at Jablje and winter wheat at Rakican). Refer to Tables 3 and 4 for strategies. Yield: harvested yield; Costs: all costs related to the crop inclusive of seeds, fertiliser, plant protection products and expenses related to machinery for treatments; Gross margin: income from yield subtracting the costs.

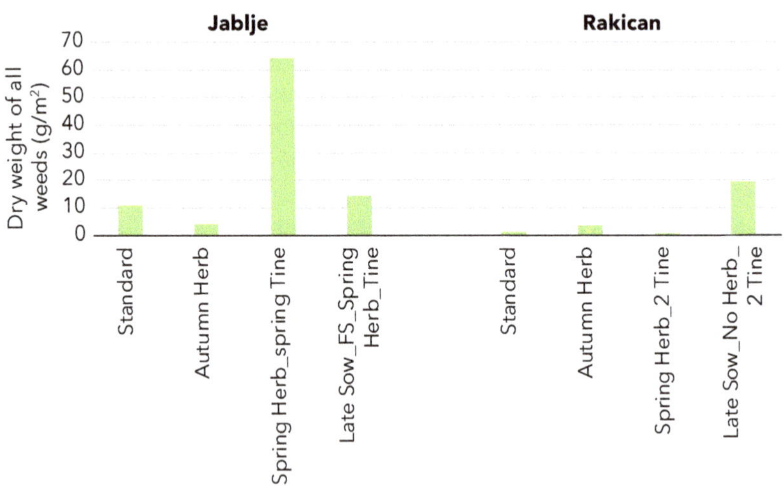

Figure 8 Weed biomass of the strategies sampled in June 2019 as total weed biomass at both locations (refer to Tables 3 and 4 for strategies).

this will lead to a change in site of action of the applied herbicides. Generally, weed control was satisfactory in most strategies.

6.2.2 Denmark

A Danish trial was established at Slimminge, on the island of Zealand, as a demonstration trial on sandy loam soil. The plots were approximately 80 m long and 10 m wide with no spatial replication, but sampling was repeated four times along a transect in the plots. This trial layout was chosen to increase the demonstration value of the trial. Differences among the strategies could be evaluated visually by farmers and advisors invited to open field days. In general, delayed sowing is recommended for winter cereals to minimise grass weed problems; however, grass weeds were not the main problem in this field. Furthermore, an increasing number of farmers are adopting reduced soil tillage practices and the two factors were integrated into the alternative strategies. The effect of delayed sowing on broadleaved weeds is also interesting, as more and more winter cereal is established later than usual. Weed hoeing, either alone or in combination with spraying in the crop row (band spraying), was implemented to achieve a reduction in herbicide input (Table 5). In strategies 2–5, different weed control strategies were combined following conventional ploughing. With the delayed sowing, a false seedbed was established in strategies 3 and 5, where emerged weeds were controlled via seedbed preparation at sowing. In the two directly sown strategies, 6 and 7, late sowing after controlling weeds with glyphosate was the main tactic. In strategy 7, spring wheat was established in November, which is a strategy some farmers use if experiencing severe grass weed problems. When sowing in November, it is assumed that the effect on grass weed emergence is comparable to a traditional spring crop but the crop gets a head start in spring and is expected to yield higher than spring-sown spring wheat. The herbicide application in the crop row was the same as in the standard strategy, but by decreasing the sprayed area to approximately 10 cm across the crop row, herbicide use was reduced by roughly 60%. The equipment required for spraying cereal crop rows is not standard so an experimental plot sprayer was equipped with double rows of nozzles (25 cm apart) to facilitate the application. A low speed and boom height was used for application. The row spacing of 25 cm was used based on the available equipment for inter-row hoeing, but smaller row spacing has been shown to produce equal results with weed hoeing in relation to the yield for spring cereals (Melander et al., 2018).

The yields in the alternative strategies were lower than that obtained in the standard strategy, which had a yield of 6.9 t/ha. Compared with average yields in the area, the yields of all strategies were generally low in 2020. The combination of weed hoeing and herbicide band application provided the

Table 5 Strategies in the Danish winter wheat demonstration trial (Slimminge) in 2020. Seeding density for winter wheat was 133 kg/ha at the normal sowing time (variety Benchmark). For strategy 7 with spring wheat, the seeding density was 210 kg/ha (variety *Alondra*). The standard strategy included weed monitoring in spring to decide on spring herbicide application. A decision support system (DSS) was used that offers herbicide solutions based on the weed composition and number of weeds

	Standard_HerbAu	NoHerb_Hoe	NoHerb_Late_Hoe	Band_Hoe	Band_Late_Hoe	NoTill_Late	Notill-SpringWheat
	Strategy 1	Strategy 2	Strategy 3	Strategy 4	Strategy 5	Strategy 6	Strategy 7
	Reference/standard	Ploughing No herbicides	Ploughing No herbicides	Ploughing bandspraying and wide rows	Ploughing bandspraying and wide rows	No-till	No-till, spring wheat
2019–2020	Winter wheat	Winter wheat	Winter wheat	Winter wheat	Winter wheat	Winter wheat Cover crop	Spring wheat Cover crop
Soil tillage	Ploughed 16/9 harrow 17/9	Ploughed and harrow same timing as Strategy 1	Ploughed same timing as Strategy 1, harrow 10/10	Ploughed and harrow same timing as Strategy 1	Ploughed same timing as Strategy 1, harrow 10/10	Cover crop sowed 16/8 Direct sowing	Cover crop sowed 16/8 Direct sowing
Sowing time	Normal sowing time 17/9	Normal sowing time 17/9	Late sowing 10/10	Normal sowing time 17/9	Late sowing 10/10	Late sowing 3/10	Very late sowing 22/11
Seeding density	Standard 133 kg/ha	Higher density in row 133 kg/ha	Higher density in row + adjusted to later sowing 205 kg/ha	Higher density in row 133 kg/ha	Higher density in row + adjusted to later sowing 205 kg/ha	Adjusted to later sowing 177 kg/ha	Adjusted to later sowing 210 kg/ha
Row width	Standard row 12 cm	Wide rows 25 cm	Wide rows 25 cm	Wide rows 25 cm	Wide rows 25 cm	Wide rows 20 cm	Wide rows 20 cm

Herbicides	Standard herbicide application autumn 800 g prosulfocarb + 50 g diflufenican per ha Need based spring based on DSS (CPO) 3 g florasulam + 3.75 g halauxifen pr ha	No herbicides	No herbicides	Bandspraying with normal spraying boom in low height 3/10: 800 g prosulfocarb + 50 g diflufenican in band No herb in spring	Bandspraying with normal spraying boom in low height 24/10: 800 g prosulfocarb + 50 g diflufenican in band No herb in spring	Cover crop termination with 1080 g glyphosate per ha: 26/9 No herb in spring	Cover crop terrmination with 1080 g glyphosate per ha: 26/9 No herb in spring
Mechanical weeding	No	Autumn: 10/10 interrow hoeing + tine harrow Spring: 22/4 interrow hoeing + tine harrow	Autumn: 25/10 interrow hoeing + tine harrow Spring: 22/4 interrow hoeing + tine harrow	Autumn: 10/10 interrow hoeing Spring: 22/4 interrow hoeing + tine harrow	Autumn: 25/10 interrow hoeing Spring: 22/4 interrow hoeing + tine harrow	No	No

Source: Crop Protection Online, Sønderskov et al. (2016).

highest yields of the ploughed strategies, while the directly drilled strategy with winter wheat had the highest yield of all the alternative strategies. The spring wheat in strategy 7 cannot be directly compared with the other strategies, and the average yield per hectare of spring wheat compared with winter wheat was generally lower.

All of the alternative strategies had a lower cost than the standard strategy, which was €990/ha (Fig. 9). With delayed sowing, the seeding density was increased to ensure the same number of heads per hectare after establishment. This increased the cost, which is evident in the difference between strategies 2 and 3 and between strategies 4 and 5, respectively. Even though the alternative strategies had lower costs, the gross margins were €10 and €99 lower per ha than for the standard strategy because of the lower yields. In fact, the gross margins of all strategies were negative in 2020, due to low yields. The weed biomass of the band-sprayed strategies, 4 and 5, was lower than the unsprayed strategies, 2 and 3 (Fig. 10). Directly sown strategy 6 had a lower weed biomass compared with the similar ploughed strategy 3 and the yield was higher. This is noteworthy as the directly sown strategies represented first-year results after transition to no-till where no benefits can be expected from decreased soil disturbance.

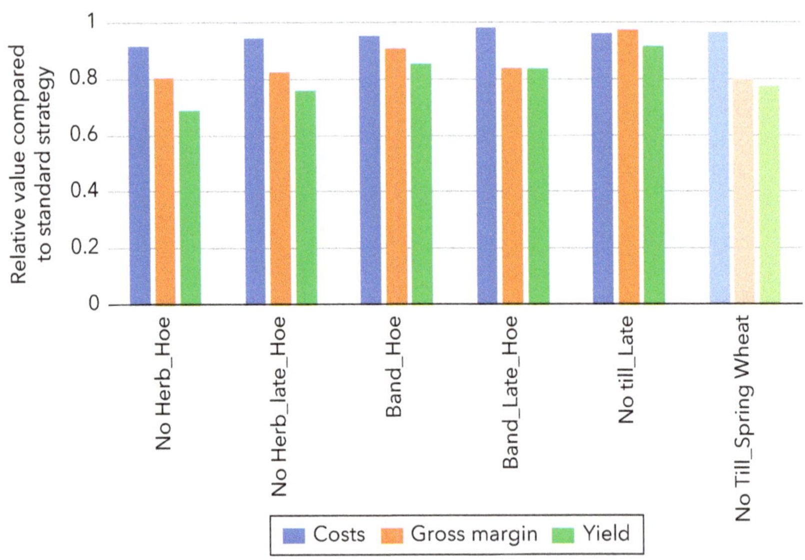

Figure 9 Relative index for costs, gross margin and yield compared to the standard strategy. Note that strategy 7 (NoTill_SpringWheat) cannot be directly compared with the other strategies in terms of yield and economic outcome. Strategies correspond to strategies 2–7 in Table 5. Yield: harvested yield; Costs: all costs related to the crop inclusive of seeds, fertiliser, plant protection products and expenses related to machinery for treatments; Gross margin: income from yield subtracting the costs.

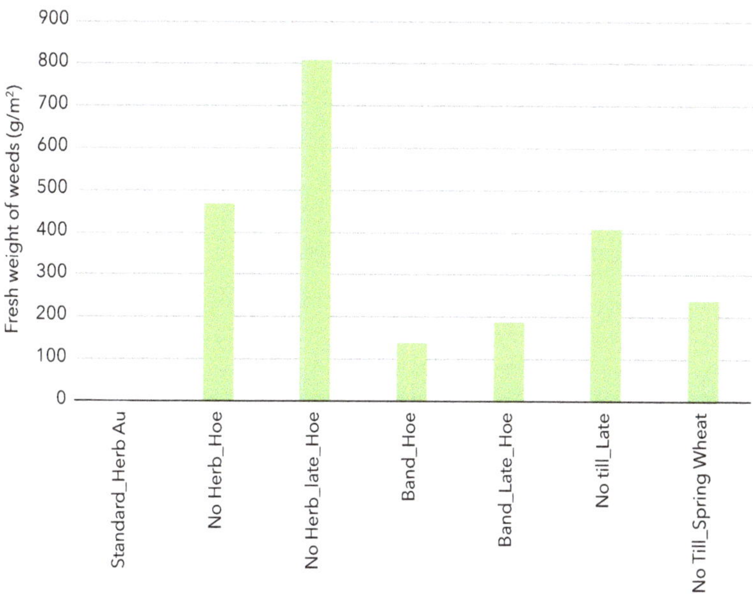

Figure 10 Weed biomass of the strategies sampled in June 2020 as total weed fresh weight biomass. The weed species were mainly broadleaved weeds such as *Papaver rhoeas*, *Viola arvensis*, *Veronica arvensis* and *Geranium* spp. The only grass weed was *Poa annua*. Strategies correspond to strategies 1–7 in Table 5.

The lower weed biomass in strategies 4 and 5, compared with strategies 2 and 3, indicated that a large number of weed plants emerged within the row, and there was a high additional efficacy of adding band-spraying to the inter-row weed hoeing in the ploughed strategies. An effect of leaving the soil surface undisturbed for direct sowing was evident in strategies 6 and 7, where an intermediate effect on weed infestation was observed. The strategies with delayed sowing provided lower weed control than sowing at normal timing. Delayed sowing primarily targets winter annual grass weeds with an emergence pattern similar to winter cereals. Since there were no grass weeds in this trial field, the results reflect the effect of these strategies on broadleaved weeds, and they indicate that delayed sowing is not necessarily recommendable in fields with no grass weed problem. Strategy 7, with spring wheat sown very late in autumn, was observed to have fewer weeds than strategy 6, where winter wheat was drilled directly, but it resulted in a lower gross margin due to lower yields of spring wheat.

This demonstration trial highlighted that alternative strategies are feasible, but they require further development and optimisation. A better understanding needs to be obtained regarding how the efficacy of alternative strategies depends on the specific local weed community, the soil type and the weather conditions in a specific year. The band-spraying equipment used here was developed for experimental purposes, and the advances in the development of equipment for band spraying in narrow-row crops will make such strategies more accessible for farmers. Automated carriers for spraying equipment could be a way forward to implement band spraying combined with inter-row weed hoeing.

6.3 Overall summary of the case studies

The demonstration trials with alternative weed management strategies highlighted that the results were highly variable and, compared with a standard herbicide strategy, the alternative strategies were harder to manage and require adaption to local conditions, for example, weed population composition and suitable conditions for mechanical weed control. There were some potentially effective strategies among the tested combinations, and the most successful strategies can be further explored and optimised for local conditions. Some of the alternative strategies, such as band spraying, will benefit from technological development.

7 Where to look for further information

More information is available on the IWMPRAISE website - Integrated Weed Management: PRActical Implementation and Solutions for Europe (iwmpraise.eu). This project has received funding from the European Union's Horizon 2020 research and innovation programme under grant agreement n° 727321. Each country has also made available information on IPM methods, adapted to local situations:

- For France: www.ecophytopic.fr/tr/programmes-de-recherche/europe/iwm-praise.
- For Denmark: https://www.landbrugsinfo.dk/public/3/2/9/planter_projekter_iwmpraise_dk_udgivelser.
- For Slovenia: https://www.ivr.si/raziskave-in-razvoj/iwmpraise/.
- For the UK : www.iwm-uk.co.uk.
- For Italy: http://www.venetoagricoltura.org/progetti/iwmpraise/.

ARVALIS - Institut du vegetal is an applied agricultural research organization dedicated to arable crops : cereals, maize, sorghum, potatoes, fodder crops, flax and tobacco. It was founded by farmers and dedicates its expertise to the creation of production systems that combine economic competitiveness, adaptation to changing markets conditions and environment protection. It considers technological innovation as a major tool to enable producers and agri-companies to respond to societal challenges.

Aahrus University in Flakkebjerg carry out basic, strategic and applied research in issues regarding the interaction between climate, soil, plants, animals and people in agro-ecosystems with a focus on promoting health, sustainability and environmentally friendly production of food, feed, energy and bio-based products.

The Agricultural Institute of Slovenia is the leading research institute in the field of agriculture in Slovenia. It comprehensively deals with the issues of modern agriculture and is expanding its activities into the fields of environmental protection and ecology. The institute performs fundamental, applied and development research and specialist tasks in agriculture, publishes the results of scientific research work as well as professional and supervision work, performs tasks based on authorisations and accreditations and checks the quality of agricultural products and products used in agriculture.

NIAB (National Institute of Agricultural Botany) is an independent, science-based crop research organisation, working across plant science, crop evaluation and agronomy, and ensuring these advances are transferred effectively onto farm. NIAB work to improve agricultural and horticultural crop production, bringing together the specialist knowledge, skills and facilities required to understand the performance and quality of agricultural crop varieties and seeds.

Institute of Life Sciences in Pisa is a leading institute that provides a multifaceted and challenging scientific environment to a broad spectrum of students: undergraduates, postgraduates and PhD. Courses and research activities span from classical and molecular human and plant biology to preclinical and clinical sciences, plant biotechnology, food quality and nutraceutics, agroecology and agrobiodiversity, and novel sustainable agricultural systems. Strong emphasis is placed on technological innovation. Research carried out in the macro-area Agricultural Sciences and plant Biotechnology addresses two main domains: plant sciences and agronomy, with a focus on various aspects of plant biology, food and energy crops, agrobiodiversity, and agroecosystem management.

8 References

Adetunji, A. T., Ncube, B., Mulidzi, R. and Lewu, F. B. (2020). Management impact and benefit of cover crops on soil quality: A review. *Soil and Tillage Research* 204, 104717.

Aharon, S., Fadida-Myers, A., Nashef, K., Ben-David, R., Lati, R. N. and Peleg, Z. (2021). Genetic improvement of wheat early vigor promote weed-competitiveness under Mediterranean climate. *Plant Science* 303, 110785.

Andrew, I. K. S. and Storkey, J. (2017). Using simulation models to investigate the cumulative effects of sowing rate, sowing date and cultivar choice on weed competition. *Crop Protection* 95, 109–115.

Andrew, I. K. S., Storkey, J. and Sparkes, D. L. (2015). A review of the potential for competitive cereal cultivars as a tool in integrated weed management. *Weed Research* 55(3), 239-248.

Barberi, P. and Lo Cascio, B. (2001). Long-term tillage and crop rotation effects on weed seedbank size and composition. *Weed Research* 41(4), 325-340.

Barralis, G., Chadoeuf, R. and Lonchamp, J. P. (1988). Longévité des semences de mauvaises herbes annuelles dans un sol cultivé. *Weed Research* 28(6), 407-418.

Barzman, M., Bàrberi, P., Birch, A. N. E., Boonekamp, P., Dachbrodt-Saaydeh, S., Graf, B., Hommel, B., Jensen, J. E., Kiss, J., Kudsk, P., Lamichhane, J. R., Messéan, A., Moonen, A., Ratnadass, A., Ricci, P., Sarah, J. and Sattin, M. (2015). Eight principles of integrated pest management. *Agronomy for Sustainable Development* 35(4), 1199-1215.

Been, T., Berti, A., Evans, N., Gouache, D., Gutsche, V., Jensen, J. E. and Rydahl, P. (2009). *Review of New Technologies Critical to Effective Implementation of Decision Support Systems (DSS's) and Farm Management Systems (FMS's)*. Aarhus University.

Benvenuti, S. and Macchia, M. (2006). Seedbank reduction after different stale seedbed techniques in organic agricultural systems. *Italian Journal of Agronomy* 1(1), 11-22.

Blanco-Canqui, H., Shaver, T. M., Lindquist, J. L., Shapiro, C. A., Elmore, R. W., Francis, C. A. and Hergert, G. W. (2015). Cover crops and ecosystem services: insights from studies in temperate soils. *Agronomy Journal* 107(6), 2449-2474.

Bonin, L. and Labreuche, J. (2007). Gestion des adventices à l'interculture: Intérêt du type de déchaumage. In: *AFPP*. Proceedings of the - XXème Conférence Du Columa, Journées Internationales sur la Lutte Contre Les Mauvaises Herbes 2007, Dijon, France.

Chikowo, R., Faloya, V., Petit, S. and Munier-Jolain, N. M. (2009). Integrated Weed Management systems allow reduced reliance on herbicides and long-term weed control. *Agriculture, Ecosystems and Environment* 132(3-4), 237-242.

Cirujeda, A., Pardo, G., Marí, A. I., Aibar, J., Pallavicini, Y., González-Andújar, J. L., Recasens, J. and Solé-Senan, X. O. (2019). The structural classification of field boundaries in Mediterranean arable cropping systems allows the prediction of weed abundances in the boundary and in the adjacent crop. *Weed Research* 59(4), 300-311.

Colbach, N., Granger, S. and Meziere, D. (2013). Using a sensitivity analysis of a weed dynamics model to develop sustainable cropping systems. II. Long-term effect of past crops and management techniques on weed infestation. *The Journal of Agricultural Science* 151(2), 247-267.

Colbach, N. and Mézière, D. (2013). Using a sensitivity analysis of a weed dynamics model to develop sustainable cropping systems. I. Annual interactions between crop management techniques and biophysical field state variables. *The Journal of Agricultural Science* 151(2), 229-245.

De Vita, P., Colecchia, S. A., Pecorella, I. and Saia, S. (2017). Reduced inter-row distance improves yield and competition against weeds in a semi-dwarf durum wheat variety. *European Journal of Agronomy* 85, 69-77.

Dittmann, B. (2012). Final results concerning the effects of long-term ploughless farming on weed infestation in the Güterfelde organic crop rotation trial ("Ökologische Fruchtfolge Güterfelde"). *Julius-Kühn-Archiv* 2(434), 708-711.

Dorn, B., Jossi, W. and van der Heijden, M. G. A. (2015). Weed suppression by cover crops: Comparative on-farm experiments under integrated and organic conservation tillage. *Weed Research* 55(6), 586-597.

Fontaine, L., Bernicot, M. H., Rolland, B. and Poiret, L. (2009). Des variétés rustiques concurrentes des adventices pour l'agriculture durable, en particulier l'agriculture biologique. *Innovations agronomiques* 4, 115-124.

Gerhards, R., Kollenda, B., Machleb, J., Möller, K., Butz, A., Reiser, D. and Griegentrog, H. W. (2020). Camera-guided weed hoeing in winter cereals with narrow row distance. *Gesunde Pflanzen* 72(4), 403-411.

Guzzomi, A. L., Ryan, M., Saunders, C. and Walsh, M. J. (2017). Reducing the integrated Harrington Seed Destructor power requirements through chaff fraction optimization. *Applied Engineering in Agriculture* 33(5), 695-701.

Hakansson, S. (2003). *Weeds and Weed Management on Arable Land an Ecological Approach*. Cambridge, UK: CABI Publishing.

Hansen, P. K., Kristensen, K. and Willas, J. (2008). A weed suppressive index for spring barley (Hordeum vulgare) varieties. *Weed Research* 48(3), 225-236.

Jensen, P. K. (2009). Longevity of seeds of four annual grass and two dicotyledon weed species as related to placement in the soil and straw disposal technique. *Weed Research* 49(6), 592-601.

Jensen, P. K. (2010a). Longevity of seeds of Poa trivialis and Vulpia myuros as affected by simulated soil tillage practices and straw disposal technique. *Grass and Forage Science* 65(1), 76-84.

Jensen, P. K. (2010b). Longevity of seeds of Poa pratensis and Lolium perenne as affected by simulated soil tillage practices and its implications for contamination of herbage seed crops. *Grass and Forage Science* 65(1), 85-91.

Johnson, W. C. and Mullinix, B. G. (1995). Weed management in peanut using stale seedbed techniques. *Weed Science* 43(2), 293-297.

Korres, N. E. and Froud-Williams, R. J. (2002). Effects of winter wheat cultivars and seed rate on the biological characteristics of naturally occurring weed flora. *Weed Research* 42(6), 417-428.

Kudsk, P. (2008). Optimising herbicide dose: A straightforward approach to reduce the risk of side effects of herbicides. *The Environmentalist* 28(1), 49-55.

Kudsk, P., Sønderskov, M., Bonin, L., Gonzalez-Andujar, J. L., Jensen, J. E., Melander, B., Moonen, C., Riemens, M., Sattin, M., Schaffner, U. and Storkey, J. (2020). IWMPRAISE-an EU Horizon 2020 project providing integrated weed management solutions to European farmers. *Outlooks on Pest Management* 31(4), 152-159.

Kumar, V., Obour, A., Jha, P., Liu, R., Manuchehri, M. R., Dille, J. A., Holman, J. and Stahlman, P. W. (2020). Integrating cover crops for weed management in the semiarid US Great Plains: Opportunities and challenges. *Weed Science* 68(4), 311-323.

Leblanc, M. L. and Cloutier, D. C. (1996). Effect of the stale seedbed technique on annual weed emergence in maize. In: *AFPP Proceedings Xe Colloque International sur la biologie des Mauvaises Herbes* (pp. 29-34), Dijon, France.

Lefebvre, M., De Cuyper, K., Loix, E. and Viaggi, D. (2014). European farmers' intention to invest in 2014-2020: Survey results. *Gomez-y-Paloma, S* 4(6. 1). Farm household lifecycle.

Lemerle, D., Verbeek, B. and Diffey, S. (2006). Influences of field pea (Pisum sativum) density on grain yield and competitiveness with annual ryegrass (Lolium rigidum) in south-eastern Australia. *Australian Journal of Experimental Agriculture* 46(11), 1465-1472.

Leoni, F., Carlesi, S., Lazzaro, M. and Moonen, A. C. (2019). Selection of the best adapted legume types for relay intercropping in durum wheat. In: *European conference on Crop Diversification*, Budapest, Hungary.

Lutman, P. J. W., Moss, S. R., Cook, S. and Welham, S. J. (2013). A review of the effects of crop agronomy on the management of A lopecurus myosuroides. *Weed Research* 53(5), 299-313.

Marshall, E. J. P. (1989). Distribution patterns of plants associated with arable field edges. *Journal of Applied Ecology* 26(1), 247-257.

Mazoyer, M. and Roudart, L. (1997). *Histoire des agricultures du monde: Du néolithique à la crise contemporaine, 505 p*. Paris: Editions du Seuil.

McCollough, M. R., Gallandt, E. R., Darby, H. M. and Molloy, T. (2020). Band sowing with hoeing in organic grains: I. Comparisons with alternative weed management practices in spring barley. *Weed Science* 68(3), 285-293.

Melander, B., Holst, N., Rasmussen, I. A. and Hansen, P. K. (2012). Direct control of perennial weeds between crops-Implications for organic farming. *Crop Protection* 40, 36-42.

Melander, B., Jabran, K., De Notaris, C., Znova, L., Green, O. and Olesen, J. E. (2018). Inter-row hoeing for weed control in organic spring cereals-Influence of inter-row spacing and nitrogen rate. *European Journal of Agronomy* 101, 49-56.

Melander, B., Rasmussen, I. A. and Barberi, P. (2005). Integrating physical and cultural methods of weed control - Examples from European research. *Weed Science* 53(3), 369-381.

Meynard, J. M., Charrier, F., Fares, M., Le Bail, M., Magrini, M. B., Charlier, A. and Messéan, A. (2018). Socio-technical lock-in hinders crop diversification in France. *Agronomy for Sustainable Development* 38(5), 54.

Mkenda, P. A., Ndakidemi, P. A., Mbega, E., Stevenson, P. C., Arnold, S. E. J., Gurr, G. M. and Belmain, S. R. (2019). Multiple ecosystem services from field margin vegetation for ecological sustainability in agriculture: Scientific evidence and knowledge gaps. *PeerJ* 7, e8091.

Mohler, C. L. and Galford, A. E. (1997). Weed seedling emergence and seed survival: Separating the effects of seed position and soil modification by tillage. *Weed Research* 37(3), 147-155.

Moss, S. (2019). Integrated weed management (IWM): Why are farmers reluctant to adopt non-chemical alternatives to herbicides? *Pest Management Science* 75(5), 1205-1211.

Mwendwa, J. M., Brown, W. B., Weidenhamer, J. D., Weston, P. A., Quinn, J. C., Wu, H. and Weston, L. A. (2020). Evaluation of commercial wheat cultivars for canopy architecture, early vigour, weed suppression and yield. *Agronomy* 10(7), 983.

Oerke, E.-C. (2006). Crop losses to pests. *The Journal of Agricultural Science* 144(1), 31-43.

Olsen, J., Kristensen, L. and Weiner, J. (2005a). Effects of density and spatial pattern of winter wheat on suppression of different weed species. *Weed Science* 53(5), 690-694.

Olsen, J., Kristensen, L., Weiner, J. and Griepentrog, H. W. (2005b). Increased density and spatial uniformity increase weed suppression by spring wheat. *Weed Research* 45(4), 316-321.

Oveisi, M., Kaleibar, B. P., Mashhadi, H. R., Müller-Schärer, H., Bagheri, A., Amani, M., Elahinejad, M. and Masoumi, D. (2021). Bean cultivar mixture allows reduced herbicide dose while maintaining high yield: A step towards more eco-friendly weed management. *European Journal of Agronomy* 122, 126173.

Padel, S., Lampkin, N. and Foster, C. (1999). Influence of policy support on the development of organic farming in the European Union. *International Planning Studies* 4(3), 303-315.

Parsons, D. J., Benjamin, L. R., Clarke, J., Ginsburg, D., Mayes, A., Milne, A. E. and Wilkinson, D. J. (2009). Weed manager–A model-based decision support system for weed management in arable crops. *Computers and Electronics in Agriculture* 65(2), 155-167.

Peruzzi, A., Martelloni, L., Frasconi, C., Fontanelli, M., Pirchio, M. and Raffaelli, M. (2017). Machines for non-chemical intra-row weed control in narrow and wide-row crops: A review. *Journal of Agricultural Engineering* 48(2), 57-70.

Peterson, M. A., Collavo, A., Ovejero, R., Shivrain, V. and Walsh, M. J. (2018). The challenge of herbicide resistance around the world: A current summary. *Pest Management Science* 74(10), 2246-2259.

Roberts, H. A. and Dawkins, P. A. (1967). Effect of cultivation on the numbers of viable weed seeds in soil. *Weed Research* 7(4), 290-301.

Ryan, M. R., Smith, R. G., Mirsky, S. B., Mortensen, D. A. and Seidel, R. (2010). Management filters and species traits: Weed community assembly in long-term organic and conventional systems. *Weed Science* 58(3), 265-277.

Schipanski, M. E., Barbercheck, M., Douglas, M. R., Finney, D. M., Haider, K., Kaye, J. P., Kemanian, A. R., Mortensen, D. A., Ryan, M. R., Tooker, J. and White, C. (2014). A framework for evaluating ecosystem services provided by cover crops in agroecosystems. *Agricultural Systems* 125, 12-22.

Snyder, L. D., Gómez, M. I. and Power, A. G. (2020). Crop varietal mixtures as a strategy to support insect pest control, yield, economic, and nutritional services. *Frontiers in Sustainable Food Systems* 4, 60.

Somerville, G. J., Sønderskov, M., Mathiassen, S. K. and Metcalfe, H. (2020). Spatial modelling of within-field weed populations: A review. *Agronomy* 10(7), 1044.

Sønderskov, M., Rydahl, P., Bøjer, O. M., Jensen, J. E. and Kudsk, P. (2016). Crop protection online–Weeds: A case study for agricultural decision support systems. In: *Integrated Series in Information Systems* (pp. 303-320). Cham: Springer.

Sønderskov, M., Somerville, G. J., Lacoste, M., Jensen, J. E. and Holst, N. (2020). DK-RIM: assisting integrated management of Lolium multiflorum, Italian ryegrass. *Agronomy* 10(6), 856.

Soni, N., Nissen, S. J., Westra, P., Norsworthy, J. K., Walsh, M. J. and Gaines, T. A. (2020). Seed retention of winter annual grass weeds at winter wheat harvest maturity shows potential for harvest weed seed control. *Weed Technology* 34(2), 266-271.

Swanton, C. J., Shrestha, A., Knezevic, S. Z., Roy, R. C. and Ball-Coelho, B. R. (2000). Influence of tillage type on vertical weed seedbank distribution in a sandy soil. *Canadian Journal of Plant Science* 80(2), 455-457.

Van Vooren, L., Reubens, B., Ampoorter, E., Broekx, S., Pardon, P., Van Waes, C. and Verheyen, K. (2018). Monitoring the impact of hedgerows and grass strips on the performance of multiple ecosystem service indicators. *Environmental Management* 62(2), 241-259.

Vrignon-Brenas, S., Celette, F., Piquet-Pissaloux, A. and David, C. (2016). Biotic and abiotic factors impacting establishment and growth of relay intercropped forage legumes. *European Journal of Agronomy* 81, 169-177.

Vullioud, P., Delabays, N., Frei, P. and Mercier, E. (2006). Résultats de 35 ans de culture sans labour à Changins III. Mauvaises herbes, *maladies fongiques et ravageurs. Revue Suisse d'Agriculture* 38(2), 81–87.

Weiner, J., Griepentrog, H. W. and Kristensen, L. (2001). Suppression of weeds by spring wheat *Triticum aestivum* increases with crop density and spatial uniformity. *Journal of Applied Ecology* 38(4), 784–790.

Werle, R., Burr, C. and Blanco-Canqui, H. (2017). Cereal rye cover crop suppresses winter annual weeds. *Canadian Journal of Plant Science* 98(2), 498–500.

Wicks, G. A., Nordquist, P. T., Baenziger, P. S., Klein, R. N., Hammons, R. H. and Watkins, J. E. (2004). Winter wheat cultivar characteristics affect annual weed suppression. *Weed Technology* 18(4), 988–998.

Zeller, A. K., Zeller, Y. I. and Gerhards, R. (2021). A long-term study of crop rotations, herbicide strategies and tillage practices: Effects on *Alopecurus myosuroides* Huds. abundance and contribution margins of the cropping systems. *Crop Protection* 145, 105613.

Chapter 13

Integrated weed management in grasslands

Urs Schaffner, CABI, Switzerland; Heinz Müller-Schärer, University of Fribourg, Switzerland; and Andreas Lüscher, Agroscope, Switzerland

1 Introduction

Grasslands, defined here as ecosystems in which graminoids, forbs and shrubs form a relatively continuous herbaceous layer of vegetation (Veldman et al. 2015), cover some 40% of the Earth's land surface, extending over large areas on all continents except Antarctica. The development of grasslands, their species composition and challenges in their sustainable management for production and conservation vary considerably across regions (Olson et al. 2001). In many parts of the world, such as in North America, Central Asia and Sub-Saharan Africa, primary grasslands dominate. These 'old-growth' grasslands (Veldman et al. 2015) often occur where tree growth is limited by shallow soils, low soil moisture availability, low temperature, frequent fires or herbivory by large grazers. While Eastern European steppes are considered to be a climax vegetation, large parts of temperate grasslands in western and central Europe are associated with human activity, and their origin and maintenance are mostly linked to forest clearing and subsequent management such as mowing, grazing by domestic livestock or fire. Due to their anthropogenic origin, these grasslands are called secondary grasslands (Bredenkamp et al. 2002).

While primary and secondary grasslands in temperate zones are often important in terms of biodiversity, the pressure to increase animal production has

http://dx.doi.org/10.19103/AS.2021.0098.15

led to grassland intensification by increasing cutting and/or grazing frequency, reseeding, herbicide and fertilizer application, resulting in high-intensification grasslands and a corresponding loss of diversity. In contrast, it is estimated that some 40% of the subtropical and tropical grasslands and savannas in Sub-Saharan Africa are degraded due to overgrazing, fire, climate change or other factors, usually resulting in a reduction of herbaceous vegetation cover and diversity (Le et al. 2016).

Problems with undesirable, weedy plant species are known from all grassland types, but the type of weeds, the nature of the problem and thus options for their management vary considerably (Plate 1). For example, the vast majority of grassland weeds in the Prairies in North America are invasive non-native plants (INNPs), both grasses and forbs that have been brought in as seed contaminants with forage grass species from Eurasia. Weeds in arid and semi-arid savannas of Sub-Saharan Africa vary widely, from space-filling annuals to woody or other non-palatable perennials; they are mostly introduced but are sometimes also native species. In contrast, most grassland weeds in Europe are native plant species that are toxic or unpalatable to livestock; these may benefit from nutrient input and from vegetation gaps due to trampling or other mechanical disturbance.

Until recently, the reliance on herbicides has been high in intensive grasslands in most regions of Europe. In low-yield grasslands in, for example, the northwestern United States or Australia, classical biological control by importing antagonists from the weed's native range has been successfully used for more than 50 years to control INNPs in a relatively cheap and sustainable way. However, the concept of integrated weed management (IWM), that is, the

Plate 1 Examples of grassland weeds. a: Tansy ragwort, *Jacobaea vulgaris*, in Oregon, USA; b: *Prosopis juliflora* in Afar region, Ethiopia; c: *Rumex obtusifolius* in Kt Zürich, Switzerland; and d: *Leucanthemum vulgare* in Kosciuszko National Park, New South Wales, Australia. Photo credits: a: Marianna Szucs, b: René Eschen, c: Julie Klötzli, d: Andrew McConnachie.

combined use of complementary weed management practices, such as grazing, herbicide application, land fallowing or biological control (FAO 2021), remains largely understudied. The evolution of herbicide resistance, environmental concerns regarding the large-scale application of herbicides in grasslands and the fact that weed management strategies based on single control options often fail to manage weeds at the landscape level are likely to foster a truly IWM approach in grasslands across the globe, although the focus on particular components of weed management options may differ among regions and ecosystems. In particular, an IWM approach is considered critical for managing herbicide resistance in weeds (Norsworthy et al. 2012) and may offer options of low-cost, environmentally friendly and sustainable weed management in low-yield grasslands or in protected grasslands.

Here, we describe the current status of IWM for grasslands by adopting a conceptual approach proposed by Kudsk et al. (2020). Its focus is on management practices available to influence transitions

1 from the soil seed bank to seedling establishment;
2 from the seedling stage to the mature plant; and
3 from the mature plant to the soil seed bank.

The latter includes export and import of propagules from and to the grassland as well as selecting well-adapted species/variety/genotype assemblages, when establishing the grassland community (Fig. 1). We thus provide a conceptual approach to illustrate how management practices available in IWM affect different transitions in a weed's life cycle and then provide examples of how weed management practices have been integrated so far. As weed management in grasslands differs considerably among geographic regions and among the type of weed species, we discuss examples of integration of weed management practices from across the globe. We end with an outlook for possible ways to promote increased uptake of IWM in grasslands.

2 The weed management toolbox for grasslands: prevention, cultural, physical, chemical and biological control

While the conceptual approach of IWM outlined in the study of Kudsk et al. (2020) is applicable to all major agricultural systems, weed management in perennial grasslands differs from weed management in annual cropping systems in a number of aspects. First, grasslands tend to form a perennial competitive environment, which, if well managed, only offers a few microsites for weeds to recruit from invading seeds or the soil seed bank (Fig. 1, transition

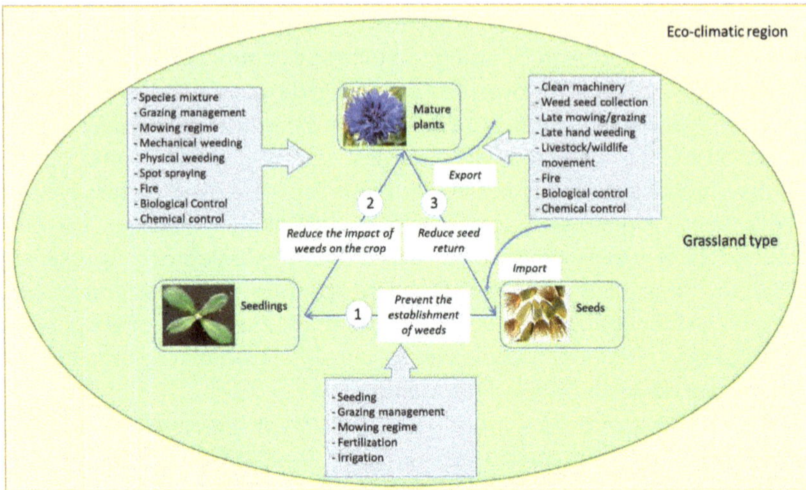

Figure 1 Integrated weed management (IWM) framework for grasslands, consisting of tools that 1) limit seedling establishment in grassland from the soil seed bank or subterranean vegetative organs, 2) limit competition for resources such as light, nutrients and water by removing weeds or reducing their competitive impact, and/or 3) limit return of seeds or vegetative organs to the soil seed/vegetative organ bank or their export to or input from other grasslands. Suitable tools may depend on grassland type and eco-climatic region (adapted from Kudsk et al. 2020).

1). In addition, this competitive environment can affect the whole life cycle of weeds, reducing their growth rate and size (Fig. 1, transition 2) and consequently their seed production or fitness (Fig. 1, transition 3). Secondly, weed problems in grasslands are often caused by one or a few problematic species, while annual cropping systems are confronted with a diverse community of weedy species (Müller-Schärer et al. 2018). Thirdly, grasslands, particularly in semi-arid and arid regions, tend to generate relatively low short-term economic benefits, which sometimes are lower than the costs of chemical weed control (Griffith and Lacey 1991). Thus, weed management strategies in grasslands may be built on different management practices than those in annual cropping systems. In the following sections, we briefly describe individual weed management practices as potential components of IWM, that is, prevention and cultural, physical, chemical and biological control measures.

2.1 Prevention

We understand prevention as any measure that prevents the transfer of weed propagules (primarily seeds) to areas where the weed has not yet established. In the context of INNPs, prevention measures may be implemented at the national border (e.g. control of goods or passengers at the port of entry).

In the case of native weeds or INNPs already established locally, prevention measures include activities that help avoid the transfer of weed propagules from an invaded to an uninvaded grassland. Practices that can prevent invasion of uninvaded grassland include controlled moving of livestock, cleaning of machinery or the use of weed seed-free fodder or seeding material. As most weeds are spread by seed, prevention can be described as the spatial component of the transition from mature plants to the soil seed bank (Fig. 1). Moreover, weed seeds may also be deliberately introduced if the weed species are sold as ornamentals or as components of commercial seed mixtures (Reichard and White 2001).

2.2 Cultural control

The aim of cultural control practices is to establish or maintain a competitive, well-managed sward, an essential component of weed management in grassland. Weed management must therefore be closely linked with adapted grazing management, as overgrazed grassland with an open sward is likely to be more susceptible to weed invasion compared with unstocked or well-managed grassland. On the other hand, undergrazing strongly increases selective foraging by the animal, resulting in a competitive advantage of less-grazed weedy species over heavily grazed forage species. Accordingly, Suter et al. (2007) observed 12 times higher relative risk of *Jacobaea vulgaris* Gaertn. (syn. *Senecio jacobaea* L.) problems in pastures with low stocking rates as compared to well-managed rotational grazing or cutting management.

Besides the prevention measures described above, cultural control measures primarily attempt to reduce the transition from the soil seed bank to the seedling establishment or seedling survival (Fig. 1), but grazing can also be used to reduce growth and seed set of established weeds (e.g. Samuel et al. 2004). Cultural measures include rotational grazing, stabling livestock during wet days to reduce trampling damage, fire, overseeding and restoration of diverse grasslands. In general, multi-species, well-managed swards consisting of species with complementary functional traits have higher biomass production and prevent the establishment of unsown species more effectively than species-poor swards (Connolly et al. 2018; Suter et al. 2017). Fire, which can be used to reduce the transition from seedlings or saplings to adult plants, and reduce the survival and fitness of adult plants, is often used to manage invasive grasses or trees. Other cultural control measures, such as targeted grazing of weeds, also aim to reduce the transition from the seedling to the adult stage as well as to seed set; for example, sheep grazing is used to reduce densities or standing biomass of leafy spurge (*Euphorbia esula* L.), an important grassland weed in the Northwestern United States (Masin et al. 2018).

2.3 Physical control

Physical control measures include manual, mechanical and thermal (e.g., flaming, hot water) practices for weed control in grassland. Manual control comprises the uprooting of plants by hand pulling or using, for example, a spate, a hoe or a garden fork, or by removing the above-ground parts of a plant with an axe or a machete. It may also include ring- and strip-barking of woody weeds. Mechanical control may involve the use of machinery, for example, bulldozers or tractors, and involves, among others, chaining of larger plants, stick-raking or blade ploughing. Mechanical control is often used to remove dense stands of woody weeds but can be expensive and may lead to disturbance of the grassland sward, thereby increasing its susceptibility to re-invasion by the same or other weeds from the soil seed bank.

2.4 Chemical control

Chemical control is the use of naturally occurring or synthesized herbicides that alter the metabolic processes of a weed, so the plant is either killed or suppressed. Post-emergence herbicides, which are applied to weeds after they have emerged, are most frequently used to manage grasslands. Dense infestations with herbaceous weeds are often treated with foliar applications, and low densities are treated with spot spraying of herbicides that selectively control broadleaf species; the advantage of spot spraying is the reduced damage to non-target species, but the cost of application can be high. Chemical control of invasive alien tree species invading grasslands is usually based on foliar application, cut-stump treatment, basal bark treatment or stem injection; the latter two treatments allow selective application of non-selective herbicides with little risk to other plants growing nearby. Herbicides are labelled to indicate which weeds are susceptible to the herbicide, the habitats in which they may be applied, and the appropriate application method.

A major difficulty of herbicide application in grassland is the multi-species nature of the non-targeted grassland sward. This makes it difficult to find a herbicide that is selective enough to only/mainly affect the weed species. It is thus difficult to chemically control unwanted grass species in grasslands that are generally grass dominated. In addition, treating dicot weeds may kill also dicot forage plants, resulting in gaps in the sward that facilitate the establishment of new weed species from the soil seed bank. Finally, herbicide treatments against dicot weed species often also kill leguminous species, which are highly advantageous in grassland systems (Lüscher et al. 2014; Suter et al. 2021). Repeated use of the same herbicide or other herbicides with the same mode of action will favour the development of resistant weed populations, a driving force for the adoption of IWM.

2.5 Biological control

Three methods of biological weed control can be distinguished based on targeted weed, origin of the control agent and the amount of initial inoculum used (Müller-Schärer and Schaffner 2008). These three methods are

- classical biological control;
- inundative biological control; and
- conservation biological control.

Classical biological control (CBC; also called importation biological control) aims to control invasive non-native weeds by the introduction of specialist control organisms, usually insects, mites or fungal pathogens, from the weed's native range. The inundative or bioherbicide method uses periodic releases of an abundant supply of a native or exotic control agent over the entire weed population to be controlled. Such biological agents, generally, are manufactured and registered as biological control products. The third approach, which is called the conservation or system management approach, aims to enhance the effectiveness of resident natural enemies by manipulating their environment to increase their survival or performance.

Grasslands rank among those habitat types with the longest and most successful history of classical biological weed control against non-native weeds (Winston et al. 2014), particularly in regions Europeans emigrated to between the sixteenth and nineteenth century. Several myco-herbicides have been developed against weeds, including members of the genus *Taraxacum*, *Isatis* and woody invasive alien species, but this inundative biological control approach has been hardly applied in grasslands (Table 1; Triolet et al. 2020, Hasan et al. 2021). Similarly, the use of commercial products consisting of herbivorous insects to manage grassland weeds has only been tested in a few cases so far (Vitelli 2000; Hahn et al. 2016).

3 Integrated weed management practices in grasslands

As has been repeatedly emphasized, in the case of grasslands, there is a need to manage the whole plant community rather than just manage individual weed species or populations (Dietl 1982; Grice and Brown 1996). The challenge of weed management in grasslands is to be effective, provide minimal negative environmental impacts and be economically sustainable. While there are examples where CBC of INNPs achieves all of these goals at the landscape scale, sustainable weed management in grasslands often requires an integrated management approach that combines management practices related to one or several transitions of the conceptual model as shown in Fig. 1.

Table 1 Bioherbicides for use against weeds in grasslands and lawns (Triolet et al. 2020; Hasan et al. 2021)

Agent	Target	Commercial name	Country developed	Year developed/ first registered
Acremonium diospyri (Crand.) W.Gams	*Diospyros virginiana L.*		USA	1960
Alternaria destruens E.G.Simmons, strain 059	*Cuscuta* spp.	Smoulder®	USA	2005
Xanthomonas campestris (Pammel) Dowson, pv. poae	*Poa annua* L.	Camperico™	Japan	1997
Puccinia thlaspeos C.Shub., *'strain woad'*	*Isatis tinctoria* L.	Woad Warrior®	USA	2002
Sclerotinia minor Jagger, strain IMI 344141	*Taraxacum officinale* L.	Sarritor®	Canada	2007
Streptomyces acidiscabies strain RL-1101, non-viable cells	Broadleaf weeds, e.g. *Taraxacum officinale*, *Senecio* spp., *Plantago* spp.	Opportune™	USA	2012
Cylindrobasidium laeve (Pers.) Chamuris	*Acacia* spp.	Stumpout®	South Africa	2008
Colletotrichum acutatum J.H. Simmonds	*Hakea sericea* Schrad. & J.C.Wendl	Hakatak	South Africa	1999 (not registered)
Pinus radiata D.Don, oil	*Ochna serrulata* Walp.	BioWeed™	Australia	?
Cymobopogon sp., oil	Broadleaf weeds and weedy grasses	GreenMatch EX	USA	?

Promoting an appropriate combination of individual management practices to tackle weed problems in grasslands (and in other habitats) often requires developing context-dependent solutions. For example, in their native range, the build-up of high population densities of *J. vulgaris* in grasslands can largely be prevented by implementing cultural management practices that avoid sward damage from continuous extensive grazing on grassland with low nitrogen fertilization (Suter et al. 2007). In contrast, in their invasive range in North America, cultural management needs to be combined with biological control to achieve long-term control of this weed, as the resident community in the invaded range appears to be less competitive than that in the native range (see above; McEvoy et al. 1993). In a study comparing different combinations of management practices against three different INNPs, Huwer et al. (2005) found a trend towards a more favourable pasture state in all cases when at least two practices were combined in an IWM system. However, the results suggested that the order in which the IWM components should be applied depended on the initial perennial grass content at the study sites. Thus, to assist farmers to maintain healthy pasture systems, the IWM approach must be sufficiently flexible so that selection of practices and the order of the IWM components can be arranged depending on initial grassland conditions and biogeographic and eco-climatic settings (Fig. 1).

The development of IWM strategies should also be based on the management objectives of the invaded grasslands, for example, whether the grasslands should be primarily managed for forage production, wildlife habitat improvement, restoration of native vegetation complexes, or recreational land maintenance (DiTomaso et al. 2006; Firn et al. 2013). To increase forage production and reduce densities of invasive forbs, IWM management in the Northwestern United States sometimes includes overseeding with perennial, competitive European grass species (Miller 2016), a practice that should be avoided in areas managed for wildlife habitat improvement or restoration of native grasslands.

4 Integrating weed management practices: case studies

In the following sections, we discuss strategies to integrate weed management practices in grasslands using case study examples listed in Table 2. We then end the chapter with an outlook on future developments and challenges related to sustainable weed management in grasslands.

4.1 Tackling multiple transitions in the weed's life cycle

A possible way to integrate weed management practices consists of combining a practice that reduces the establishment of seeds from the soil seed

Published by Burleigh Dodds Science Publishing Limited, 2022.

Table 2 Examples of integrated weed management of grassland weeds

Target weed	Region	Type of integration	Agents and substances	Effect	References
Rhaponticum repens (L.) Hidalgo					
Carduus nutans L.	North America	BC/plant competition	*Rhinocyllus conicus* Frölich		Tipping (2001)
Bromus tectorum L.	North America	Grazing/fire	Cattle		Diamond et al. (2012)
	North America	CC/grazing	Sheep		Lehnhoff et al. (2019)
	North America	BC/CC	*Pyrenophora semeniperda* (Brittleb. & D.B. Adam) Shoemaker; imazapic		Ehlert et al. (2014)
Cirsium arvense (L.) Scop	Europe	BC/cutting	*Puccinia punctiformis*		Kluth et al. (2003)
	Europe	BC/CC	*Hadroplontus litura*, *Pseudomonas syringae pv. Tagetis*; glyphosate	synergistic	Sciegienka et al. (2011)
		BC/plant competition	*Cassida rubiginosa* Muller		Ang et al. (1994)
	North America	CC/fertilization	Picloram and 2,4-d	+	Grekul and Bork (2007)
Centaurea s stoebe L.	North America	BC/plant competition	*Agapeta zoegana* (L.); native vegetation		Müller-Schärer (1991)
Centaurea solstitialis L.	North America	CC/fire	Clopyralid		DiTomaso et al. (2006)
Echium plantagineum L.,	Australia	BC/CC/grazing/plant competition	*Mogulones larvatus* (Schultze)	multiplicative	Huwer et al. (2005)
Taraxacum officinale (L.) Weber ex F.H.Wigg.	North America	BC/CC	*Sclerotinia minor* Jagger; 2,4-D		Schnick et al. (2002)
Various invasive alien plant species	North America	Grazing/fire	Cattle		Delaney et al. (2016)
Rosa bracteate J.C. Wendl.	North America	Mowing/CC	Various herbicides		Enloe et al. (2013)
Euphorbia esula L.	North America	BC/fire	*Aphthona nigriscutis* Foudras;	+	Fellows and Newton (1999)
	North America	BC/CC	*A. nigriscutis*		Nelson and Lym (2003)

	North America	BC/grazing	*Aphthona spp.*; sheep	+	Samuel et al. (2004)
	North America	BC/CC	*Hyles euphorbiae* L.; 2,4-D	ns	Lym (1998)
	North America	BC/CC/reseeding	*Aphthona spp.*, imazapic, grasses		Richardson et al. (2008), Joshi (2008)
Ranunculus acris L.	New Zealand	Mowing/BC	*Sclerotinia sclerotiorum* (Lib.) de Bary		Green et al. (1998)
Jacobaea vulgaris L.	North America	BC/plant competition	*Longitarsus jacobaeae* (Waterhouse)		McEvoy et al. (1993)
Onopordum spp.	Australia	BC/CC/plant competition	*Larinus latus* Herbst. *Lixus cardui* Olivier		Huwer et al. (2005)
Opuntia stricta (Haw.) Haw.	South Africa	BC/CC	*Cactoblastis cactorum* (Berg)		Hoffmann et al. (1998)
Mimosa pigra L.	Australia	BC/CC/physical control/ fire	*Acanthoscelides puniceus* Johnson, *Carmenta mimosa* Eichlin & Passoa, *Chlamisus mimosae* Karren, *Coelocephalapion pigrae* Kissinger, *Neurostrota gunniella* (Busck)		Paynter and Flanagan (2004)
Hypericum perforatum L.	Australia	BC/fire	*Aphis chloris* Koch, *Chrysolina quadrigemina* (Suffrian)		Briese (1996)
		BC/plant competition	*Aculus hyperici* Liro	additive	Willis et al. (1998)
Rumex crispus L.	Europe	BC/BC	*Gastrophysa viridula* Degeer; *Uromyces rumicis* (Schum.) Wint.	+	Hatcher et al. (1994)
Rumex obtusifolius L.	Europe	BC/CC	*Gastrophysa viridula* Degeer; asulam	ns	Speight and Whittaker (1987)
	Europe	BC/BC	*Gastrophysa viridula* Degeer; *Uromyces rumicis* (Schum.) Wint.	ns	Hatcher et al. (1994)
		Mowing/fertilizer	Cattle slurry, NPK fertilizer	ns, -	Niggli et al. (1993); Hopkins and Johnson (2002)
Isatis tinctoria L.	USA	BC/CC	*Puccinia thlaspeos* C. Shub.		Phatak et al. (1983)

BC = biological control; CC = chemical control.

bank (transition 1 in Fig. 1) with a practice that kills plants before they start setting seeds (transitions 2 or 3; Fig. 1). McEvoy et al. (1993) could show that establishing a competitive sward to reduce both seedling establishment in combination with biological control by the leaf beetle *Longitarsus jacobaeae* Waterhouse 1858, which kills established plants, had the highest impact on the population dynamics of the European plant *J. vulgaris* in the invaded range in Oregon, United States. Similarly, Grekul and Bork (2007) found a strong synergistic effect of fertilization on the herbicide treatments for *Cirsium arvense* (L.) Scop. control, which was at least partly attributed to enhanced competition from the increase in grass vigour and biomass of the fertilized forage sward. To manage the invasive yellow starthistle, *Centaurea solstitialis* L., DiTomaso et al. (2006) first applied prescribed burning to kill established plants; the efficacy of prescribed burning was significantly increased when it was followed by chemical control, probably due to a decreased recruitment from the remaining soil seed bank.

A combination of practices reducing the performance of established plants and reducing the input of seeds into the soil seed bank also appears promising. In short-lived weeds, this may be achieved by combining practices that target the transitions from the seedling to the adult stage and the transition from the adult plant to the soil seed bank (Huwer et al. 2005). In long-lived weeds, particularly in woody species, both practices may target the adult plants, that is their survival and their reproductive output. For example, mechanical removal of established trees and releases of biological control agents led to the successful control of the tree *Hakea sericea* Schrad. & J.C.Wendl. (Proteaceae), which invaded fynbos and grassland ecosystems in South Africa. The seed-feeding biological control agents reduced seed output by more than 95%, which significantly reduced the weed's population growth rates (Le Maitre et al. 2008). Modelling analyses conducted by Buckley et al. (2004) indicated that the most successful strategy for suppressing the invasive tree *Mimosa pigra* L. involved a combination of herbicide application, mechanical control, burning, a reduction of small-scale disturbances and the use of insect biological control agents.

4.2 Vertical and horizontal integration of weed management practices

The integration of weed management practices can be viewed as a vertical integration of various management practices against a single weed species or as a horizontal integration across different weed species in one crop (Müller-Schärer and Collins 2012). In grassland, horizontal integration mainly involves practices that aim to establish or maintain competitive vegetation that offers as few microsites for weed recruitment or growth as possible. Practices for horizontal integration thus include grazing and mowing practices or practices to prevent the import of seeds through machinery or livestock. Vertical integration of weed management practices can be implemented by separating the individual practices spatially or temporally, depending on weed densities or location relative to the invasion front or by fully integrating the different practices locally. For example, Chalak-Haghighi et al. (2008) suggested a combination of chemical control and intensified grazing at low density of *C. arvense*, while mowing in late summer plus chemical control and targeted grazing management techniques at high density of *C. arvense*. Grice et al. (2011) proposed a spatially explicit management strategy against the INNP *Hymenachne amplexicaulis* (Rudge) Nees in Australia that considers the stage of invasion and the assets to be protected. They developed a map that distinguished zones with different management objectives and thus different sets of suitable management practices. The study by Grice et al. (2011) provides a nice example of a spatially explicit management strategy against an INNP, focusing on better local integration of management practices. Paynter and Flanagan (2004) showed that the impact of biological control on the invasive tree *M. pigra* can be maximized by integrating it with other management practices locally, rather than by separating the practices spatially or temporally.

4.3 Integrating grazing and mowing practices in integrated weed management

As overgrazing is one of the main factors promoting problematic weeds in grasslands, integrating grazing/mowing management practices is often key for long-term sustainable weed management in grasslands. For example, Suter and Lüscher (2011, 2012) found that herbicides and mowing once a year reduced the density of *Jacobaea aquatica* (Hill) G. Gaertn. & al. in Swiss grasslands by almost 90% in the short term. However, after 3 years, weed densities had recovered again if no site-adapted mowing or grazing management was implemented, as gap formation in the vegetation and increased availability of light on bare soil facilitate weed recruitment from the soil seed bank. This study exemplifies the primary importance of the soil seed bank in a weed species' life cycle. *J. aquatica* is biannual, meaning that in a stable population every year 50% of the plants die and are replaced by new plants recruited from the soil seed bank. If a herbicide is applied and kills all established weed plants, some 50% of the original population will re-establish in the first year and another 50% in the second year, resulting in the loss of the herbicide effect.

The example of control of *J. aquatica* highlights two important issues:

- For an effective IWP, it is important to know the weed's biology; and
- Preventing the build-up of a large soil seed bank is a key factor for successful and sustainable weed control.

Adjusting livestock stocking rates to or below the carrying capacity and implementing cultural grazing practices such as rotational grazing should be considered in numerous weed management projects on grassland; they also help restoring or conserving ecosystem services such as soil organic carbon stocks or flood protection (Baer et al. 2015). Targeted grazing practices can also increase the efficacy of other weed management practices. For example, leaf beetles of the genus *Aphthona* have been shown to be more effective when used in sequence with livestock grazing than either strategy used alone (Samuel et al. 2004). When contrasting herbicide treatments with environmentally more sustainable management practices, Pywell et al. (2010) concluded that lenient grazing in spring and autumn was sufficient to give long-term control of *C. arvense* in lowland and upland grasslands in the United Kingdom; herbicide wiping was the most effective control measure, but effects were lost rapidly. These examples illustrate the importance of implementing an appropriate mowing frequency or grazing rate as part of IWM in grasslands. One should consider, though, that mowing and grazing may have differential effects on the spacing and genetic structure of grassland weeds and thus affect prospects of other management practices (e.g. biological control; Kleijn and Steinger 2002).

4.4 Weeding with invertebrates and pathogens in combination with other management practices

Specialist invertebrate herbivores or pathogens have been repeatedly used in IWM of INNPs, in combination with either chemical control, physical control, prescribed burning or grazing (Fig. 1; Table 2). For example, Paynter and Flanagan (2004) found that herbicide application, bulldozing and fire alone were not effective in the management of the woody INNP *M. pigra*, but they enhanced the impact of invertebrate CBC agents (Buckley et al. 2004). Importantly, integrating CBC with other management can also significantly reduce management costs (Paynter and Flanagan 2004). While CBC is based on the deliberate introduction of specialist natural enemies to control INNPs, the use of native pathogens to control INNPs has also been considered. For example, Ehlert et al. (2014) proposed a two-pronged approach to control the INNP *Bromus tectorum* L. combining inoculation with the soil-borne generalist fungal pathogen *Pyrenophora semeniperda* (Brittlebank and Adam) Shoemaker with post-emergent application of the herbicide imazapic to limit the invasion of this weed in grasslands of western North America.

In contrast to integrated management of pests, the use of native herbivores has rarely been considered in IWM, neither in inundative nor conservation biological control. *Rumex obtusifolius* L. and other European dock species are problematic grassland weeds in their native range as well as in the introduced range in Australia (Scott and Sagliocco 1991a). For biological control of invasive

Rumex species in Australia, two closely related European clearwing moths, *Pyropteron chrysidiforme* (Esper) and *P. doryliforme* (Ochsenheimer), were examined as potential control agents (Scott and Sagliocco 1991a,b). Ultimately, *P. doryliforme* was released in Australia where it significantly decreased densities of invasive *Rumex* populations (Strickland et al. 2012).

Based on this successful CBC project, a research project has been initiated to assess the feasibility of using *P. chrysidiforme* for inundative BC in the native European range of the insect and the target weed. Based on a field experiment assessing different application techniques, Hahn et al. (2016) proposed that mass releases of *P. chrysidiforme* may be a valuable approach to control *R. obtusifolius* in the native range by biological means. However, the considerable variation in infestation and subsequent impacts detected under experimental field conditions call for long-term studies to assess the full potential and efficacy of *P. chrysidiforme* for inundative BC of *R. obtusifolius*. Intuitively, the development of commercial inundative BC products using native invertebrate herbivores or pathogens (Kluth et al. 2003) and their integration in IWM holds considerable promise, but the proof of concept has yet to be established.

5 Future trends

IWM of weeds in grasslands is based on a good understanding of the biology and population dynamics of the target weed, particularly of site-specific transitions between stages of the weed's life cycle where particular management techniques can be effective (Fig. 1). Furthermore, the composition of the grassland, soil nutrient status and top-down pressure by natural enemies must be considered in IWM. The examples in Table 2, and discussed earlier, illustrate that targeted integration of different weed management practices can help to successfully reduce reliance on herbicides and result in more environmental friendly and sustainable management practices when tackling weed problems in grasslands across the globe. However, because the most suitable IWM strategies are context-dependent, developing new strategies often requires well-designed field experiments, which run for sufficiently long periods to allow community responses to develop. Moreover, successful implementation of IWM in grasslands requires careful planning that includes capacity building among stakeholders, prevention programs and dissemination of validated strategies (Liebman et al. 2016).

In long-term perennial systems like grassland, understanding the competitive ability of the grassland sward relative to that of the weed species is a key factor for long-term success in weed management. The reason for weed dominance in a grassland is that growth conditions may give the weed a strong competitive ability relative to the grassland sward. In such a situation, any intervention that affects the weed only for a short time span – independent whether the measure is of chemical, physical or biological manner - will only treat the symptom but not the fundamental cause of infestation. Several studies show that weeds come back immediately after a short-term control measure stops. To sustainably manage weeds in grasslands, the challenge is to adapt growth conditions in a way that the competitive ability of the grassland sward is increased or that of the weed species reduced or both.

Species composition and productivity of grasslands are linked to climatic conditions. Weed management strategies should therefore take climate change into account, potential warming and the increasing prevalence of extreme weather conditions due to climate change. It has been suggested that such changes may affect population dynamics of weeds by affecting physiological seed dormancy, germination and emergence pattern, morphological characteristics (e.g., tougher plants) and resulting in reduced herbicide efficiency, and thus, their competitive ability and, in turn, the grassland community (Ziska 2016). Climate change is also expected to directly influence herbicide effects via changing herbicide uptake, translocation and metabolism (Varanasi et al. 2016) and to break herbicide selectivity (Jursík et al. 2020). Few studies have investigated such climate change effects, although they create a need for adapted control strategies as part of mitigation planning (Sun et al. 2021). Rapid increases in herbicide resistance have further highlighted the ability of weeds to undergo rapid evolutionary change. This also has been rarely studied so far but most likely does occur with consequences for the distribution, community composition and herbicide efficacy. It remains to be further explored how the recipient communities may also be affected by climate change, either directly (e.g., drought stress) or indirectly (e.g., change in land use), which in turn will affect their susceptibility to or impact on weeds (e.g., Sandel and Dangremond 2012). As a key prediction and observation of climate change is a shift in species ranges, a resilient weed management strategy should also take changes in the composition of desirable grassland and weed species into account (Catford et al. 2019).

The expected accelerated evolution of herbicide resistance under climate change, increased herbicide regulations (e.g., the ban of numerous acting ingredients of herbicides) and a reduction in the discovery of new active ingredients of herbicides are further moving the field from herbicide-dominated weed management to IWM.

6 Acknowledgements

The authors would like to thank Per Kudsk for inviting them to write this book chapter. IWMPRAISE has received funding from the European Union's Horizon 2020 research and innovation programme under grant agreement No. 727321. Urs Schaffner was supported by CABI with core financial support from its member countries (see https://www.cabi.org/about-cabi/who-we-work-with/key-donors/).

7 References

Alder, C. G. 2012. Evaluating integrated weed management: Russian knapweed control with goat grazing and aminopyralid. Master thesis. Utah State University.

Ang, B. N., Kok, L. T., Holtzman, G. I. and Wolf, D. D. 1994. Competitive growth of Canada thistle, tall fescue, and crownvetch in the presence of a thistle defoliator, *Cassida rubiginosa* Muller (Coleoptera: Chrysomelidae). *Biol. Control* 4(3):277-284.

Baer, S. G., Bach, E. M., Meyer, C. K., Du Preez, C. C. and Six, J. 2015. Belowground ecosystem recovery during grassland restoration: South African Highveld compared to US tallgrass prairie. *Ecosystems* 18(3):390-403.

Bredenkamp, G. J., Spada, F. and Kazmierczak, E. 2002. On the origin of northern and southern hemisphere grasslands. *Plant Ecol.* 163:209-229.

Briese, D. 1996. Biological control of weeds and fire management in protected natural areas: are they compatible strategies? *Biol. Conserv.* 77:135-141.

Buckley, Y. M., Rees, M., Paynter, Q. and Lonsdale, M. 2004. Modelling integrated weed management of an invasive shrub in tropical Australia. *J. Appl. Ecol.* 41(3):547-560.

Catford, J. A., Jones, L. P., Gibson, D. J. and Newman, J. A. 2019. Grassland invasion in a changing climate. In: Gibson, D. J. and Newman, J. A. (Eds). *Grasslands and Climate Change*. Cambridge University Press (Catholic University of Pusan), Cambridge, UK, pp.149-171.

Chalak-Haghighi, M., Van Ierland, E. C., Bourdôt, G. W. and Leathwick, D. 2008. Management strategies for an invasive weed: a dynamic programming approach for Californian thistle in New Zealand. *N. Z. J. Agric. Res.* 51(4):409-424.

Connolly, J., Sebastià, M. T., Kirwan, L., Finn, J. A., Llurba, R., Suter, M., Collins, R. P., Porqueddu, C., Helgadóttir, Á., Baadshaug, O. H., Bélanger, G., Black, A., Brophy, C., Čop, J., Dalmannsdóttir, S., Delgado, I., Elgersma, A., Fothergill, M., Frankow-Lindberg, B. E., Ghesquiere, A., Golinski, P., Grieu, P., Gustavsson, A. M., Höglind, M., Huguenin-Elie, O., Jørgensen, M., Kadziuliene, Z., Lunnan, T., Nykanen-Kurki, P., Ribas, A., Taube, F., Thumm, U., De Vliegher, A. and Lüscher, A. 2018. Weed suppression greatly increased by plant diversity in intensively managed grasslands: a continental-scale experiment. *J. Appl. Ecol.* 55(2):852-862.

Delaney, J. T., Moranz, R. A., Debinski, D. M., Engle, D. M. and Miller, J. R. 2016. Exotic-dominated grasslands show signs of recovery with cattle grazing and fire. *PLoS ONE* 11(11):e0165758.

Diamond, J. M., Call, C. A. and Devoe, N. 2012. Effects of targeted grazing and prescribed burning on community and seed dynamics of a downy brome (Bromus tectorum)-dominated landscape. Inv. Plant Sci. Manag. 5:259-269.

Dietl, W. 1982. Oekologie und Wachstum von Futterpflanzen und Unkräutern des Graslandes. Schweiz. Landwirtschaft. Forsch. 21:85-110.

DiTomaso, J. M., Kyser, G. B., Miller, J. R., Garcia, S., Smith, R. F., Nader, G., Connor, J. M. and Orloff, S. B. 2006. Integrating prescribed burning and clopyralid for the management of yellow starthistle (*Centaurea solstitialis*). *Weed Sci.* 54(4):757-767.

Ehlert, K. A., Mangold, J. M. and Engel, R. E. 2014. Integrating the herbicide imazapic and the fungal pathogen *Pyrenophora semeniperda* to control *Bromus tectorum*. *Weed Res.* 54(4):418-424.

Enloe, S. F., Kline, W. N., Aulakh, J. S., Bethke, R. K., Gladney, J. B. and Lauer, D. K. 2013. Macartney rose (*Rosa bracteata*) response to herbicide and mowing treatments. *Invasive plant sci manag* 6(2):260-267.

FAO 2021. What is integrated weed management? Available at: http://www.fao.org/agriculture/crops/thematic-sitemap/theme/spi/scpi-home/managing-ecosystems/integrated-weed-management/iwm-what/en/.

Fellows, D. P. and Newton, W. E. 1999. Prescribed fire effects on biological control of leafy spurge. *J. Range Manag.* 52(5):489–493.

Firn, J., Price, J. N. and Whalley, R. D. 2013. Using strategically applied grazing to manage invasive alien plants in novel grasslands. *Ecol. Processes* 2:1–11.

Green, S., Gaunt, R. E., Bourdôt, G. W. and Field, R. J. 1998. Influence of phenology, defoliation, and *Sclerotinia sclerotiorum* on regrowth potential of *Ranunculus acris*. *New Zeal. J. Agric. Res.* 41:125–133.

Grekul, C. W. and Bork, E. W. 2007. Fertilization augments Canada thistle (*Cirsium arvense* L. Scop) control in temperate pastures with herbicides (*Cirsium arvense* (L.) Scop). *Crop Prot.* 26(4):668–676.

Grice, A. C. and Brown, J. R. 1996. An overview of the current status of weed management in Australia. In: *Proceedings of the 11th Australian Weeds Conference* (vol. 30), pp.195–204.

Grice, A. C., Clarkson, J. R. and Calvert, M. 2011. Geographic differentiation of management objectives for invasive species: a case study of Hymenachne amplexicaulis in Australia. *Environ. Sci. Policy* 14(8): 986–997.

Griffith, D. and Lacey, J. R. 1991. Economic evaluation of spotted knapweed [Centaurea maculosa] control using picloram. *J. Range Manag.* 44(1):43–47.

Hahn, M. A., Schaffner, U., Häfliger, P. and Lüscher, A. 2016. Establishment and early impact of the native biological control candidate *Pyropteron chrysidiforme* on the native weed *Rumex obtusifolius* in Europe. *BioControl* 61(2):221–232.

Hasan, M., Ahmad-Hamdani, M. S., Rosli, A. M. and Hamdan, H. 2021. Bioherbicides: an eco-friendly tool for sustainable weed management. *Plants (Basel)* 10(6): 1212.

Hatcher, P. E., Paul, N. D., Ayres, P. G. and Whittaker, J. B. 1994. The effect of an insect herbivore and a rust fungus individually, and combined in sequence, on the growth of two *Rumex* species. *New Phytol.* 128(1):71–78.

Hoffmann, J. H., Moran, V. C. and Zeller, D. A. 1998. Long-term population studies and the development of an integrated management programme for control of *Opuntia stricta* in Kruger National Park, South Africa. *J. Appl. Ecol.* 35(1):156–160.

Hopkins, A. and Johnson, R. H. 2002. Effect of different manuring and defoliation patterns on broad-leaved dock (*Rumex obtusifolius*) in grassland. *Ann. Appl. Biol.* 140(3):255–262.

Huwer, R. K., Briese, D. T., Dowling, P. M., Kemp, D. R., Lonsdale, W. M., Michalk, D. L., Neave, M. J., Sheppard, A. W. and Woodburn, T. L. 2005. Can an integrated management approach provide a basis for longterm prevention of weed dominance in Australian pasture systems? *Weed Res.* 45(3):175–192.

Joshi, A. 2008. Integrating flea beetles (*Aphthona* spp.) with herbicide and grasses for leafy spurge (*Euphorbia esula*) management. *Weed Technol.* 22(3):523–529.

Jursík, M., Kočárek, M., Kolářová, M. and Tichý, L. 2020. Effect of different soil and weather conditions on efficacy, selectivity and dissipation of herbicides in sunflower. *Plant Soil Environ.* 66(9):468–476.

Kleijn, D. and Steinger, T. 2002. Contrasting effects of grazing and hay cutting on the spatial and genetic population structure of *Veratrum album*, an unpalatable, long-lived, clonal plant species. *J. Ecol.* 90(2):360–370.

Kluth, S., Kruess, A. and Tscharntke, T. 2003. Influence of mechanical cutting and pathogen application on the performance and nutrient storage of *Cirsium arvense*. *J. Appl. Ecol.* 40(2):334-343.

Kudsk, P., Sønderskov, M., Bonin, L., Gonzalez-Andujar, J. L., Jensen, J. E., Melander, B., Moonen, C., Riemens, M., Sattin, M., Schaffner, U. and Storkey, J. 2020. IWMPRAISE-an EU Horizon 2020 project providing integrated weed management solutions to European farmers. *Outlooks Pest Manag.* 31(4):152-159.

Le, Q. B., Nkonya, E. and Mirzabaev, A. 2016. Biomass productivity-based mapping of global land degradation hotspots. In: Nkonya, E., Mirzabaev, A. and von Braun, J. (Eds). *Economics of Land Degradation and Improvement-A Global Assessment for Sustainable Development*. Springer, New York, pp.55-84.

Lehnhoff, E. A., Rew, L. J., Mangold, J. M., Seipel, T. and Ragen, D. 2019. Integrated management of cheatgrass (*Bromus tectorum*) with sheep grazing and herbicide. *Agronomy* 9(6):315.

Le Maitre, D. C., Krug, R. M., Hoffmann, J. H., Gordon, A. J. and Mgidi, T. N. 2008. *Hakea sericea*: development of a model of the impacts of biological control on population dynamics and rates of spread of an invasive species. *Ecol. Modell.* 212(3-4):342-358.

Liebman, M., Baraibar, B., Buckley, Y., Childs, D., Christensen, S., Cousens, R., Eizenberg, H., Heijting, S., Loddo, D., Merotto Jr., A., Renton, M. and Riemens, M. 2016. Ecologically sustainable weed management: how do we get from proof-of-concept to adoption? *Ecol. Appl.* 26(5):1352-1369.

Lüscher, A., Mueller-Harvey, I., Soussana, J. F., Rees, R. M. and Peyraud, J. L. 2014. Potential of legume-based grassland-livestock systems in Europe: a review. Grass For. Sci. 69:206-228.

Lym, R. G. 1998. The biology and integrated management of leafy spurge (Euphorbia esula) on North Dakota rangeland. Weed Technol. 12:367-373.

Masin, E., Nelson, C. R. and Valliant, M. T. 2018. Can sheep control invasive forbs without compromising efforts to restore native plants? *Rangel Ecol. Manag.* 71(2):185-188.

McEvoy, P. B., Rudd, N. T., Cox, C. S. and Huso, M. 1993. Disturbance, competition, and herbivory effects on ragwort *Senecio jacobaea* populations. *Ecol. Monogr.* 63(1):55-75.

Miller, T. W. 2016. Integrated strategies for management of perennial weeds. *Invasive plant sci manag* 9(2):148-158.

Müller-Schärer, H. 1991. The impact of root herbivory as a function of plant density and competition: survival, growth and fecundity of *Centaurea maculosa* (Compositae) in field plots. *J. Appl. Ecol.* 28:759-776.

Müller-Schärer, H. and Collins, A. R. 2012. Integrated weed management. In: Jorgensen, S. E. (Ed.), *Encyclopedia of Environmental Management*. Taylor and Francis, New York, pp. 439-447.

Müller-Schärer, H. and Schaffner, U. 2008. Classical biological control: exploiting enemy escape to manage plant invasions. *Biol. Invasions* 10(6):859-874.

Müller-Schärer, H., Sun, Y., Chauvel, B., Karrer, G., Kazinczi, G., Kudsk, P., Oude, A. G. J. M. L., Schaffner, U., Skjoth, C. A., Smith, M., Vurro, M., de Weger, L. A. and Lommen, S. T. E. 2018. Cross-fertilizing weed science and plant invasion science to improve efficient management: a European challenge. *Basic Appl. Ecol.* 33:1-13.

Nelson, J. A. and Lym, R. G. 2003. Interactive effects of *Aphthona nigriscutis* and picloram plus 2, 4-D in leafy spurge (*Euphorbia esula*). *Weed Sci.* 51(1):118-124.

Niggli, U., Nösberger, J. and Lehmann, J. 1993. Effects of nitrogen fertilization and cutting frequency on the competitive ability and the regrowth capacity of Rumex obtusifolius L. in several grass swards. *Weed Res.* 33(2):131-137.

Norsworthy, J. K., Ward, S. M., Shaw, D. R., Llewellyn, R. S., Nichols, R. L., Webster, T. M., Bradley, K. W., Frisvold, G., Powles, S. B., Burgos, N. R. and Witt, W. W. and Barrett, M. 2012. Reducing the risks of herbicide resistance: best management practices and recommendations. Weed Sci. 60:31-62.

Olson, D. M., Dinerstein, E., Wikramanayake, E. D., Burgess, N. D., Powell, G. V., Underwood, E. C., D'amico, J. A., Itoua, I., Strand, H. E., Morrison, J. C. and Loucks, C. J. 2001. Terrestrial ecoregions of the world: a new map of life on earth. *BioScience* 51(11), 933-938.

Paynter, Q. and Flanagan, G. J. 2004. Integrating herbicide and mechanical control treatments with fire and biological control to manage an invasive wetland shrub. Mimosa *Pigra. J. Appl. Ecol.* 41:615-629.

Phatak, S. C., Sumner, D. R., Wells, H. D., Bell, D. K. and Glaze, N. C. 1983. Biological control of yellow nutsedge with the indigenous rust fungus *Puccinia canaliculata. Science* 219(4591):1446-1447.

Pywell, R. F., Hayes, M. J., Tallowin, J. B., Walker, K. J., Meek, W. R., Carvell, C., Warman, L. A. and Bullock, J. M. 2010. Minimizing environmental impacts of grassland weed management: can *Cirsium arvense* be controlled without herbicides? *Grass Sci.* 65(2):159-174.

Reichard, S. H. and White, P. 2001. Horticulture as a pathway of invasive plant introductions in the United States: most invasive plants have been introduced for horticultural use by nurseries, botanical gardens, and individuals. *BioScience* 51:103-113.

Richardson, L. A., Juricek, C. J., Lym, R. G., Kirby, D. R. and Tober, D. A. 2008. Integrated leafy spurge (*Euphorbia esula*) control using imazapic, *Aphthona* spp. biological control agents, and seeded native grasses. *Inv. Plant Sci. Manag.* 1(3):255-264.

Samuel, L. W., Sedivec, K. K., Dahl, J. D., Jonson, L. L. and Faller, T. C. 2004. Effects of multi-species grazing and biocontrol on leafy spurge-infested rangeland using rotational grazing. In: *Proceedings of the Sociaal Range Managed*. Littleton, CO, p.326.

Sandel, B. and Dangremond, E. M. 2012. Climate change and the invasion of California by grasses. *Glob. Change Biol.* 18(1):277-289.

Schnick, P. J., Stewart-Wade, S. M. and Boland, G. J. 2002. 2,4-D and *Sclerotinia minor* to control common dandelion. *Weed Sci.* 50(2):173-178.

Sciegienka, J. K., Keren, E. N. and Menalled, F. D. 2011. Interactions between two biological control agents and an herbicide for Canada thistle (*Cirsium arvense*) suppression. *Invasive plant sci manag* 4(1):151-158.

Scott, J. K. and Sagliocco, J. -L. 1991a. *Chamaesphecia doryliformis* (Lep.: Sesiidae), a second root borer for the control of *Rumex* spp. (Polygonaceae) in Australia. *Entomophaga* 36(2):245-251.

Scott, J. K. and Sagliocco, J. -L. 1991b. *Bembecia chrysidiformis* (Lep.: Sesiidae), a potential control agent for *Rumex* spp. (Polygonaceae) in Australia. *Entomophaga* 36(2):235-244.

Speight, R. I. and Whittaker, J. B. 1987. Interactions between the chrysomelid beetle *Gastrophysa viridula*, the weed *Rumex obtusifolius* and the herbicide asulam. *J. Appl. Ecol.* 24(1):119-129.

Strickland, G. R., Fogliani, R. and Scott, J. K. 2012. *Rumex* spp.–docks. In: Julien, M., McFadyen, R. and Cullen, J. (Eds) *Biological Control of Weeds in Australia*. CSIRO Publishing, Collingwood, pp.140–148.

Sun, Y., Kaleibar, B. P., Oveisi, M. and Müller-Schärer, H. 2021. Addressing climate change: what can plant invasion science and weed science learn from each other? *Front. Agron.* 2:626005. doi: 10.3389/fagro.2020.626005

Suter, M., Siegrist-Maag, S., Connolly, J. and Lüscher, A. 2007. Can the occurrence of *Senecio jacobaea* be influenced by management practice? *Weed Res.* 47(3):262–269.

Suter, M. and Lüscher, A. 2011. Measures for the control of *Senecio aquaticus* in managed grassland. *Weed Res.* 51(6):601–611.

Suter, M. and Lüscher, A. 2012. Rapid and high seed germination and large soil seed bank of *Senecio aquaticus* in managed grassland. *Sci. World J.* Available at: https://doi.org/doi:10.1100/2012/7238082012:723808.

Suter, M., Hofer, D. and Lüscher, A. 2017. Weed suppression enhanced by increasing functional trait dispersion and resource capture in for-age ley mixtures. *Agric. Ecos Environ.* 240:329–339.

Suter, M., Huguenin-Elie, O. and Lüscher, A. 2021. Multispecies for multifunctions: combining four complementary species enhances multifunctionality of sown grassland. *Sci. Rep.* 11(1):3835.

Tipping, P. W. 2001. Canada thistle (*Cirsium arvense*) control with hexazinone in crown vetch (*Coronilla varia*). *Weed Technol.* 15(3):559–563.

Triolet, M., Guillemin, J. P., Andre, O. and Steinberg, C. 2020. Fungal-based bioherbicides for weed control: a myth or a reality? *Weed Res.* 60(1):60–77.

Varanasi, A., Prasad, P. V. V. and Jugulam, M. 2016. Impact of climate change factors on weeds and herbicide efficacy. *Adv. Agron.* 135:107–146.

Veldman, J. W., Buisson, E., Durigan, G., Fernandes, G. W., Le Stradic, S., Mahy, G., Negreiros, D., Overbeck, G. E., Veldman, R. G., Zaloumis, N. P., Putz, F. E. and Bond, W. J. 2015. Toward an old-growth concept for grasslands, savannas, and woodlands. *Front. Ecol. Environ.* 13(3):154–162.

Vitelli, J. S. 2000. Options for effective weed management. *Trop. Grassl.* 34:280–294.

Willis, A. J., Groves, R. H. and Ash, J. E. 1998. Interactions between plant competition and herbivory on the growth of *Hypericum* species: a comparison of glasshouse and field results. *Aust. J. Bot.* 46(6):707–721.

Winston, R. L., Schwarzländer, M., Hinz, H. L., Day, M. D., Cock, M. J. W. and Julien, M. H. 2014. Biological control of weeds: a world catalogue of agents and their target weeds, 5th edn. *FHTET-2014-04*. USDA Forest Service, Forest Health Technology Enterprise Team, Morgantown.

Ziska, L. H. 2016. The role of climate change and increasing atmospheric carbon dioxide on weed management: herbicide efficacy. *Agric. Ecosys Environ.* 231:304–309.

Chapter 14

Integrated weed management in perennial woody crops

Verónica Pedraza and José Luis González-Andújar, IAS-CSIC, Spain; Victoire Huet and Paul Tuteirihia, NIAB EMR, UK; and Julien Lecourt, Pôle Scientifique des vignobles Bernard Magrez, France

1 Introduction

Tree crops are the most common perennial crops, covering an area of more than 82 million ha worldwide (FAO, 2018). More specifically, orchards are one of the most common and successful forms of perennial agriculture (Roberts, 2017) and Europe plays an important role, cultivating 15% of the total area (12.5 million ha) (FAO, 2018).

Perennial woody crops span a broad variety of species in Europe, with olive groves (5.1 million ha) and vineyards (3.6 million ha) representing the largest cultivated area (FAO, 2018). In addition, their productions are of great economic importance particularly for smaller producers. Olive orchards are mainly found in the Mediterranean region, in countries like Spain, Italy, Greece and Portugal (FAO, 2018). However, grapes are cultivated across the continent, from the Southern Mediterranean countries (Italy, Spain or Portugal) to Northern Europe (the UK and France). Nevertheless, both crops share many physiological and agronomic features that allow for similar soil and weed management strategies.

Both the perennial crops are self-pollinating and need mild winters and warm summers for successful flowering and fruiting. The fruits follow a similar processing procedure. Moreover, they are planted in a similar spatial arrangement - in rows, with intra-row and inter-row spacing across the field. This cropping pattern allows plants to achieve good light interception (IOC,

http://dx.doi.org/10.19103/AS.2021.0098.16

2007), and to draw water and nutrients from the soil by the roots of their woody trunks, in addition to maintaining soil quality and productivity.

The effective management of annual and perennial weed species plays a key role, since weeds cause substantial yield losses across all crops (Oerke, 2006). In these cropping systems, weeds also reduce tree growth by competing for water, nutrients, light and rooting space (IWMPRAISE, 2018). Competition is more severe during the first four to five years when root growth is limited. Therefore, competition for water is a critical issue in rainfed areas of Southern Europe with scarce water availability, while frost damage due to ground cover is a significant problem in the Northern part of Europe.

Traditionally, conventional agriculture has achieved weed control through multiple applications of herbicides. In fact, the reliance on herbicides is very high and herbicides are the single most used group of pesticides. However, in recent years, three factors have been driving a need to change weed control practices in conventional farming systems: 1) the rapidly increasing problem of herbicide resistance, 2) the expectation that many of the currently used herbicides will be withdrawn from the market and 3) the negative effects of herbicides on farmland biodiversity and hence on the associated ecosystem services (Storkey et al., 2012).

This has triggered interest in integrated weed management (IWM), which allows farmers to use alternative weed management approaches, all of them focused on reducing the reliance on herbicides by replacing them, wholly or partly, with non-chemical methods (Kudsk et al., 2020). In tree crops, conservation tillage methods and no-till systems are commonly used, combining cultural, mechanical, biological and chemical practices. Conservation tillage systems involve minimum mechanical soil disturbance, while no-tillage implies no soil movement. Both systems are less dependent on herbicides due to the weed-suppressing effects of a permanent soil organic cover on the surface, with a minimum of 30% of the soil cover being recommended by the principles of conservation agriculture. This ground coverage is usually achieved by establishing cover crops, incorporating mulch, leaving wood residues from pruning or by all three methods together. These strategies can help to reduce not only the need for chemical weed control but also soil erosion (Cosentino et al., 2015), in addition to improving soil structure and fertility in the long term (Kassam et al., 2012) and increasing overall sustainability of the farming system (Pedraza, 2018). In fact, the inclusion of spontaneous or sown cover crops in the cropping system provides multiple benefits to the agroecosystem (Hartwig and Ammon, 2002), from weed and pest control to soil protection, depending on the cover crop species and their adaptability to local environmental conditions.

Given the different geographic, climatic and agronomic conditions existing between perennial woody crops grown, for example, in Spain and the UK, the most appropriate IWM strategy should comprise several of these practices in keeping with the features and needs of each olive orchard or vineyard. In this chapter, two case studies on IWM in perennial tree crops will be presented: olive groves in Spain and vineyards in the UK.

2 Case study: olive orchards in Spain

2.1 Introduction

Olive groves occupy over 10.5 million hectares worldwide, 95% of which are in the Mediterranean region (FAO, 2018). Spain has been cultivating olive trees for centuries. In fact, Spain is the country with the largest olive growing area in the world (2.7 million hectares of olive groves), which is mainly concentrated in Andalusia (1.6 million ha), the southernmost region of Spain (MAPA, 2019b). Next in importance are the olive groves located in the central-south areas of the country, Castilla La Mancha and Extremadura, although with much less regional weight (436 000 ha and 287 000 ha, respectively) (MAPA, 2019a).

Given the broad geographical area that olive orchards cover, soil and weed management decisions are significantly influenced by location, climatic conditions, soil, topography and grower preferences (Huqi et al., 2009). Moreover, as a key crop, new management and cultivation techniques and new technologies are constantly being introduced, resulting in different olive cultivation systems.

Soil management techniques in olive cropping systems have always aimed to promote high profitability and quality production (Saavedra et al., 2016b). Improvements, however, are only possible where the orographic conditions of the farming area are suitable for olive growing. As a result, Spain's traditional olive cultivation systems currently co-exist with high-density ones (Fig. 1). The traditional systems grow olive trees under rainfed conditions, with one to three trunks per tree, widely spaced (7-12 × 5-10 m) and an average density of 80–120 trees/ha. Almost half of the olive farms are located in unfavourable areas with slopes steeper than 15% and about 72 000 ha of the area sloped at a gradient >30% (e.g. hillsides, mountains) where farmers carry out their work with non-mechanized means (AEMO, 2012). However, high-density systems cultivate single-trunk olive trees in favourable areas and use integral mechanized means. These orchards are intensive (high density, with 200-600 trees/ha and a wide inter-row spacing of 6-7 m) or super-intensive (very high density with 1000-2000 trees/ha with tree spacing of 4 × 1.5 m) and most of them employ localised irrigation systems.

Despite the existing differences in crop establishment and features, machinery use and subsequent yields, olive groves employ similar weed management strategies. The most used soil management systems in Spain are reduced tillage (40% of the total area), spontaneous cover crops (28%)

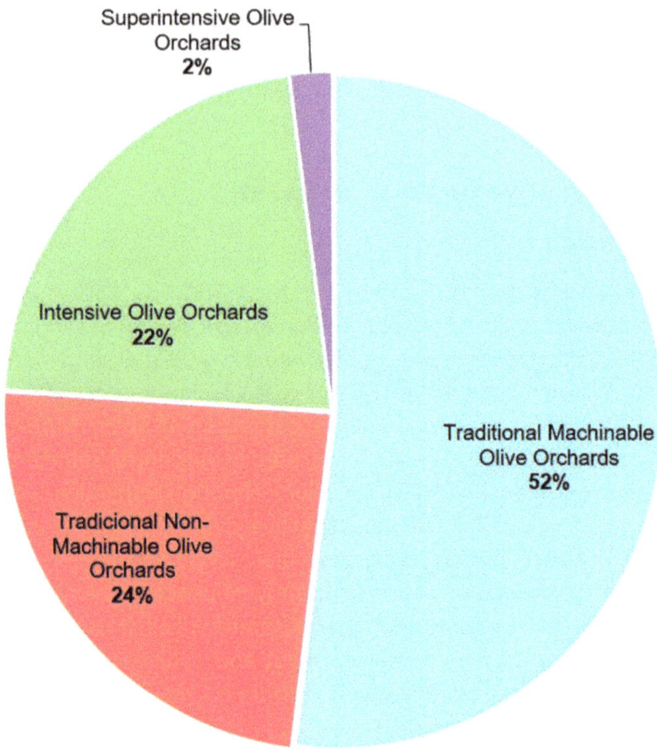

Figure 1 Distribution of the olive cultivation systems in Spain. Source: AEMO (2012).

and no-tillage with application of herbicides (12%) (MAPA, 2019a). Often, a combination of these practices is used on farms, since most of the olive orchards have two clearly distinctive areas: the bare soil beneath the olive trees, which facilitates harvesting, and the area along the lanes (intra-row and inter-row spacing), where soil compaction and higher susceptibility to runoff and erosion will influence the system chosen (Castro, 1993; CAP, 2006).

Farmers typically manage weeds through repeated tillage and/or herbicide application with the primary goal of reducing weed competition for water and mineral resources (Saavedra et al., 2015a). However, the combination of a Mediterranean-type climate, sloping areas and management practices with scarce herbaceous vegetation cover has led to severe problems of water availability and soil erosion, accompanied by soil fertility depletion, biodiversity loss and environmental degradation.

In this context, the Common Agricultural Policy (CAP) environmental programmes and cross-compliance principles in Europe have been aimed at improving and achieving more balanced and sustainable practices, such as adding inert/plant residue mulches or establishing living cover crops to promote biodiversity and to prevent soil erosion during the rainy season (Pedraza, 2018). Moreover, current knowledge reveals great potential for using cover crops as a preventive method in IWM strategies in temperate areas since they are able to suppress 70–95% of weeds in the fall-to-spring season of arable crops. The cover crop residues can also reduce weed emergence during early development of the following crop by presenting a physical barrier and releasing allelopathic compounds into the soil solution (Gerhards and Schappert, 2020). Consequently, the adoption of covered soil techniques has significantly increased in the last ten years, going from 434 828 ha in 2006 to 835 262 ha in 2019. However, these practices need to be adapted to the local conditions of the farm, and their successful establishment in olive orchards requires careful management and control to reduce not only the likelihood of pests and diseases appearing (Martinelli et al., 2017) but also to control weeds by reducing herbicide use (Abu-Irmaileh and Abu-Rayyan, 2004).

The fact that there is no practice without drawbacks highlights the importance of IWM in olive orchards (Pedraza et al., 2019), and it is designed to reduce the negative impacts on soil and production while maintaining beneficial flora at an affordable and manageable level (Huqi et al., 2009). Therefore, proper IWM should take into account not only the efficacy of weed control but also how the practices affect the weed population, the olive cropping system and the agro-ecosystem (Fracchiolla et al., 2016).

2.2 Commonly used soil management systems in Spain

Nowadays, growers have different soil and weed management tools available to achieve weed control objectives. The best strategy for employing these tools, however, will vary between years and farms according to local conditions. However, all IWM strategies consist of the weed management tools listed in Table 1.

2.3 Tillage

Tillage entails moving the soil with the main aim of managing weeds and facilitating infiltration (Fig. 2). Tillage operations continue to be the most used soil management system by olive growers in Spain, although this practice causes the greatest soil losses (Gómez et al., 2009). Since late 2000, the public policies under the CAP regulations (cross-compliance requirements) have resulted in a reduction of tillage operations and ploughing depth. Conventional tillage involves inversion tillage operations at a depth greater than 20 cm, while reduced tillage includes non-inversion tillage systems at 10–15 cm depth (Pedraza, 2018).

Table 1 Outline of the weed management options according to soil management technique in the olive groves of Spain

Soil management technique			Weed control
Tillage adapted to the cross-compliance regulations: shallow tillage (<20 cm depth) with lower labour frequency and avoiding mouldboard ploughs and disc harrows, especially in olive farms located on slopes steeper than 10-15%			-Annual and biennial weed control by shallow tillage - Effective control of flora highly adapted to no-tillage practices -Not always effective against perennials
No tillage with chemical control			Pre-emergence and post-emergence herbicides
Inert cover with plant residue mulch (olive leaves, wood chips, pruning residues or cereal straw)			-Physical barrier against weed development -Allelopathic substances produced by mulch decomposition ensure partial weed control -Additional weed control is required by herbicides
Living cover crop	**Weeds (Spontaneous flora)**		Competition with other weeds
		Chemical mowing	Contact and translocated authorised herbicides
		Mechanical mowing at ground level and left on the soil as a mulch	Brush cutter and shredder
		Grazing	Livestock introduction
		Chopped and incorporated by tillage	Brush cutter, shredder and rotary tiller for incorporation
	Plants cultivated under controlled growth conditions (Gramineous, leguminous and cruciferous species)		Competition with the spontaneous flora
		Chemical mowing	Contact and translocated authorised herbicides
		Mechanical mowing	Brush cutter and shredder
		Grazing	Livestock introduction
		Chopped and incorporated by tillage	Brush cutter, shredder and rotary tiller for incorporation

Source: Own elaboration based on Saavedra et al. (2015b) and Pastor et al. (2001).

Nowadays, reduced tillage is used in 40% of the total olive growing area (1 093 280 ha) while conventional tillage is only used in 10% of the total area (286 768 ha) (MAPA, 2019a). In fact, reduced tillage is the most used technique in rainfed olive groves (784 870 ha), which represents 72% of the total use, and the second most used practice in irrigated olive groves (308 409 ha).

Figure 2 Tillage management of olive groves.

Thus, mechanical traction is commonly used in a crossed pattern to the slope direction, avoiding mouldboard ploughs and disc harrows.

Olive farms located on slopes steeper than 15% can be tilled with chain-tractors operating along the direction of the steepest slope. However, where slopes are more than 45%, tillage operations are excluded by the minimum-security working regulations (Saavedra et al., 2015b). Moreover, tillage is not advisable beneath the olive trees, because soil compaction and infiltration problems do not usually occur in this area (Castro, 1993; IOC, 2007). Therefore, it is mainly used along the lanes, in the intra-row and inter-row spacing.

Additionally, some regions of Spain have specific olive IP regulations which limit tillage operations. For example, the olive IP regulation of Andalusia, in the south of Spain, forbids tillage practices in the direction of the slope or mechanical traction on slopes ≥10%. In this latter case, terracing, strip cropping and cover cropping are traditionally practiced. Furthermore, there are some exceptions where a shallow tillage (<20 cm) is allowed, for example, on compacted soils, on Vertisol soils with expandable mineral clays and for incorporation of organic amendments in soils infested with resistant creeping or perennial plants not controlled by herbicides or mechanical methods (i.e. mowing or clearing) (IFAPA, 2011b).

This soil management technique helps to control annual and biennial weeds, but it is not always effective against perennials. This mainly affects reduced-tillage systems, where shallow practices can lead to a rapid increase in the superficial soil seed bank and problems with perennial weeds (Abu-Irmaileh and Abu-Rayyan, 2004). However, tillage is very useful for controlling flora which are highly adapted to no-tillage practices, which is difficult to manage by other means (e.g. *Conyza* spp).

2.4 No tillage with chemical control

No tillage and reduced tillage are practices that form an integral part of conservation agriculture. Conservation agriculture attempts to minimise disruption of the soil's structure, composition and natural biodiversity, thereby reducing erosion and degradation, as well as fuel consumption (Holland, 2004). No tillage, and thus, no soil movement occurs in the intra-row and inter-row spacing (Fig. 3), and the ground is kept weed-free by applying herbicides (Pedraza, 2018).

No tillage is the third most used technique in olive groves in Spain, being used on 338 196 ha (12% of the total cultivated area), of which 67% is rainfed and 33% is irrigated (MAPA, 2019a).

Initially, herbicides were used to keep the soil in olive groves completely bare of vegetation. In recent years, the recommended practice has been to keep bare soil only beneath the trees, especially in those olive varieties with a tendency to detach, while maintaining covered soil between the tree rows (Saavedra et al., 2015a). Moreover, a combination of tillage and no tillage methods is a very common practice in olive groves of Spain. The

Figure 3 No-tillage management of olive groves. Source: INTIA.

proportion of soil with tillage or no-tillage management depends on the plantation type (traditional or intensive/super-intensive olive groves with traditional spacing or high-/very high-density patterns, respectively) and the olive tree spacing (planting pattern). For example, tillage inter-row and intra-row management with herbicide application beneath the olive trees are used in olive-growing regions with a broad-scale pattern and in most of the traditional systems. However, in areas with less tree spacing, such as some intensive and super-intensive olive systems, the herbicide application can encompass all the intra-row management combined with a tillage inter-row management.

There is some controversy in research studies about the effectiveness of this system compared with tillage in olive orchards on erosion control and water balance (Pastor et al., 2001; Saavedra, 2015). Nevertheless, herbicide use became a common practice in Spanish olive groves and has rapidly increased in the last 20 years, with the use of pre-emergence herbicides such as diflufenican and oxyfluorfen. Even so, glyphosate is still widely used.

The herbicides currently authorised in Spain are regulated according to the Official Register for Phytosanitary Products and Materials of the Ministry of Agriculture and Fisheries, Food and Environment (MAPAMA, 2017). In the case of areas located in regions of Spain with specific olive integrated production (IP), herbicide use is restricted to those authorised by the integrated production norms (Specific Regulation of olive IP).

Herbicides allow effective control of the majority of the olive orchard weed flora. However, one should take into account the serious problems caused by the appearance of resistant weed biotypes selected by the overuse of herbicides (Saavedra and Pastor, 2002). For example, there is a widespread occurrence of multiple herbicide resistance in annual ryegrass (*Lolium rigidum*) and *Conyza* spp. in Spain. For that reason, weed control with herbicides requires a good knowledge of the species to be treated, their cycles and the optimum time of treatment. In addition, a careful choice of active substance and application rate is crucial to achieve good management and prevent not only resistance but also problems of pollution.

2.5 Inert cover with plant residue mulches

Application of plant residue mulches consists of leaving the soil untilled and covered with olive leaves, wood chips, pruning residues or other plant residues such as cereal straw. According to conservation agriculture practices, a minimum of 30% soil cover is needed (Pedraza, 2018).

This type of cover is only used on 65475 ha of olive growing area in Spain, which only accounts for 2% of the total cultivated area (MAPA, 2019a). However, this technique is widely used in the southernmost region of Spain, as a supplement to tillage and no tillage practices with chemical control, in the inter-row spacing of traditional olive grove systems (Fig. 4).

Every year, olive orchards generate a significant volume of pruning wood residues after harvesting, which, once they have been chopped up and scattered on the soil surface by crushers or choppers, are excellent as plant cover that supplies organic matter, contributes to erosion control and reduces phytosanitary contamination hazards (Saavedra et al., 2015b).

These olive pruning remains, in addition to enabling the farmer to use the residues generated on his farm, can help to ensure partial weed control, of both of the autumn-winter and spring- emerging species, due to the physical barrier they constitute. Therefore, it is not necessary to incorporate them into the soil by tillage, ensuring a long-term decomposition and soil protection. The effect depends to a great extent on the quantity of debris, and it is considered that for the effect to contribute significantly to weed management, the quantity of residues should be around 7–8 kg/m^2 (Alcántara et al., 2009).

Weed control should be completed with herbicides, due to the difficulty of controlling weeds by grazing or mechanical methods. Nonetheless, these organic mulches help to reduce the use of herbicides and improve the clay-humus complex by increasing absorption and promoting degradation. They also diminish the transport of herbicide-containing sediment and water (Saavedra et al., 2015b).

Figure 4 Using pruning wood residues to cover soil along inter-row spacing.

The great advantage of these mulches is that they supply the soil with organic matter (Márquez-García et al., 2014), but there is a risk of introducing pathogens like *Verticillium dahliae* through leaves and pruning residues from sick trees. The prevention of infections by this soil-borne fungus is a priority for orchard survival. Consequently, if infection occurs, the residues should be removed and should never be incorporated into the soil (Saavedra et al., 2015b).

2.6 Spontaneous cover crops

This system involves the use of the spontaneous flora growing on the farm as living cover crops, keeping them alive for a specific period over the whole surface or in strips (Fig. 5). Properly managed living cover crops between the rows of olive trees increase water infiltration rate and improve soil water balance by reducing runoff, promoting soil moisture conservation if cover crop density and biomass are not excessive, competing with other undesired weeds and protecting the soil from the direct impact of rain and erosion (Saavedra et al., 2015b). In fact, to tackle the growing problem of soil erosion, EU cross-compliance regulations have implemented a mandatory requirement of maintaining a vegetation cover of at least 1-m width in the lanes of olive groves located on areas with mean slopes of ≥10% (FEGA, 2014).

These cover crops must be allowed to emerge spontaneously in autumn and winter, during the cold rainy period when water is available, and they should be killed off in the early spring under Mediterranean conditions, when the water balance starts to become negative, to maximise soil protection and minimise soil water use by the cover crop (Alcántara et al., 2011a). They are especially recommended on soils where steep orography makes sowing of cover crops complicated, such as in traditional mountain olive groves, on arable land with great weed species diversity that can provide a dense protective cover crop and in organic olive orchards where herbicides are not used (IFAPA, 2011a).

Spontaneous cover crops are the second most used technique in the olive groves of Spain, being implemented on 761 648 ha (28% of the total cultivated area) (MAPA, 2019a). Moreover, spontaneous cover crops are the most used technique in the irrigated olive groves of Spain (316 443 ha, which represents 38% of the total area) and the second most used practice in rainfed olive groves (445 205 ha, which constitutes 23% of the total rainfed olive growing area).

Despite their widespread use, the great diversity of species (weeds with short or long cycle; annual, biennial and perennial, dicotyledonous or monocotyledonous; erect or creeping) hampers their management. Therefore, only one group of weed species is usually selected in spontaneous cover crops, and the rest of the weeds are usually removed with selective herbicides to eliminate the most competitive species.

Natural grasses such as false barley (*Hordeum murinum*), compact brome (*Bromus madritensis*), annual ryegrass (*Lolium rigidum*) and annual meadow grass (*Poa annua*), among others, are found frequently in olive orchards (Saavedra et al., 2016a). Spontaneous cruciferous species like field mustard (*Sinapis arvensis*) and white rocket (*Diplotaxis erucoides*), among others, are also frequent in olive orchards (Saavedra, 2015).

These spontaneous species will have to establish again in the subsequent years but the soil seedbank has a limited duration. Consequently, the strategy is to leave narrow bands (from 0.5 m) or patches of uncontrolled cover crop to produce seeds and ensure their establishment in the subsequent year by 'self-seeding'. Therefore, only part of the cover crop should be mown, to allow for seed production and regeneration. Moreover, to prevent flora inversions in the seeding strips, they should be located in different positions every year, facilitating plant cover uniformity in successive years (Saavedra et al., 2016a).

Once properly installed, mechanical mowing is effective against spontaneous crucifers given their limited regrowth ability, but it does not fully control natural grasses. Hence, tillage or herbicides are better control measures in spontaneous grass cover crop systems. In olive groves grown on mountainous areas with livestock, grazing is a recommended technique. Furthermore, cover crops can be chopped and incorporated into the soil by tillage, although this can lead to the loss of soil moisture (Saavedra et al., 2015b). This loss of water

Figure 5 Spontaneous grass cover crops in strips.

may lead to a decrease in yield compared with other control systems. However, this may be partly offset by the effective and timely control of the living cover crop, which will minimise evaporation losses. Notwithstanding, cover crop killing by tillage is also limited by the cross-compliance regulations (FEGA, 2014) .

2.7 Cultivated cover crops

Sown cover crops are highly recommended for application in the middle of the orchard lanes for a certain period, as they improve the physical, chemical and biological properties of soils, similar to spontaneous cover crops. Sometimes, the establishment of spontaneous cover crop is complicated, especially if the farm is managed with a bare soil system based on no tillage with herbicide applications where the weed seedbank is depleted or residues of herbicides are present in the soil. Moreover, on farms where intensive tillage use was practiced, the spontaneous flora will be composed of short-cycle species with poor development that will hardly form a suitable cover crop (Saavedra et al., 2015b). Therefore, in these cases, it is advisable to sow cover crops every year (Fig. 6).

Autumn-sown cover crops are the least used technique in Spain, being adopted on only 8139 ha, which represents 0.3% of the total olive-growing area (MAPA, 2019a). Indeed, the costs of seeds and sowing can explain the lower interest among farmers in sown cover crops compared to spontaneous cover crops. The soil must be carefully prepared to maximise emergence. Therefore, due to the uneven terrain in olive groves, a shallow tillage is recommended (vibro-cultivator) followed by a roller pass after cover crop sowing. They are usually sown by centrifugal broadcaster or by hand (according to the farm tillage system) and seeds need to be buried with a shallow tillage (cultivator or vibro-cultivator followed by a roller pass after cover crop sowing). It is not possible to establish a cover crop in steep-sloped areas (>15–20%).

Fertilisation is very important to allow early cover crop growth and to compete with undesirable weed species, reducing the need for subsequent chemical or non-chemical control. The recommendation for cover crops in areas with an average rainfall of 500–600 mm is to apply a minimum of 50 kg of nitrogen (N) per hectare (IFAPA, 2011b).

The benefits from sown cover crops will only be achieved if the selected species are adapted to local environmental conditions and are appropriate for achieving the agro-ecological targets defined by the farmer. It is crucial to choose species with a short phenological cycle, fast growth during the winter, shallow root system, short height and high biomass production and well-adapted to olive-growing conditions (Saavedra et al., 2015b). However, soil conditions change and ecological succession takes place, and it may be advisable to change the type of cover crop, in order to establish a rotation and to alternate the management systems over the years.

The most commonly cultivated cover crop species are grasses, legumes, crucifers or a legume-grass mixture (Saavedra et al., 2015a). Grass cover crops such as barley (*Hordeum vulgare* L.), oat (*Avena sativa*) or native annual grass (*Brachypodium distachyon)* are advantageous for erosion control as they provide effective soil protection and high biomass production (Saavedra et al., 2016a). However, grasses are difficult to establish in compacted soils, and their development is insufficient to protect the soil or to compete with weeds (Alcántara et al., 2011a).

The main advantage of legume cover crops such as faba bean (*Vicia faba*), white lupin (*Lupinus albus*) or vetch (*Vicia sativa*) is their ability to fix atmospheric N, which can be transferred to the main crop or remain available for the subsequent crop. Nevertheless, because of the low C/N ratio, legume plant material is decomposed quickly, reducing the protective effect against evaporation from the soil surface (Pedraza, 2018). For this reason, intercropping of winter grasses with annual legumes is another interesting option tested in olive groves with livestock. The vetch (*Vicia sativa* L.)-oat (*Avena sativa* L.) mixture is traditionally the most used in Southern Europe, and many studies have reported that these species are the most appropriate for a cover crop mixture in the Mediterranean regions (Pedraza et al., 2017). They provide an effect intermediate to grasses and legumes in terms of ground cover, soil protection and soil fertility. Moreover, this system can produce fodder with a good yield and quality for grazing livestock or feed production.

Figure 6 Sown crucifer cover crop along the inter-row spacing. Source: INTIA.

Several cruciferous species such as white mustard (*Sinapis alba subsp. mairei*), rocket (*Eruca vesicaria*), Ethiopian mustard (*Brassica carinata*) or radish (*Raphanus sativus*) are being introduced as alternatives to grass cover crops, mainly because of their tap root that makes them very promising for overcoming soil compaction (Saavedra et al., 2016b). Moreover, they have a high potential for controlling soil-borne diseases, weeds and nematodes due to their high content of allelopathic glucosinolates. In fact, leaving some residues on the soil surface reduce and slow down the emergence of the spring-summer-cycle weeds (Alcántara et al., 2011b), and when they are mown and chopped, they have been found to have a positive biofumigant effect, reducing *Verticillium dahliae* inoculum in the soil.

Additionally, there are a few cases where species of the *Asteraceae* family have been sown (such as *Anthemis arvensis*) as cover crops in olive groves with many rabbits, where other palatable species are quickly consumed (e.g. grasses) (Carpio et al., 2017).

Cultivated cover crops compete with the spontaneous flora and make weed control easier, helping to reduce herbicide use and tillage operations. However, such covers beneath the olive trees can become competitive and difficult to manage (Alcántara et al., 2011a). For that reason, it is important to keep this area weed-free to facilitate the operations of harvesting, pruning and weed management. Weeds are usually allowed to grow during the autumn and winter and then controlled in late winter or early spring by herbicides or mowing. The optimal living cover crop can reach a height up to 50–80 cm and occupy around 50 % of the surface in the middle of the orchard lanes (Saavedra et al., 2015b). The optimum mowing date depends on climatic factors, which precludes universal recommendations on this aspect. Nevertheless, an early sowing date (September) and, therefore, an early mowing date (February) seem advisable to reduce competition for water and facilitate its retention by the soil (Gómez and Soriano, 2020; Pedraza, 2018). Therefore, cover cropping is a valuable IWM tool to target weeds through competitive, ecological, biochemical and physical pressure during the fall-to-spring period (Gerhards and Schappert, 2020).

2.8 Conclusion and future trends

Given the economic and agronomic importance of the olive crop in the Mediterranean region, it could be expected that different soil and weed management practices are used by farmers in an attempt to optimise the management system. The successful establishment of these systems in olive orchards requires careful management to maintain productivity and maximise the water availability for the olive trees, the most limiting growth factor in this area.

Although tillage continues to be the most-used soil management system in inter-row spacing, soil conservation is crucial in the semiarid regions where soil cover is not frequent but necessary for erosion control. For that reason, tillage reduction, continuous soil cover by crop residues and cover crops are increasingly used in olive groves. Moreover, combining the use of non-chemical weed control methods with herbicides could lead to an improved olive grove biodiversity in the long run. The goal is not to eliminate all weeds but to keep them at a density that is economical and manageable, without negatively affecting olive production.

Considering the multiple ecological services and well-proven effects of each technique, farmers need to manage soil and weeds together with the olive crop and crop residues to effectively ensure that these management techniques provide the desired agro-environmental effects.

Future trends in research should be aimed at improving these IWM strategies, considering that different practices can be chosen according to the distinctive areas within the same farm (soil beneath the olive trees, intra-row and inter-row spacing). For that reason, it is not possible to establish a single valid integrated management system for all olive groves, not even for a specific farm, nor for all years. The selection of the techniques included in the IWM strategy will depend on several factors, such as the soil type, olive crop features, water availability, topography, the main goals desired by the farmers on their farm and the adaptability of the techniques to the local conditions.

3 Case study: vineyards in the UK

3.1 Introduction

Globally, 7.4 million ha is planted with vines but 50% of this area is contained in just five countries. Three of these countries are in Europe, and the area dedicated to vines in the continent is now 3.2 million ha. In 2016, European restrictions on new plantings were put in place to help to stabilise the area covered by vineyards (OIV, 2019).

From an area of just 1687 ha in 2015 (Eurostat, 2017), viticulture in the UK has more than doubled, to 3500 ha in 2020 (Skelton, 2020). The UK is classified as a cool-climate production area and the relatively low temperatures, limited light intensities and frequent and sometimes heavy rain in a typical UK growing season mean that all aspects of vine planting and training must be optimised (Jones, 2018).

When planting a vineyard, the chosen vine density is usually a compromise between vigour, crop load and production costs (Reynier, 2011). In UK vineyards, the spacing between rows varies from 2 m to 2.4 m and the in-row spacing is usually 1.2–1.4 m (Skelton, 2009). This translates into a vine density of 3427–4166 vines/ha. The height of the trunk is usually around 1 m to reduce the risk of frost damage and the vines are usually trained to a vertical shoot position with a Guyot pruning system.

The rapid expansion in UK viticulture has created many new challenges, one of which is the need to develop new weed management strategies for established and new vineyards that are tailored to the UK climate. In this study, Bunting's (1960) definition of weeds is used – any plant rendered undesirable by its occurrence in a cultivated area dedicated to another plant.

Weeds are known to compete with vines for nutrition and water (ENDURE, 2010) and excessive weediness can affect establishment and vigour. The resulting nutrient deficiencies can reduce yields and berry quality, especially if limited nitrogen availability leads to reductions in Yeast Available Nitrogen (YAN), which is vital for fermentation. Consequently, weed management strategies such as the use of herbicides and tillage are commonplace in UK vineyards, in an effort to minimise the effects of weeds on yield and berry quality.

Perennial weeds tend to dominate in perennial crops when there is little or no tillage. Perennial weeds are also more likely to be tolerant to herbicides, since both vegetative and sexual reproduction can occur (Navas, 1991).

3.2 Commonly used soil management systems in the UK

The NIAB EMR conducted interviews with eight growers to better understand which weed management systems are most often deployed in the UK. Their responses suggest that many vineyards use cover-cropped inter-rows for a variety of reasons, including a perceived positive effect on soil quality, the alleviation of soil compaction and the ease of access and use of machinery. Most interviewees favoured spontaneous cover crops over sown crops. For these and other reasons, it is rare for inter-rows to be left as bare ground in the UK. However, inter-rows can be left bare every other row to limit the impact of a vigorous cover crop on the vine.

Herbicides are the preferred method of weed control within the rows of vines, and glyphosate is commonly used to control perennial weeds such as thistles, nettle and grass. The frequency of applications in UK vineyards is unknown, but an average of two applications per year is used in UK tree fruit orchards. However, concerns over the environmental impact of herbicides have seen the use of mechanical weeding increase in UK vineyards, and the majority of growers now own a mechanical weeder for under-vine use. Interestingly, this increased reliance on mechanical weeders has not yet spread to UK orchards.

3.3 Materials and methods of the NIAB IWM experiment

The IWM experimental vineyard covers 0.15 ha and is located at NIAB EMR, New Road East Malling, West Malling, Kent ME19 6BJ (Fig. 7). The topsoil is a sandy loam with a variable depth of 45–75 cm, which lies above a 5-cm layer of clay with a low stoniness. The subsoil is a cracked ragstone. The soil is free-draining and dries relatively quickly, which increases competition for available water between vines and weeds.

The IWM vineyard was planted in 2018 with 96 vines of Chardonnay clone grafted onto 3309Couderc rootstocks. The row spacing is 2.4 m and in-row spacing is 1.1 m, giving a density of 3,600 vines/ha. The vines are trained to a vertical shoot position with a single Guyot pruning system of ten to twelve buds per vine. The vineyard is managed conventionally. Due to the logistics associated with the use of mechanical weeders, treatments were applied per row instead of per randomised plot. The experimental design is summarised in Fig. 8.

Weeding treatments began in April and ended in September/October. Four treatments were applied (Fig. 9 and Table 2):

- Non-treated control (NTC);
- Inter-vine blade cultivator (Blade);
- Finger hoe + finger disc (Disc); and
- Herbicide.

The herbicides used in this study were fluazifop-p-butyl, diquat dibromide or carfentrazone-ethyl, depending on the accreditation at the time. Three applications were made each year. A LiDAR (Light Detection And Ranging) (SICK) system was used at flowering and veraison in 2019, and at flowering and bunch closure in 2020, to visualise canopy parameters through remote sensing by a laser. The Dualex (Force A) was used at veraison in 2019, and at fruit set

Figure 7 Location of NIAB EMR in the UK, and of the site of the Integrated Weed Management (IWM) vineyard (shown in red).

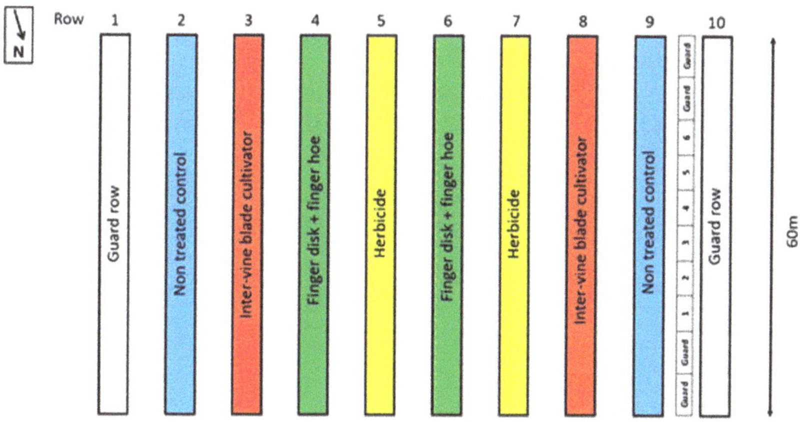

Figure 8 The experimental design for the IWM experiment at NIAB EMR, East Malling. White rectangles represent guard rows. The non-treated control treatment was used in rows in the blue rectangles, the inter-vine blade cultivator treatment in the red rectangles, the finger hoe + finger disc treatment in the green rows, and the herbicide treatment in the yellow rectangles. Each row was divided into ten plots, with the two plots at each end serving as guard plots. Plots one to six were used in this experiment, and each plot contained five vines. Source: NIAB EMR.

Figure 9 The four treatments applied at the IWM experimental vineyard. A) Non-treated control treatment, B) Herbicide treatment, C) Inter-vine blade cultivator treatment and D) Finger disc + finger hoe treatment. Source: NIAB EMR (2020), photos taken on 01/06/2020.

and veraison in 2020. This device interpolates an index of nitrogen balance (NBI) with a light sensor (Dualex, 2019).

Petiole analysis was conducted at veraison on 30 September 2020. The petiole sampled was taken from the leaf opposite to the lowest bunch, and if

Table 2 Summary of the different weeding treatments pictured and described according to the action taken, the mode of action and the timing of application

Weeding modalities	Picture	Action taken	Mode of action	Timing
Non-treated control		A one-meter strip of weed is cut down using a strimmer (STIHL)	Cuts weed stem	Weeds are 20-25 cm high
Herbicides		A one-meter strip is kept weed-free using herbicide applied with a knapsack (CP3)	Kills weeds	Weed coverage reaches 10-15%
Inter-vine blade cultivator		A one-meter strip is kept weed-free using an inter vine-blade cultivator (CLEMENS)	Cuts weed roots and lifts the weeds	Weed coverage reaches 10-15%
Finger Disc + finger hoe		A one-meter strip is kept weed-free using a combination of a finger disc (CLEMENS) and a finger hoe (CLEMENS)	The finger disc crumbles the soil around weeds and buries them, the finger hoe digs out the weeds closer to the plant.	Weed coverage reaches 10-15%

Source: NIAB EMR.

this leaf was damaged or missing, the petiole from the leaf above was taken. A pooled sample of 100 g of petiole tissue was analysed for each treatment. Yield was averaged from the sum of the five vines of each plot, and samples for quality analysis were collected randomly from the three middle vines of each plot. Berry quality attributes were quantified using an Oenofoss (FOSS) on pooled samples of hundred berries gathered at harvest from the different treatments. A total of 48 samples were analysed, with twelve samples for each treatment.

Statistical analyses were carried out using Rstudio software. To determine whether the differences between treatments were statistically significant, analysis of variance (ANOVA) tests were carried out and the least significant difference (lsd) values for $P < .05$ were calculated.

3.4 Influence of weed management on canopy development

No treatment differences in canopy height were found at flowering in 2019, but row volumes were significantly higher from the blade treatment compared to the non-treated control. At veraison in 2019, the canopy height was significantly greater in all treatments when compared to non-treated control values by an average of 20%. The canopy row volumes were also greater in the three weed management treatments, and values were increased by 67% with the use of blades and discs, and by 83% by using herbicides (see Table 3).

In 2020, treatment differences in canopy height were apparent at flowering. Compared to non-treated control values, canopy height was significantly greater under the herbicide treatment and significantly lower under the disc treatment. All treatments significantly increased canopy volume compared to the non-treated control. At bunch closure, canopy height was significantly increased under all treatments, being on average 14% higher than non-treated control values, and row volumes were on average 48% greater when a weeding management strategy was used (Table 3).

Vigour was reduced later in the season in the non-treated control vines, presumably due to increased competition from weeds for water and nutrients. This response was observed in 2019 and 2020. The improved rooting depth could have contributed to the higher row volume values in 2020. The beneficial effects of the blade treatment on canopy height in both years could have stemmed from the effect of mulching, which is an inherent result of this weed management method. The NBI calculated from the Dualex measurements indicated a significantly higher level of nutrition under the herbicide, blade and disc treatments in 2019, when compared to non-treated controls. However, the NBI was significantly higher under the herbicide treatment compared to the non-treated control (see Table 4).

Table 3 Evolution of canopy development depending on weed management. 2019 and 2020 were assessed respectively at flowering and veraison and flowering and bunch closure. Two canopy parameters were taken into account, canopy height and row volume. They are expressed in mean value and percentage when compared to the non-treated control when the difference was significant. a-c: The letters describe the groups' significance compared to one another. When groups did not share a letter, the difference was significant ($P < .05$)

		2019						2020					
		Flowering			Veraison			Flowering			Bunch closure		
		Mean value		NTC (%)	Mean value		NTC (%)	Mean value		NTC (%)	Mean value		NTC (%)
Canopy height (*m*)	Herbicide	1.58	a		1.98	a	+19	1.48	a	+4	1.56	a	+14
	Disc	1.45	a		1.98	a	+19	1.36	c	−4	1.53	a	+12
	Blade	1.55	a		2.01	a	+21	1.38	bc		1.59	a	+16
	NTC	1.53	a		1.66	b		1.42	b		1.37	b	
Row Volume (m^3)	Herbicide	0.07	b		0.11	a	+83	0.23	a	+28	0.27	a	+50
	Disc	0.07	b		0.10	ab	+67	0.22	a	+22	0.27	a	+50
	Blade	0.08	a	+14	0.10	b	+67	0.23	a	+28	0.26	a	+44
	NTC	0.07	b		0.06	c		0.18	b		0.18	b	

Source: NIAB EMR (2020).

Table 4 Changes in foliar nitrogen concentration under the weed management treatments. This parameter was assessed at veraison in 2019 and at fruit set and veraison in 2020. Data are mean values, and the percentage difference from non-treated control values is given when the difference was significant. The letters a, b and c are used to denote significant differences between groups – groups not sharing a letter, the difference was significant ($P < .05$)

		2019			2020					
		Veraison			Fruit set			Veraison		
		Mean value		NTC (%)	Mean value		NTC (%)	Mean value		NTC (%)
NBI	Herbicide	10.42	a	+49	9.45	a		13.35	a	+55
	Disc	8.85	b	+27	9.66	a		13.06	a	+51
	Blade	9.25	b	+33	9.47	a		13.73	a	+59
	NTC	6.97	c		7.99	a		8.63	b	

Source: NIAB EMR (2020).

The NBI calculated from the 2020 data indicated that there was no difference in nutrition at fruit set, but at veraison, the uptake of nitrogen was significantly higher under the three weed management treatments compared to non-treated control values. The NBI values were greater under all treatments in 2020 than in 2019 (Table 4).

The differences in NBI suggest that the vines' root systems were more established in 2020 than in 2019, which, in turn, may have led to smaller treatment differences at veraison. The NBI values correlate with the canopy data and indicate that vine vigour in the non-treated controls was limited by increased competition for nitrogen and water. The NBI was also greater under the blade treatment compared to the discs in 2019, and was higher than that for the herbicide treatment in 2020. These results support the notion that the blades facilitated deeper rooting and enhanced vigour via a mulching effect.

Results from the petiole analyses showed that at veraison, concentrations of sodium, zinc, iron and boron were higher in the non-treated controls, but potassium concentrations were lower compared to the weed management treatments. In samples taken from the herbicide treatment, potassium concentrations were the highest, but those of calcium, manganese and iron were lower, and magnesium and phosphorus were deficient. In samples from the blade treatment, the concentrations of nitrogen, calcium, copper and iron were higher than in other treatments, but phosphorus was deficient. the concentrations of nitrogen and potassium were higher under the disc treatment, but the concentrations of sulphur, copper and iron were lower than in other treatments.

It should be noted that the results from the petiole analyses should be interpreted with caution, since samples were pooled within treatments and were collected on a single date with no replication.

Soil nutrients are made available to the plant through dilution in water (IFV Occitanie, 2020). Their leaching is prevented by the cation exchange in the clay-humic complex, which is five to ten times more important in humus than clay. Nutrient uptake is strongly influenced by the depth of rooting and humus content controlled by the addition of mulches.

In the herbicide treatment, the measured calcium concentrations in the petioles were lower. Although calcium is always present in the soil as calcium salts, this low value suggests a likely lack of humus due to the absence of mulching in the herbicide treatment. This might also help to explain the lower uptake of phosphate, manganese and iron. The herbicide treatment is unlikely to have encouraged deep, penetrative vine root growth (Gaviglio, 2020).

The mulching effect of the blade treatment presumably facilitated the higher uptake of nitrogen, calcium, copper and iron. It is also likely that the roots would have grown deeper, leading to a greater exploration of larger and deeper soil volumes (Gaviglio, 2020). The impaired uptake of phosphate

Table 5 Summary of the influence of weed managements on vigour and its expected effects on rooting depth, mulching effect and competition

	Vigour	Rooting depth	Mulching effect	Competition
Herbicide	Better	shallow	none	low
Disc	Good	moderate	high	moderate
Blade	Best	deep	high	moderate
NTC	Poor	deep	low	high

Source: NIAB EMR (2020).

was perhaps due to a mulching-induced effect on the lowering of rhizosphere pH. Nutrient availability under the blade treatment may have been adversely affected by impaired water absorption through the surface crust.

The mulching effect, inherent in the blade management approach, should be expected to increase the availability of nitrogen and potassium. However, it is possible that the vine roots would only be mildly disturbed by the blade, leading to a shallower root system that is less able to exploit soil reserves of sulphur, copper and iron in deeper soil horizons.

Mulching in the non-treated control would be expected to be minimal, but the lack of root disturbance would encourage good uptake of sodium, zinc, iron and boron. Although the presence of weed roots in shallow soil horizons in the non-treated control plots may encourage a deeper vine root system, as suggested by l'Institut Français de la Vigne et du Vin (2014), potassium uptake would be expected to be lower due to the absence of mulching.

An optimal weed management would bestow high vigour, deep rooting depth, high mulching effect and low weed competition upon the vine. Our results suggest that the blade weeding management is the most promising weeding system of those tested and higher yields and improved berry quality could be expected as a result. The non-treated control performed poorly in most measured parameters and it is likely that lower yields and quality would result if weeds in vineyards were left unmanaged (Table 5).

3.5 Influence of weed management on yield and berry quality

Following three years of vine establishment, 2020 was the first year of berry production. Our results show that the impacts of the three weed management strategies on yields were significant (Fig. 10).

Yields from the non-treated controls were 72% lower than in the three treatments. Higher yields were obtained under the blade treatment, and lower yields from the herbicide treatment, although these differences were not statistically significant.

All measured quality parameters were lowest in the non-treated control samples, although not all treatment differences were statistically significant

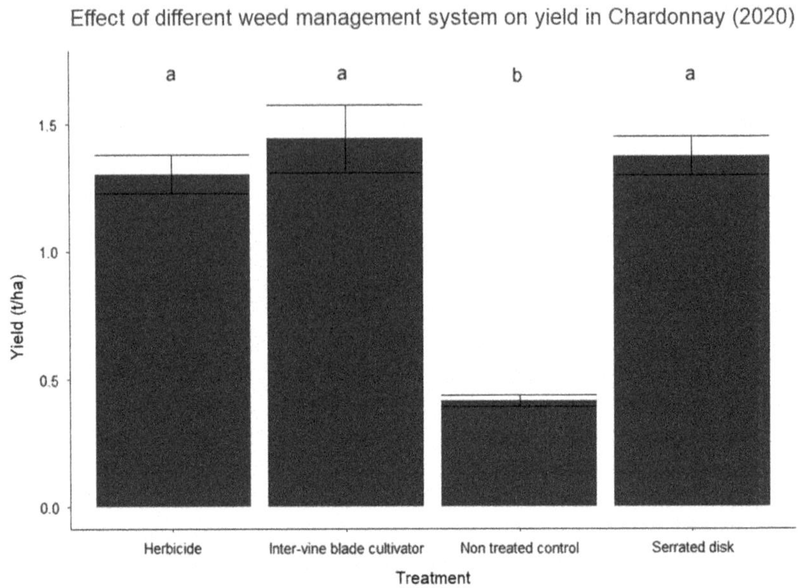

Figure 10 The effect of the four weed management strategies on yields. Yields were measured at harvest. (a-b) The letters describe the groups' significance compared to one another. When groups do not share a letter, the difference was significant ($P < .05$). Source: NIAB EMR (2020).

(Table 6). The concentrations of sugar (TSS) and organic acids were similar between treatments but the concentrations of tartaric acid and malic acid were significantly higher in the herbicide and disc treatments. Yeast Assimilable Nitrogen (YAN) was also significantly improved by the disc treatment (Table 6).

Berry maturity or ripeness is often indicated by higher sugar and tartaric acid and lower malic acid concentrations (Deloire, 2007). In the non-treated controls, the lowered quality parameters were presumably due to the lack of vine vigour (Table 6). The herbicide treatment presented no significant differences in sugar or total acid concentrations when compared to the blade and disc treatments, but the lower tartaric acid concentration indicated that maturity was delayed. The blade treatment, although it had the best NBI at veraison, did not show values of significant difference to the non-treated control with regard to the YAN value. This is perhaps due to its higher yield, implying a higher dilution of the YAN in the blade treatment.

3.6 Summary and future trends

Our results suggest that weed management strategies would benefit the UK viticulture industry, despite the widespread belief amongst growers that rainfall in the UK is sufficient to sustain both weeds and vines. The use of blade and disc

Table 6 The effects of weed management treatments on Must quality. Quality attributes were assessed at harvest in 2020 and included Total Soluble Solids (TSS), total acid, tartaric acid, malic acid and YAN (Yeast Available Nitrogen). They are expressed in mean value and percentage when compared to the non-treated control when the difference was significant. a-b: The letters describe the groups' significance compared to one another. When groups did not share a letter, the difference was significant ($P < .05$)

	TSS °*Brix*		Total acid *g/l*		Tartaric acid *g/l*		Malic acid *g/l*		YAN	
	Mean value	NTC (%)	Mean value	NTC (%)	Mean value	NTC (%)	Mean value	NTC (%)	Mean value	NTC (%)
Herbicide	20.0 A	6%	5.6 a	12%	8.4 b		6.6 a	22%	91.55 b	
Disc	19.9 a	5%	5.6 a	12%	9.0 ab		6.8 a	26%	125.81 a	53%
Blade	20.2 a	7%	5.6 a	12%	9.7 a	17%	6.2 ab		98.90 ab	
NTC	18.9 b		5.0 b		8.3 b		5.4 b		82.12 b	

Source: NIAB EMR (2020).

treatments in our experimental vineyard were good alternatives to herbicides once the vines had established and were given the time for their root structures to adapt to the mechanical weeding. Root system development and rooting depth are often enhanced in plots where any mechanical weeder is used, and while this has been proven in France (Gaviglio, 2020), further work is needed to understand root responses to mechanical weeders in UK vineyards.

Although viticulture in the UK is still a relatively new venture, WineGB has recently launched a sustainability scheme to encourage the sector to serve as an example (Foss, 2020). This scheme is based on a minimal use of pesticides and fertilisers, on protecting biodiversity in vineyards and on reducing the carbon footprint through better water management and use of renewable energies. Collectively, these initiatives will create a greater demand for efficient vineyard weed management systems.

4 Where to look for further information

The following guides and organisations provide a good overview of soil and weed management for olive orchards in Spain:

- Saavedra, M., Hidalgo Moya, J. J., et al. (2015), *Guía de cubiertas vegetales en olivar*, Sevilla (España): Instituto de Investigación y Formación Agraria y Pesquera, Junta de Andalucía, Consejería de Agricultura, Pesca y Desarrollo Rural (Agricultura. Formación).
- IFAPA (2011), *Producción Integrada de Olivar*, Sevilla (España): Consejería de Agricultura y Pesca, Junta de Andalucía, Consejería de Agricultura y Pesca, Instituto de Investigación y Formación Agraria y Pesquera (Agricultura. Formación).
- Sociedad Española de Malherbología (https://semh.net/).
- SERVIFAPA Olivar. Instituto de Investigación y Formación Agraria y Pesquera. Consejería de Agricultura, Ganadería, Pesca y Desarrollo Sostenible (https://www.juntadeandalucia.es/agriculturaypesca/ifapa/servifapa/ buscador?fulltext=olivar&f[0]=ambito%3AOlivar).

Additional vineyard-oriented research can be provided by:

- The Agroscope in Changin provides excellent grower-oriented research on tillage, whether on vineyards in the UK or horticulture in general (https://www.revuevitiarbohorti.ch/archives/).
- The Institut Français de la Vigne et du Vin also has a section dedicated to reducing chemical input in vineyards (https://www.vignevin.com/reductions-intrants/ and https://www.vignevin.com/environnement/).
- Vine and wine open access journal: Oeno-one website (https://oeno-one.eu/).

- Additional research on integrated weed management strategies can be found on the IWMPRAISE website: Integrated Weed Management: PRActical Implementation and Solutions for Europe (https://iwmpraise.eu/). Key research in this area can be found at the following website: vineyards in UK (http://iwm-uk.co.uk/) and olive orchards in Spain (https://www.uco.es/agroecologia/iwmpraise/).

5 Acknowledgements

This research was made possible with funding provided by the European Union's Horizon 2020 research and innovation programme under grant agreement Nº 727321 through the IWMPRAISE Project *'Integrated Weed Management: PRActical Implementation and Solutions for Europe'*.

The Spanish authors would like to thank the Navarre Institute of Transfer and Innovation in Agri-food sector (INTIA) for providing graphical material, and Fatemeh Mkhani for helping in the search for information.

The British authors would also like to acknowledge Clemens GmbH & Co. KG, and especially Johannes Krütten for providing the mechanical weeders used in this project, and Zoe Clarke for her help with field work.

6 References

Abu-Irmaileh, B. and Abu-Rayyan, A. (2004). In-row preplant manure composting reduces weed populations, *HortSci.* 39(6), 1456–1460.

AEMO (2012). *Aproximación a los costes del cultivo del olivo*, Asociación Española de Municipios del Olivo (AEMO), Córdoba (Spain), 54 pp.

Alcántara, C., Carbonell, R., Ordóñez, R., Vega, V., Hidalgo, J., Hidalgo, J. C. and Saavedra, M. (2009). Control de malas hierbas con cobertura de restos de poda de olivo, *XII Congresso da SEMh/XIX Congresso da ALAM/II Congresso da IBCM*, 367–370.

Alcántara, C., Pujadas, A. and Saavedra, M. (2011a). Management of cruciferous cover crops by mowing for soil and water conservation in southern Spain, *Agric. Water Manag.* 98(6), 1071–1080. doi: 10.1016/j.agwat.2011.01.016.

Alcántara, C., Pujadas, A. and Saavedra, M. (2011b). Management of *Sinapis alba* subsp. mairei winter cover crop residues for summer weed control in southern Spain, *Crop Prot.* 30(9), 1239–1244. doi: 10.1016/j.cropro.2011.04.007.

Bunting, A. H. (1960). Some reflections on the ecology of weeds. *Biology of Weeds, Symp. Brit. ecol. Soc.* 1, 11–26.

CAP (2006). Manual de buenas prácticas agrarias en los diferentes sistemas productivos del olivar andaluz, Programa de mejora de la calidad de la producción de aceite de oliva y aceitunas de mesa, Consejería de Agricultura y Pesca, Junta de Andalucía, Sevilla (España), 60 pp.

Carpio, A. J., Soriano, M.-A., Guerrero-Casado, J., Prada, L. M., Tortosa, F. S., Lora, Á. and Gómez, J. A. (2017). Evaluation of an unpalatable species (*Anthemis arvensis* L.) as an alternative cover crop in olive groves under high grazing pressure by rabbits (*Anthemis arvensis* L.), *Agric. Ecosyst. Environ.* 246, 48–54. doi: 10.1016/j.agee.2017.05.028.

Castro, J. (1993). Control de la erosión en cultivos leñosos con cubiertas vegetales vivas, Tesis Doctoral, Departamento de Agronomía, Universidad de Córdoba.

Cosentino, S. L., Copani, V., Scalici, G., Scordia, D. and Testa, G. (2015). Soil erosion mitigation by perennial species under mediterranean environment, *Bioenergy Res.* 8(4), 1538–1547. doi: 10.1007/s12155-015-9690-2.

Deloire, A. (2007). Fiche N°7 : Métabolismes Primaires rimairAcides Organiques–Fiche Document, Montpellier SupAgro. Available at: http://obs.viti.herault.fr/viti/fiche_doc.jsp?idaction=null&id_doc=1047&idtheme=011101; accessed December 20th, 2020.

Dualex (2019). Dualex Force A. Force A. Available at: https://www.force-a.com/fr/produits/dualex; accessed December 20th, 2020.

ENDURE (2010). Grapevine Case Study Guide Number 1. Available at: http://www.endure-network.eu/about_endure/all_the_news/five_leaflets_explore_grapevine_alternatives; accessed January 15th, 2021.

Eurostat (2017). Vineyards in the EU–Statistics. Available at: https://ec.europa.eu/eurostat/statistics-explained/index.php/Vineyards_in_the_EU_-_statistics#Quality_wines_make_up_83_.25_of_the_area_under_vines; accessed December 19th, 2020.

FAO (2018). FAOSTAT database, Food and Agriculture Organization of the United Nations (FAO). Available at: http://www.fao.org/faostat/en/#data/QC; accessed November 6th, 2020.

FEGA (2014). *Manual para el cumplimiento de la condicionalidad olivar, Fondo Español de Garantía Agraria*, Ministerio de Agricultura, Alimentación y Medio Ambiente, 38 pp.

Foss, C. (2020). WineGB now has its own sustainability scheme, *WineGB* 4, 22–25.

Fracchiolla, M., Terzi, M., Frabboni, L., Caramia, D., Lasorella, C., De Giorgio, D., Montemurro, P. and Cazzato, E. (2016). Influence of different soil management practices on ground-flora vegetation in an almond orchard, *Renew. Agric. Food Syst.* 31(4), 300–308. doi: 10.1017/S1742170515000241.

Gaviglio, C. (2020). Désherbage mécanique des sols et maîtrise technique | Institut Français de la Vigne et du Vin, Institut Français de la Vigne et du Vin. Available at: https://www.vignevin.com/article/desherbage-mecanique-des-sols-et-maitrise-technique/; accessed January 30th, 2020.

Gerhards, R. and Schappert, A. (2020). Advancing cover cropping in temperate integrated weed management, *Pest Manag. Sci.* 76(1), 42–46. doi: 10.1002/ps.5639.

Gomez, J. A., Sobrinho, T. A., Giraldez, J. V. and Fereres, E. (2009). Soil management effects on runoff, erosion and soil properties in an olive grove of Southern Spain, *Soil Till. Res.* 102(1), 5–13. doi: 10.1016/j.still.2008.05.005.

Gomez, J. A. and Soriano, M.-A. (2020). Evaluation of the suitability of three autochthonous herbaceous species as cover crops under Mediterranean conditions through the calibration and validation of a temperature-based phenology model, *Agric. Ecosyst. Environ.* 291. doi: 10.1016/j.agee.2019.106788, 106788.

Hartwig, N. L. and Ammon, H. U. (2002). Cover crops and living mulches, *Weed Sci.* 50(6), 688–699.

Holland, J. M. (2004). The environmental consequences of adopting conservation tillage in Europe: reviewing the evidence, *Agric. Ecosyst. Environ.* 103(1), 1–25. doi: 10.1016/j.agee.2003.12.018.

Huqi, B., Dhima, K., Vasilakoglou, I., Keco, R. and Salaku, F. (2009). Weed flora and weed management in established olive groves in Albania, *Weed Biol. Manag.* 9(4), 276-285. doi: 10.1111/j.1445-6664.2009.00351.x.

IFAPA (2011a). *Erosión y cubiertas vegetales en olivar adehesado*, Sevilla (España): Instituto de Investigación y Formación Agraria y Pesquera, Junta de Andalucía, Consejería de Agricultura, Pesca y Desarrollo Rural (Agricultura. Formación).

IFAPA (2011b). *Producción Integrada de Olivar*, Sevilla (España): Consejería de Agricultura y Pesca, Junta de Andalucía, Consejería de Agricultura y Pesca, Instituto de Investigación y Formación Agraria y Pesquera (Agricultura. Formación).

IFV Occitanie (2020). La matière organique, IFV Occitanie. Available at: https://www.vignevin-occitanie.com/fiches-pratiques/la-matiere-organique/; accessed November 26th, 2020.

Institut Français de la Vigne et du Vin (2014). L'enherbement sous le rang : une nouvelle alternative au désherbage chimique de la ligne des ceps ?, *lesentretiensvignevinL-R* 2. 20–23.

IOC (2007). Production techniques in olive growing, International Olive Council, Madrid (Spain). Available at: http://www.internationaloliveoil.org/; accessed September 20th, 2018.

IWMPRAISE (2018). IWM strategies for on-farm experiments in perennial woody crops, IAS-CSIC (Spain), INTIA (Spain) and NIAB(UK): IWMPRAISE Project. Available at: https://ec.europa.eu/research/participants/documents/downloadPublic?documentIds=080166e5b9d08304&appId=PPGMS; accessed November 10th, 2018.

Jones, G. V. (2018). The climate component of Terroir, *Elements* 14(3), 167–172. doi: 10.2138/gselements.14.3.167.

Kassam, A., Friedrich, T., Derosch, R., Lahmar, R., Mrabet, R., Basch, G., González-Sánchez, E. J. and Serrai, R. (2012). Conservation agriculture in the dry Mediterranean climate, *Field Crops Res.* 132, 7–17. doi: 10.1016/j.fcr.2012.02.023.

Kudsk, P., Sønderskov, M., Bonin, L., Gonzalez-Andujar, J. L., Jensen, J. E., Melander, B., Moonen, C., Riemens, M. M., Sattin, M., Schaffner, U. and Storkey, J. (2020). IWMPRAISE - an EU horizon 2020 project providing integrated weed management solutions to European farmers, *Outlooks Pest Manag.* 31(4), 152–159. doi: 10.1564/v31_aug_02.

MAPA (2019a). Análisis de las técnicas de mantenimiento de suelo y de los métodos de siembra en España, Informe sectorial, Encuesta sobre Superficies y Rendimientos Cultivos (ESYRCE), Ministerio de Agricultura, Pesca y Alimentación. Available at: https://www.mapa.gob.es/es/estadistica/temas/estadisticas-agrarias/agricultura/esyrce/informes-sectoriales/; accessed October 15th, 2020.

MAPA (2019b). Encuesta sobre superficies y rendimientos cultivos (ESYRCE), encuesta de Marco de áreas de España, ministerio de agricultura, Pesca y alimentación. Available at: https://www.mapa.gob.es/es/estadistica/temas/estadisticas-agrarias/agricultura/esyrce/; accessed October 21st, 2020.

MAPAMA (2017). Registro de productos fitosanitarios, ministerio de agricultura y Pesca, alimentación y medio ambiente. Available at: http://www.mapama.gob.es/es/agricultura/temas/sanidad-vegetal/productos-fitosanitarios/fitos.asp; accessed July 17th, 2017.

Márquez-García, J., Repullo-Ruiberriz de Torres, M. A., Carbonell-Bojollo, R., Moreno-García, M. and Ordóñez-Fernández, R. (2014). La cubierta de restos de poda como fuente de nitrógeno, carbono y materia orgánica en el olivar, SERVIFAPA, Instituto de Investigación y Formación Agraria y Pesquera. Consejería de Agricultura, Pesca y Desarrollo Rural, 20 pp.

Martinelli, R., Monquero, P. A., Fontanetti, A., Conceição, P. M. and Azevedo, F. A. (2017). Ecological mowing: an option for sustainable weed management in young citrus orchards, *Weed Technol.* 31(2), 260–268. doi: 10.1017/wet.2017.3.

Navas, M.-L. (1991). Using plant population biology in weed research: A strategy to improve weed management, *Weed Res.* 31(4), 171–179. doi: 10.1111/j.1365-3180.1991.tb01756.x.

Oerke, E.-C. (2006). Crop losses to pests, *J. Agric. Sci.* 144(1), 31–43. doi: 10.1017/S0021859605005708.

OIV (2019). State of the World Vitivinicultural Sector in 2019, OIV, International Organisation of Vine and Wine, 15 pp.

Pastor, M., Castro, J., Humanes, M. D. and Muñoz, J. (2001). Sistemas de manejo del suelo en olivar de Andalucía. In: *Edafología*, 75–98. Available at: http://www.edafologia.net/revista/tomo8/art8t8t.htm; accessed 4 November 2020.

Pedraza, V. (2018). New strategies for sustainable management of the traditional durum wheat-sunflower rotation: Introducing cover crops and alternative crops, Universidad de Córdob. Available at: http://hdl.handle.net/10396/16735; accessed 12/7/2021.

Pedraza, V., González-Andújar, J. L., Lezaún, J. A. and Garnica, I. (2019). Proyecto IWMPRAISE: manejo integrado de malas hierbas, implementación práctica y soluciones para Europa, *XVII Congr. Soc. Española Malherbología, Vigo (España)*, 284-289.

Pedraza, V., Perea, F., Saavedra, M., Fuentes, M. and Alcántara, C. (2017). *Vicia narbonensis-Avena strigosa* mixture, a viable alternative in rainfed cropping systems under Mediterranean conditions, *Span. J. Agric. Res.* 15(4), 0905. doi: 10.5424/sjar/2017154-10882.

Reynier, A. (2011). *Manuel de viticulture: Guide technique du viticulteur*. Lavoisier.

Roberts (2017). The importance of tree crops in sustainable agriculture, New South Wales (Australia): Permaculture Research Institute. Available at: https://www.permaculturenews.org/2017/11/27/importance-tree-crops-sustainable-agriculture/; accessed November 12th, 2020.

Saavedra, M. (2015). Erosión del suelo y uso de cubiertas vegetales, SERVIFAPA, Instituto de Investigación y Formación Agraria y Pesquera, Consejería de Agricultura, Pesca y Desarrollo Rural, 55 pp.

Saavedra, M., Castillo-Llanque, F., Melgares, J. D., Hidalgo Moya, J. C. and Alcántara, C. (2015a). Características de *Sinapis alba* subsp. mairei como Cubierta Vegetal y para Biofumigación, SERVIFAPA, Instituto de Investigación y Formación Agraria y Pesquera, Consejería de Agricultura, Pesca y Desarrollo Rural, 27 pp.

Saavedra, M., Hidalgo Moya, J. J., Pérez, D. and Hidalgo Moya, J. C. (2015b). *Guía de cubiertas vegetales en olivar,* Sevilla (España): Instituto de Investigación y Formación Agraria y Pesquera, Junta de Andalucía, Consejería de Agricultura, Pesca y Desarrollo Rural (Agricultura. Formación).

Saavedra, M. and Pastor, M. (2002). *Sistemas de cultivo en olivar. Manejo de malas hierbas y herbicidas*, Editorial Agrícola Española, Madrid.

Saavedra, M., Casanova, C., Sánchez-Jiménez, F. J. and Alcántara, C. (2016a). Características de la gramínea *Brachypodium distachyon* como cubierta vegetal,

SERVIFAPA, Instituto de Investigación y Formación Agraria y Pesquera, Consejería de Agricultura, Pesca y Desarrollo Rural, 21 pp.

Saavedra, M., Pedraza, V. and Alcántara, C. (2016b). Implantación y manejo de Sinapis alba subsp. mairei para cubierta vegetal y biofumigación, SERVIFAPA, Instituto de Investigación y Formación Agraria y Pesquera, Consejería de Agricultura, Pesca y Desarrollo Rural, 28 pp.

Skelton, S. (2009). *Viticulture*. Lulu.com.

Skelton, S. (2020). Vineyards in the UK, the Republic of Ireland and the Channel Islands. Available at: http://www.englishwine.com/vineyards.htm; accessed November 14th, 2020.

Storkey, J., Meyer, S., Still, K. S. and Leuschner, C. (2012). The impact of agricultural intensification and land-use change on the European arable flora, *Proc. Biol. Sci.* 279(1732), 1421-1429. doi: 10.1098/rspb.2011.1686.

Chapter 15

Evaluating the economics of integrated weed management

Pieter de Wolf, Saskia Houben, William Bijker and Koen Klompe, Wageningen Plant Research, The Netherlands

1 Introduction

Integrated weed management (IWM) is a strategy in which multiple tactics are combined for optimal weed control. The tactics include not only preventive options like crop rotation and row distance but also control options like herbicides as well as non-synthetic chemical use and mechanical weeding (Fig. 1). Since it depends on multiple tactics and includes many non-chemical measures, the implementation of IWM is considered to contribute to the reduction of pesticide use.

The EU Horizon2020 IWMPRAISE Project promotes the implementation of IWM by developing, testing and assessing IWM strategies that meet the societal demand for farming practices that minimize environmental risks while maintaining productivity. A core activity in this project is a pre-implementation evaluation of the economics and environmental impact of IWM strategies to see which strategies are most promising in terms of their cost-effectiveness, applicability and environmental impact. The most promising options can then be communicated to farmers and advisors. Understanding the differences in

http://dx.doi.org/10.19103/AS.2021.0098.17

Figure 1 In IWM, multiple tactics to control weeds are integrated (Riemens et al., 2022).

cost-effectiveness, applicability and environmental impact of various strategies is also valuable for policy makers, when working on policies for implementing more sustainable weed management practices in agriculture.

In this chapter, we focus on approaches to the economic evaluation of IWM. Whilst IWM does not necessarily exclude chemical weed management control, it is important to develop effective strategies that depend less on herbicides. The efficacy of chemical weed control strategies in European farming systems is threatened by the development of herbicide resistance and the possible withdrawal of specific active ingredients like glyphosate. This chapter focuses on the economic evaluation of IWM strategies that are less dependent on herbicide use or which seek to replace reliance on herbicides entirely.

Although the development and adoption of novel strategies such as IWM are often first promoted by changes in government policies and regulations, it is important to be aware from the start of their potential economic and practical consequences for practitioners. IWM strategies can have a major economic impact on crop operations, for example, by changing row distances or even the existing crop rotation. It is therefore important to assess the cost-effectiveness of IWM programmes, including taking account of local conditions such as market structure, input costs and on-farm limitations and opportunities such as existing crop rotations. These issues raise questions like, 'What are the costs and benefits of a novel strategy? Does the new strategy affect agronomic and financial risks? How does the IWM strategy and the results of a specific experiment translate to particular conditions on a specific farm?' A farmer needs to know the practical implications and economic consequences of different strategies to decide which can be sustainably implemented, whilst researchers need to be able to decide what strategies have the best potential and therefore which strategies should be developed and promoted.

There are some particular challenges when trying to assess a novel strategy. Data availability is one of the biggest challenges. Data from farm accountancy data networks, for example, are not very useful because they relate to conventional economic performance indicators. The use of data from experiments is also problematic because the economics of an experimental set-up are completely different from the economics of a working farm. For example, experiments use much smaller equipment and much more manual labour for monitoring and sampling than a typical farm. Moreover, in many cases, the design of experiments does not entirely match farming practices. Crop rotation, for example, is a primary weed control measure, but experiments often focus on single crops to reduce the complexity and duration of the experiment. In addition, if an economic evaluation can make a contribution to better (cost-effective) solutions, it has to be done at a very early stage of the research, when no experimental data are available anyway.

Any analysis should get as close to practical conditions as possible while taking into account the specific conditions in different regions. The most promising strategy can then be selected for further examination and development before being implemented in practice. This approach was used in IWMPRAISE, starting with a pre-examination of IWM options to decide which were most promising to address the local challenges that farmers faced regarding weed management.

A starting point is to compare the IWMPRAISE approach with other economic evaluations in the same field of IWM. How influential are the choices of data used, calculation methods and what are the pros and cons of different evaluation approaches? Which requirements should be met when in finding solutions that best fit practice? To find out, the chapter starts with a literature review of ways of evaluating the economics of IWM.

2 Approaches to economic evaluation

A literature review of economic evaluation methods shows that most analyses have been done using cost-benefit analyses, some combined with calculation of net present value (NPV). The economic performance of a new farm strategy is then compared to the performance of alternative strategies, often the existing strategy currently applied in practice. Some evaluations were based on multiple years and most analyses were carried out at field level by calculating input costs and output value. Social values were usually defined at the regional or national level and related to effects on health and biodiversity and the degree of contamination of groundwater and surface water. Most of the studies were done after an experiment using the data from that experiment.

In IWMPRAISE, the economic evaluation has been done at the farming system level at the scale of a normal farm. But what is a normal farm? From an agronomic perspective, it should at least resemble a cropping system that is typical for a specific region. From the economic point of view, it makes sense to define a farming system able to generate a reasonable income over the short term. This typical scale is typically higher than the statistical average of a regional farm population, in which small-size part-time farms are often numerically overrepresented.

When the scale and cropping system of this 'standardised farm' are defined, more detailed data about all inputs and outputs are required. For this farm, current weed management practices are defined in detail to establish a point of reference, while one or more IWM strategies are defined for comparative evaluation. This requires several sources: databases with farm data (e.g. for average yields), interviews with (local) experts like researchers, advisors and farmers, and data relating to inputs and yields from field experiments. Experts are crucial to estimate how a farmer could apply a novel strategy in practice in a particular system.

All information is combined to quantify the reference weed management system and the IWM variant(s). Several economic indicators are calculated, including not only the profitability of the farm but also more detailed indicators such as labour use and costs of machinery, assuming that IWM strategies will affect these indicators. From the perspective of the farmer, it is important not just to know the absolute impact (e.g. IWM will cost 10 h extra labour per ha) but also the relative impact: for a large-scale extensive arable farm, for example, total labour use could be less than 20 h per ha, so 10 h extra is a substantial increase. As pointed out by Pardo et al. (2010), it is also important to have a look at the labour distribution during the year. To explain why and how we developed our evaluation approach, we will describe the economic evaluation undertaken in IWMPRAISE of an IWM programme for winter wheat in Denmark and compare this with two other economic case studies on IWM (Pardo et al., 2010; Vasileiadis et al., 2015).

3 The case study in IWMPRAISE

Crop protection and weed management in particular are not limited to one crop but include crop rotation. An IWM programme affects both direct costs (e.g. herbicide input) and indirect costs (e.g. equipment). For both existing and novel weed management systems, data are collected as much as possible on the farm to stay as close as possible to practice. Information was also provided by local experts in a 2-day visit to the relevant farm to define the context, followed by a discussion to define realistic IWM strategies for the future.

The economic performance of IWM for winter wheat production was evaluated for a standard farm in the Sealand region in Denmark. Grass seed production is the main crop in this area in economic terms. Besides grass seed, farmers mainly grow oilseed rape, cereals and sugar beets. Looking to the future, sugar beet and winter wheat production is expected to become less competitive compared to other European production regions, in particular Eastern Europe. This could lead to a decrease and even complete disappearance of sugar beet production in Denmark, while winter wheat could partly be replaced by spring barley. For this reason, sugar beet was not included in the crop rotation of the representative farm.

In this region, not all land is suitable for grass seed production and the acreage is limited as it is always grown on contract. As a result, many farms have two crop rotations (Tables 1 and 2):

- a rotation with grass seed, oilseed rape and cereals; and
- another rotation with oilseed rape and cereals only.

The farm size required to generate a sufficient farmer income over the short term is 200 ha with rotations of 100 ha each. The grass seed crop red fescue (*Festuca rubra*) was chosen as it is a common grass seed crop in the region.

Table 1 Rotation 1 with a grass seed crop on an average farm in Sealand

Year	Crop	Total ha
1	Oilseed rape	16.7
2	Winter wheat	16.7
3	Winter wheat	16.7
4	Spring barley	16.7
5	Grass seed Red Fescue	16.7
6	Grass seed Red Fescue	16.7

Table 2 Rotation 2 without a grass seed crop on an average farm in Sealand

Year	Crop	Total ha
1	Oilseed rape	25
2	Winter wheat	25
3	Winter wheat	25
4	Spring barley	25

We assumed that winter wheat and spring barley are sold directly at harvest, grass seed yield is stored on the farm and oilseed rape is stored outside the farm to be sold later. The winter wheat straw is bailed and sold, but straw from the other crops is left in the field. The farmer does most of the work by himself, except for the straw bailing, which is outsourced to contractors while, during sowing and harvesting, extra labour is hired. For the labour requirement without contract work, we agreed with local experts to settle for 5–6 h labour per ha.

3.1 Current weed management

Weed management is challenging in both crop rotations, especially in rotation 1. Grass seed production has a very low tolerance for weed infestation, especially for grass weeds, because of the risk of contamination of the cultivated grass seeds with weed seeds. Cleaning grass seeds is costly, and yield is reduced due to the loss of grass seeds from cleaning. The aim of weed management of all crops of rotation 1 is to achieve very low weed numbers during the two growing seasons with grass seed production. Weed management in both rotations is becoming more and more difficult due to widespread herbicide resistance from blackgrass (*Alopecurus myosuroides*) which causes problems in effective weed control in cereals.

Weeds are currently managed mainly by herbicide spraying, including pre-harvest glyphosate applications in oilseed rape and following the second-year harvest of grass seeds (Table 3). Ploughing and seedbed preparation are also part of weed management strategy but are also done for other reasons. In the analysis, we split ploughing and seedbed costs between 'weed management' and 'soil structure', with 50% of the investment and operation costs attributed to weed management. Costs for the use of the tractor and spraying machine are attributed to weed management according to the hours used for weed control. The same goes for spraying a tank mix of a herbicide and another pesticide

Table 3 Time of herbicide spraying in both rotations with the current strategy

	Oilseed rape	Winter wheat	Spring barley	Grass seed year 1	Grass seed year 2
Autumn spray	3	1			
Spring spray		1	1	2	2
Glyphosate at the end of the growing season	Yes				Yes
Ploughing	No	Depending on conditions (weed and soil)	No	No	After harvest

(targeting disease or pest management): the costs are split in proportion between these applications.

3.2 Selecting integrated weed management strategies to compare with current weed management

To compare the economics of IWM, two IWM strategies were designed in accordance with the information collected from interviews with experts and farmers. There was a thorough discussion to decide the two most promising and realistic IWM strategies to apply in local conditions. These were as follows:

- IWM wider rows; and
- IWM delayed sowing.

In the IWM strategy that we called *IWM wider rows*, winter wheat and oilseed rape are sown in wider rows to allow for inter-row cultivation to manage weeds mechanically. This resulted not only in a 60% reduction of herbicide use but also in a requirement for new mechanical weeding equipment for inter-row cultivation. Inter-row cultivation was applied twice a season for both winter wheat and oilseed rape. In winter wheat, weed suppression was also supplemented by the use of a false seedbed (Table 4). Glyphosate is still part of the strategy because, according to the experts, the required quality of the grass seed crop cannot be reached without the chemical control of problematic weeds. Organic grass seed production reflects this since it requires expensive post-harvest cleaning to offset not being able to use herbicides.

The other IWM strategy, *IWM delayed sowing*, follows the principles of delayed sowing by sowing winter wheat 2 weeks later to allow time to prepare a false seedbed prior to the sowing of winter wheat. Modelling predicted a good effect on weed suppression, removing the need for herbicide spraying in the spring (Table 5).

Table 4 Times of herbicide sprays and mechanical weeding actions in the wider row strategy for both rotations

	Oilseed rape	Winter wheat	Spring barley	Grass seed year 1	Grass seed year 2
Autumn spray		1 (band spray)			
Spring spray		1 (band spray)	1	2	2
Mechanical weeding	2	2			
Glyphosate	Yes				Yes

Table 5 Amount of herbicide sprays applied in the delayed sowing strategy for both rotations

	Oilseed rape	Winter wheat	Spring barley	Grass seed year 1	Grass seed year 2
Autumn spray	3	1			
Spring spray			1	2	2
Glyphosate	Yes				Yes

4 Comparing the economics of different integrated weed management strategies

The economic and environmental impact of the two IWM strategies was analysed and compared to the current weed management system by focusing on eight indicators. Since we only discuss the economic evaluation in this chapter, we will not discuss the environmental evaluation. The economic indicators used in IWMPRAISE were as follows:

- *Direct energy use:* This is the fuel use for all field operations. A chemical spray with both herbicides and other pesticides was partly assigned to weed management based on the ratio of herbicides to other pesticides in the spray tank. A spray with only herbicides or mechanical weed control was fully assigned to weed control.
- *Farm profitability (costs per €100 revenue):* Total profit is made comparable with the other indicators by calculating the amount of costs per €100 revenue. This is a relative indicator of the profitability of a farm. All the different costs and revenues are calculated per crop and added up to get a total overview. With this indicator, the financial performance of the system can be assessed.
- *Labour costs for weed management*: An IWM strategy with less input of herbicides may result in the requirement for more labour, for example, for mechanical or hand weeding. To assess this within and across cases, we compared the labour requirements for weeding with the total farm labour requirement (in hours). The division between labour for weeding and non-weeding operations was similar to the calculation of direct energy use.
- *Total costs of weed management in relation to total costs of the farm:* The total costs include yearly costs of machinery, labour costs, herbicide costs and fuel costs.
- *Investment costs for weeding equipment:* A different type of weed management may require other machines compared to current weed management practice. This is especially the case when strategies combine chemical, mechanical and even thermal methods, in which more equipment may be required compared to a fully chemical strategy.

> These investment costs are calculated per system. This is done by using the replacement value and comparing the total replacement value of weeding machines to the total replacement value of all machines present on the farm. The mouldboard plough and the machines for seedbed preparation were assigned to be used 50% for weeding, while the tractors and sprayers were assigned as being used for weeding based on the ratio of hours used for weed control. With this, we created an overview of the extra investment that adoption of a new IWM strategy could require.

These indicators were calculated mainly on basis of local data supported by expert estimations using the business economics analysis (BEA) tool. BEA is used to design, evaluate and compare the business economic effects of different farm strategies. This is an Excel-based tool developed by Wageningen Research for arable and vegetable farming systems. Like most other studies, BEA is based on cost-benefit analysis. What differs from most studies was that we evaluated farm decisions at a crop and enterprise level, including the required investment costs, and that we combined this with expert advice to include important information from practice, as well as using information from another database.

This second database is KWIN-AGV 2018 (Wageningen University & Research, 2018), a reference with quantitative agricultural data concerning Dutch arable and field vegetable farming. It contains current balance calculations for conventional and organic practices for the majority of arable crops and field vegetables grown in The Netherlands. Because a database similar to KWIN was not available in Denmark, we discussed with Danish experts whether the Dutch data were applicable in a Danish context. Fortunately, the cultivation of crops in The Netherlands is very similar to cultivation in Denmark, and where required, we replaced Dutch data with data or estimations from Danish experts.

Perhaps the most important single source of information used in IWMPRAISE was information from experts (researchers, advisors and farmers) who were interviewed about the practical implications of implementing particular practices and about local farming conditions. This process can be characterised as 'expert elicitation', a well-established methodology for synthesising expert opinion to allow for parameterisation and quantifying uncertainty. The technique is a valuable way to support decision-making in uncertain conditions and with limited data available (Schmidt et al., 2017). Expert elicitation makes it possible to gather important information such as what happens if farmers no longer have access to certain crop protection products, for example, what happens if farmers can no longer use glyphosate in grass seed production? The aim of this process is to get as close as possible to practice, both for current practice and for the future IWM strategies. The data for the three strategies were collected primarily on site and transformed to relative values to make the different systems comparable.

The costs and profits were mostly based on local data, with expert estimations and data from KWIN-AGV 2018 where necessary. The overall evaluation consisted of model-based calculations of the economics of weed management strategies in a farm context. The main aim was to get a clear picture of the proportions of costs and profits within the enterprise and the magnitude of impact or change compared to current practice. The results were validated with a sensitivity analysis.

The results from our economic analysis show that both IWM strategies do not perform as well as current practice in economic terms (direct energy input, costs per profit, labour for weeding, total costs of weeding and the investments for weeding machinery) (Fig. 2). However, the performance of the *IWM delayed sowing* strategy is close to standard weed management in economic performance. In IWMPRAISE, we analysed the environmental performance of the three systems as well to show the extra costs in relation to environmental impact. The IWM strategies resulted in a lower environmental impact. Besides the comparison between the three systems, the calculations and indicators show the underlying factors involved, that is, which cost items cause lower performance as well as which elements of novel strategies are positive compared to the standard. In this case study, *IWM wider rows* involves

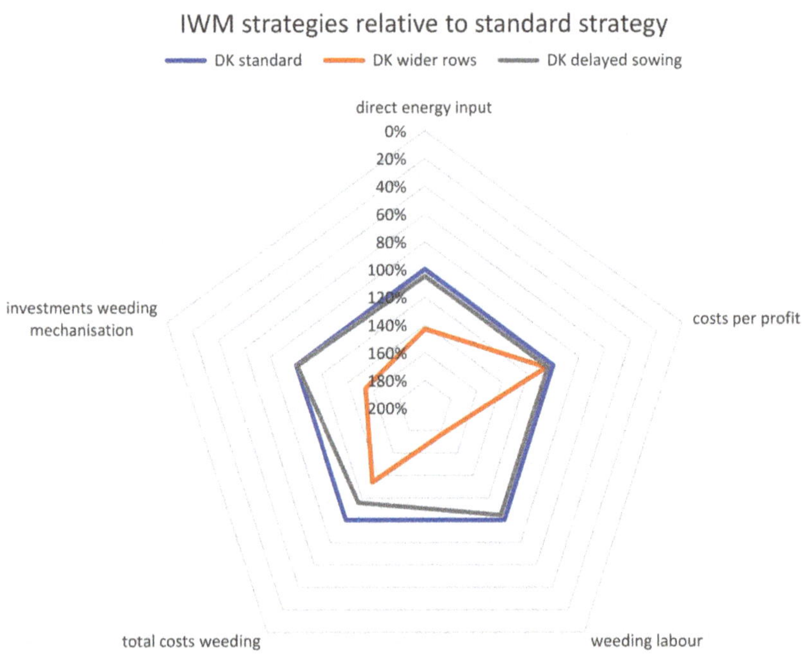

Figure 2 Performance of the two IWM strategies relative to the standard strategy, based on the economic indicators.

mechanical weeding which results in higher fuel consumption and more operations required (e.g. two mechanical weeding operations in oilseed rape with *IWM wider rows* vs. one spray action in standard). Bringing us back to the aim of this evaluation, in terms of economic performance, *IWM delayed sowing* is the most promising strategy for the future when more and more herbicides are banned.

5 Different approaches in assessing the economics of integrated weed management strategies

This section briefly describes two other economic evaluations of weed management which can be compared to the approach in IWMPRAISE (Pardo et al., 2010; Vasileiadis et al., 2015). The approaches of the three studies are compared and summarised in Tables 6 and 7.

5.1 Study by Pardo et al. (2010)

In the study by Pardo et al. (2010), the authors analysed the consequences of several IWM cropping systems both for labour requirements and economic profitability at the farm level. Different IWM cropping systems were analysed using simulations of 10 virtual farms based on a labour requirement model and input data from a 6-year cropping system experiment at a field scale, located at an experimental farm in central-eastern France (Chikowo et al., 2009, after Pardo et al., 2010). One standard weed management system (designed to maximise economic profitability without considering any environmental issues) was compared with four IWM cropping systems. Each of the 10 virtual farms used for model simulations corresponds to one experimental field. A set of decision rules was defined for each cropping system, which was then applied

- to the fields;
- to collect data to test whether the a priori objectives were reached; and
- to perform a multi-criteria assessment of the different cropping systems (Pardo et al., 2010).

The decision rules were based on expert knowledge, simulation with biotechnical models or based on experimental results, either collected locally or reported in the literature (Chikowo et al., 2009; Debaeke et al., 2009; Pardo et al., 2010).

The data collected over the 6 years of experimentation were used to estimate the profitability of the 10 virtual farms, for example, by establishing actual crop yields. Produce selling and input prices were the same as the prices

Table 6 Overview of the three approaches by means of data inputs, analysis methods and tools used

Approach	Input data	Analysis methods and tools
IWMPRAISE	- Direct energy input (fuel use, storage energy use, etc.) - Yields - Weeding labour (relative to total labour) - Total costs of weed management (relative to total costs) - Investments in machines for weed management - Costs per profit	- BEA - KWIN-AGV 2018 - Interviews with experts and farmer - Data from field experiments
Pardo et al., 2010	- Estimations from 6-year field experiment for profitability of 10 virtual farms - Actual crop yields obtained in the experiment - Sale and input prices - Net margin accounting for land rental costs, subsidies, building depreciation, farm overheads, social agrarian insurance - Cumulative time allocated to cultural operations at each time-step compared to the net availability of working time	- Data from a 6-year field experiment --> Estimation of profitability of 10 virtual farms - Ten simulated farms - Trafficable days for each type of operation based on references provided by local agricultural advisory services, climate frequency analysis and model simulations. Validated by technical advisers in the area.
Vasileiadis et al., 2015	- Weed pressure - Crop management cost - Input costs (fertiliser, pesticides, operation costs of labour, mechanical weeding, fuel) - Profit of each commercial or demonstration farm (crop yields, prices)	- Crop yields from all partners involved in experiments - Local prices for contract work (labour, machinery, fuel) - Prices that farmers paid for inputs - Grain maize prices in the years of the experimentation from Eurostat database --> Cost-benefit analysis. - Gross margin: financial yield = (yield × price) – variable costs

on the market at harvesting time in the last year of the experiment. Farmers' income was approximated by calculating the net margin, taking into account public subsidies, rental costs for land, farm overheads, social agrarian insurance and building depreciation.

5.2 Study by Vasileiadis et al. (2015)

As part of the European Pesticide Use-and-risk-REduction in European farming systems with integrated pest management Research Project, Vasileiadis et al.

(2015) evaluated the efficacy of locally selected IWM tools for direct weed control in maize in nine on-farm experiments. They compared these IWM tools with conventional weed management in three different European maize-producing regions. A comparative assessment was undertaken of the economic sustainability of the IWM tools against the conventional weed management approach.

For each farm, the most promising local IWM tools were selected based on

- outcomes of working groups of the EU project (European Network for Durable Exploitation of Crop Protection Strategies) and their results for maize systems in the specific regions (Meissle et al., 2009; Vasileiadis et al., 2011, 2013); and
- consultation with local stakeholders (e.g. farmers, researchers, extension services and agrochemical companies).

The efficacy of individual IWM tools was evaluated by determining weed pressure and through a cost-benefit analysis. Crop management costs (costs of inputs such as fertilizer and pesticides and operational costs such as labour, mechanical weeding equipment and fuel) and crop yields were collected from the experiments. Operational costs were based on local prices for contract work (labour, machinery, fuel) and costs of inputs based on the prices that farmers paid. Grain maize prices over the period of the experiment were derived from the Eurostat database.

6 Comparing different approaches in the economic evaluation of integrated weed management strategies

The approaches in the three case studies have many similarities (Table 7). Each study is based on calculations of costs and benefits to compare the financial profitability of each weed management strategy. Each takes into account the regional setting and uses field data. In the case of IWMPRAISE, however, field data are also used, but costs and profits are calculated within the context of a representative farm, based on its crop rotation and other factors that determine economic feasibility at the farm level.

6.1 The aims of economic evaluation

The aim of IWMPRAISE was to determine whether two specific IWM strategies were a feasible strategy. In addition, we examined what the proportions of different costs were, for example, what were the investments compared to the profits and savings, and how much of the total costs are attributed to weed management? Besides assessing overall feasibility, this evaluation formed a starting point for

Table 7 Overview of the data sources, the results, calculations and methods behind the results

Approach	Source of data			Results	Calculations and methods behind the results
	Experiment/farm	Database or tool	Persons		
IWMPRAISE	- Multiple field experiments - A representative farm	KWIN-AGV 2018BEA	Interviews with local experts and farmers (expert elicitation)	- Economic performance of IWM strategies compared to conventional	Model-based calculations of the costs and benefits of a representative profitable farm
				- Performance on a representative farm (enterprise level)	
				- Investments for weeding and mechanisation	
				- Labour for weeding	
				- Total costs of weeding	
				- Costs per profit	
				- Direct energy input	- Total fuel use at farm level - Fuel use for weed management --> Proportion use for weed management relative to total fuel use
Pardo et al., 2010	- Six-year cropping experiment at an experimental farm at field scale	Simulation model with 10 virtual farms based on data from the 6-year experiment		Economic performance of IWM strategies	Cost-benefit analysis
				- Profitability of 10 virtual farms	Estimated from the data collected in a 6-year experiment
				- Crop yields	Actual yields obtained in a 6-year experiment

				- Selling prices - Input prices	Actual prices at harvesting time in 2006 (last year of the experiment)
				- Net margin	Calculated accounting for public subsidies, land rental costs, building depreciation, farm overheads, social agrarian insurance --> Approximating farmers' incomes
		Climate frequency analysis and model simulations for values used for calculations	- Local agricultural advisory services for values used for calculations - Validation by expert technical advisors from the geographical area	Cumulative time for cultural operations compared to the net availability of working time	- Calculated for each time-step
Vasileiadis et al., 2015	- Nine on-farm (commercial or demonstration) experiments in three regions			- Efficacy of IWM tools for direct weed control in maize compared to the conventional approach - Comparative assessment of their economic sustainability	- Weed infestation in conventional and IWM plots - Cost-benefit analysis
				- Costs of inputs	
				- Operation costs (regional contract work prices including costs for labour, fuel, machinery)	
				- Crop yields and prices	Gross margin = financial yield (yield × price) − variable costs

improvement and adjustment of new strategies to match practical conditions. This approach helps to bridge the gap between research experiments and actual practice with all its regional characteristics and constraints.

A key advantage of the approach used in IWMPRAISE, compared to other studies which used single data from an experiment, was assessing investments at the farm scale. All costs and benefits were made, as much as possible, comparable by calculating relative values, for the following reasons:

- to highlight the balance within the farm, how much is spent on different elements (machinery/labour/investments/permanent costs/weeding, etc.);
- to compare systems across regions/countries/cropping systems/climatic zones; and
- to assess the added value and disadvantages of a novel system compared to a current system.

The time span for assessing economic impact is another important consideration. It is important to calculate NPV to compare similar investment options. Calculating NPV provides insights into the long-term economics of a new strategy, a key concern for farmers or researchers in deciding on which strategy to focus on.

6.2 Data and methods

In IWMPRAISE, data were based on field experiments, a standardised farm, databases and expert elicitation. In the study of Pardo et al. (2010), the analysis was based solely on one experiment. The data were subsequently validated by technical experts and used as input variables in a virtual farm model. Vasileiadis et al. (2015) also used data from just one experiment. The disadvantage of using the data from only one experiment is that it differs from farming practice and gives significant weight to years with either high or low harvest yields. Although the approach used in IWMPRAISE involves more complexity, by combining data from field experiments, databases and expert elicitation, a scenario can be developed that is very close to practice.

Like the IWMPRAISE study, the study of Pardo et al. (2010) stands out from other studies by connecting farm system modelling with experimental cropping system research rather than comparing systems based on cumulative labour requirements for a given crop or rotation alone, which neglects the distribution of labour over seasons and operations.

Pardo et al. (2010), as well as Vasileiadis et al. (2015), expressed the economic impact in absolute values. In IWMPRAISE, we do that by calculating relative values for the different costs and benefits to make a comparison across cases. This brings us back to the question: What do you want to learn from your analysis? If the aim is to show the ratio of costs and benefits to compare strategies, it seems more useful to calculate revenues and reduced costs because that gives a relative difference, making revenues and costs comparable between strategies and across different regions or climates. On the other hand, calculating NPV provides more insight into investment options in considering a new strategy, either as a farmer or a researcher who must decide on which strategies to focus resources.

6.3 Methods for different aims

Vasileiadis et al. (2015) rightfully stated that 'Since all the farming operations and choices, such as crop or hybrid selection, crop husbandry, fertilisation, could affect weed population dynamics, IWM should be fully evaluated not only for its direct weed control but also taking into account the whole cropping system . . . in the long term to capture any shifts in the weed community'. Since crop protection and especially weed management takes place at the crop rotation level, we agree that this should be considered in an economic analysis. Most other studies were not done at the farm level, probably due to the availability of data that are often only from field experiments which do not reflect actual farm practice.

In IWMPRAISE, we used external data to verify whether the data we used were representative of actual farm practice and we used data based on expert elicitation where this brought us closer to practice compared to data from experiments or databases. Other studies do not combine different sources of data in this way. Because we did not depend on experimental data alone, we were able to evaluate the economics of potential IWM strategies prior to doing the research, making it possible to skip research on less-promising strategies and gain insights into possible areas of improvement. The downside is that you always have to stay aware of whether data are 'hard' or 'soft' and which data are based on actual field data and which on expert elicitation? What are the assumptions and what is the possibility that actual implementation will differ from the assumptions made? The same goes for other studies since you have to consider the differences between models and practice, as well as experimental conditions that differ from those in real farms.

To meet the aim of IWMPRAISE – to get insight into the most promising strategies for further development and research – it was logical to evaluate the strategies in the way we did because it brought the evaluation close to actual farming practice and because it evaluated strategies at farm level since IWM involves the entire farm system. For other aims, however, it may make more sense to choose an approach which, for example, is based on experimental data only. Evaluation with multiple sources and expert elicitations is more complex. For example, if the evaluation is conducted for fundamental research, for example, examining the effects of plant physiological mechanisms on productivity, the farm context is not necessarily required. In this case, a strong experimental setup that excludes external factors as much as possible may be more appropriate.

Another consideration, when evaluating the feasibility of a strategy from a farmer's perspective, is the variation in representative farms since each farm is unique, for example, with its own rotation, vision and local conditions. The IWMPRAISE case involving IWM for winter wheat in Denmark involves two crop rotations which represent two systems. The choice of a particular standardised farm setup therefore also affects the result. The context and farm setup should therefore be carefully described so that the differences between strategies can be translated to other situations.

To choose an appropriate approach in an economic evaluation, the following issues need to be considered:

- pre-examination of novel strategies to decide which strategies are most promising to use in practice, for example, to select strategies that will be studied or developed further (like in IWMPRAISE);
- examination of already proven strategies for policy decision-making, for example, to reflect on what is reasonable to demand from the practitioners;
- examination of multiple operation options to show farmers the benefits and disadvantages of each in decision making; and
- regional studies where the aim is to get insight into the effects of a strategy over a time span or sector, for example, public health or economic impact at a system or sector level.

Besides the aim of evaluation, the choice of method depends on the research context and scale of analysis:

- The cost-benefit analysis at the field level (the approach used in the studies of Vasileiadis et al. (2015) and Pardo et al. (2010)) is appropriate for considering costs and benefits and notably to determine the effects of a strategy on the variable costs; and
- Calculations at the farm level, such as IWMPRAISE, are well suited to give insight into the differences between farm systems with differences in investments in new machinery for mechanical weeding. It is therefore necessary to analyse different parameters at the farm level-like crop rotation balance. A cost-benefit analysis at the farm level can also be done but has the disadvantage that it only shows the changed costs and benefits. It does not show whether the economic impact is substantial, for example, an increase of 200 €/ha is significant on a total result of 200 €/ha but relatively minor on a total result of 2000 €/ha.

To conclude, for each approach, there is something to consider. It is important to consider the aim of your evaluation, access to data and the advantages and disadvantages of adding multiple or more complex methods of analysis. For example, expert elicitation gives valuable information about the opportunities, bottlenecks and local conditions in evaluating the economics and feasibility of a system. If the aim is to examine the feasibility of novel measures or strategies from the perspective of a farmer, it is most appropriate to compare a novel IWM strategy with a standardised farm at the farm level, using expert elicitation and data that brings the evaluation close to a real farm situation, including local conditions that can affect a potential solution positively or negatively. By combining multiple data sources and expert elicitation, a strategy can be examined at an early stage and even be adjusted since you can start analysis before experimental data are collected.

7 Where to look for further information

The following guide forms a good starting point to apply economic evaluations of Integrated Pest Management (IPM) specifically:

- Fleischer, G., Jungbluth, F., Waibel, H. and Zadoks, J. C. (1999). *A Field Practitioner's Guide to Economic Evaluation of IPM* (No. 9). Institute for Economics in Horticulture (https://core.ac.uk/reader/29316240).

The following articles provide a good overview of economic evaluation methods, applied in different fields (not specific in agriculture or for IWM):

- Smith, E. G., Upadhyay, B. M., Blackshaw, R. E., Beckie, H. J., Harker, K. N. and Clayton, G. W. (2006). Economic benefits of integrated weed management systems for field crops in the Dark Brown and Black soil zones of western Canada. *Canadian Journal of Plant Science 86*(4), 1273–1279.
- Jones, R., Cacho, O. and Sinden, J. (2014). The importance of seasonal variability and tactical responses to risk on estimating the economic benefits of integrated weed management. *Agricultural Economics 35*(3), 245–256.
- Cameron, N., Wardlaw, T., Venn, T., Carnegie, A. and Lawson, S. (2018). Costs and benefits of a leaf beetle Integrated Pest Management (IPM) program II. Cost-benefit analysis. *Australian Forestry 81*(1), 53–59.
- Visalakshmi, V., Rao, G. V. R. and Rao, P. A. (2005). Integrated pest management strategy against Hubner in chickpea. *Indian Journal of Plant Protection 33*(1), 17–22.
- Gorddard, R. J., Pannell, D. J. and Hertzler, G. (1996). Economic evaluation of strategies for management of herbicide resistance. *Agricultural Systems 51*(3), 281–298.

- Vasileiadis, V. P., Otto, S., Van Dijk, W., Urek, G., Leskovšek, R., Verschwele, A., Furlan, L. and Sattin, M. (2015). On-farm evaluation of integrated weed management tools for maize production in three different agro-environments in Europe: agronomic efficacy, herbicide use reduction, and economic sustainability. *European Journal of Agronomy 63*, 71–78.
- Pardo, G., Riravololona, M. and Munier-Jolain, N. M. (2010). Using a farming system model to evaluate cropping system prototypes: Are labour constraints and economic performances hampering the adoption of Integrated Weed Management? *European Journal of Agronomy 33*(1), 24–32.

The studies in Table 8 can help to decide on a method for economic evaluation.

Table 8 An overview of studies that can be used as an example for economic evaluation

Authors	Title	Level analysis	Methods	Data	Results	Country
Smith et al. (2006)	Economic Benefits of Integrated Weed Management Systems for Field Crops in the Dark Brown and Black soil Zones of Western Canada	Farm level (ha)	• Contribution margin • Statistical analysis to compare IWM with common current practices	• Field experiment • Input: crop price, fertiliser, machinery, seed, marketing	• Means contribution margin	Canada
Jones et al. (2014)	The Importance of Seasonal Variability and Tactical Responses to Risk on Estimating the Economic Benefits of Integrated Weed Management	Farm level (ha)	• Biophysical models • NPV: to compare economic benefit with and without IWM • Stochastic dynamic programming model	• Production inputs • Crop yield and loss • Costs of control • Costs of application • Costs of production • Expert opinions to obtain parameters where no published data was found	• Cumulative probability NPV distributions	Australia

Authors	Title	Level analysis	Methods	Data	Results	Country
Cameron et al. (2018)	Costs and Benefits of a Leaf Beetle Integrated Pest Management (IPM) Program II. Cost-Benefit Analysis[a]	Regional	• Cost-benefit analysis • Cost-benefit ratio	• Costs of control • Yield and loss • Expert opinions to obtain parameters where no published data were found	• NPVs to compare IWM with common current practices for a plantation case study • Comparison of cost--benefit ratios	Australia
Visalakshmi et al. (2005)	Integrated Pest Management Strategy Against *Helicoverpa armigera* Hubner in Chickpea	Farm level (ha)	• Cost-benefit ratio	• Field Experiment • Input: crop price, fertiliser, machinery, seed, marketing	• Yield/ha • Gross income • Costs • Net income	India
Gorddard et al. (1996)	Economic Evaluation of Strategies for Management of Herbicide Resistance	Farm level (ha)	• Cost-benefit analysis • NPV • Economic regression model: MIDAS	• Costs • Yield price • Fixed costs • Variable costs	• NPV/ha • Cost benefits	Australia
Vasileiadis et al. (2015)	On-Farm Evaluation of Integrated Weed Management Tools for Maize Production in Three Different Agro-Environments in Europe: Agronomic Efficacy, Herbicide Use Reduction, and Economic Sustainability	Farm level (ha)	• Cost-benefit analysis	• Experiments farmers' fields • Costs of inputs, e.g., fertilisers and pesticides, and costs of operations, e.g., mechanical weeding, labour, fuel	• Costs/ha • Gross margin/ha	Europe: Italy, Slovenia, Southern Germany

(*Continued*)

Table 8 (*Continued*)

Authors	Title	Level analysis	Methods	Data	Results	Country
Pardo et al. (2010)	Using a Farming System Model to Evaluate Cropping System Prototypes: Are Labour Constraints and Economic Performances Hampering the Adoption of Integrated Weed Management?	Farm level (ha)	• Cost-benefit analysis	• Experiments research	• Input costs comparison • Labour comparison • Income/ha: harvest value • Costs/ha: mechanisation, other inputs, other costs • Net margin	Europe: France

[a]Study is not related to agriculture, but to forestry.

Key research in this area can be found at the following organisation:

- Wageningen Research Field crops (www.wur.nl/en/Research-Results/Research-Institutes/plant-research/Field-crops.htm).

8 References

Chikowo, R., Faloya, V., Petit, S. and Munier-Jolain, N. M. 2009. Integrated weed management systems allow reduced reliance on herbicides and long-term weed control. *Agric. Ecosyst. Environ.* 132(3-4), 237-242.

Debaeke, P., Munier-Jolain, N. M., Bertrand, M., Guichard, L., Nolot, J. M., Faloya, V. and Saulas, P. 2009. Iterative design and evaluation of rule-based cropping systems: methodology and case studies. A review. *Agron. Sustain. Dev.* 7, 235-250.

KWIN-AGV 2018. *Kwantitatieve Informatie Akkerbouw en Vollegrondsgroenteteelt, 2018.* Wageningen University & Research, Business Unit Open Teelten, publicatienummer 776, ISSN 1571-3059.

Meissle, M., Mouron, P., Musa, T., Bigler, F., Pons, X., Vasileiadis, V. P., Otto, S., Antichi, D., Kiss, J., Pálinkás, Z., Dorner, Z., Van Der Weide, R., Groten, J., Czembor, E., Adamczyk, J., Thibord, J. -B., Melander, B., Nielsen, G. C., Poulsen, R. T., Zimmermann, O., Verschwele, A. and Oldenburg, E. 2009. Pests, pesticide use and alternative options in European maize production: current status and future prospects. *J. Appl. Entomol.* 134(5), 357-375.

Pardo, G., Riravololona, M. and Munier-Jolain, N. M. 2010. Using a farming system model to evaluate cropping system prototypes: are labour constraints and economic performances hampering the adoption of integrated weed management? *Eur. J. Agron.* 33(1), 24-32.

Riemens, M., Sønderskov, M., Moonen, A., Storkey, J. and Kudsk, P. 2022. An Integrated Weed Management framework: A pan-European perspective. *Eur. J. Agron.* 133, 126443.

Schmidt, O., Gambhir, A., Staffell, I., Hawkes, A., Nelson, J. and Few, S. 2017. Future cost and performance of water electrolysis: an expert elicitation study. *Int. J. Hydrog. Energy* 42(52), 30470–30492.

Vasileiadis, V. P., Sattin, M., Otto, S., Veres, A., Pálinkás, Z., Ban, R., Pons, X., Kudsk, P., van der Weide, R., Czembor, E., Moonen, A. C. and Kiss, J. 2011. Crop protection in European maize-based cropping systems: current practices and recommendations for innovative integrated pest management. *Agric. Syst. Dev.* 104(7), 533–540.

Vasileiadis, V. P., Moonen, A. C., Sattin, M., Otto, S., Pons, X., Kudsk, P., Veres, A., Dorner, Z., van der Weide, R., Marraccini, E., Pelzer, E., Angevin, F. and Kiss, J. 2013. Sustainability of European maize-based cropping systems economic, environmantal and social assessment of current and proposed innovative IPM-based cropping system. *Eur. J. Agron.* 48, 1–11.

Vasileiadis, V. P., Otto, S., van Dijk, W., Urek, G., Leskovšek, R., Verschwele, A., Furlan, L. and Sattin, M. 2015. On-farm evaluation of integrated weed management tools for maize production in three different agro-environments in Europe: agronomic efficacy, herbicide use reduction, and economic sustainability. *Eur. J. Agron.* 63, 71–78.

Index

Ingram Content Group UK Ltd.
Milton Keynes UK
UKHW020750250723
425740UK00001B/8

9 781786 767455